WITHDRAWN
University of
Illinois Library
at Urbana-Champaign

Capillary Electrophoresis of Small Molecules and Ions

Capillary Electrophoresis of Small Molecules and Ions

Petr Jandik
and
Günther Bonn

Petr Jandik
Department of Analytical Chemistry
Johannes-Kepler-University
Altenbergerstr. 69
4040 Linz
Austria

Günther Bonn
Department of Analytical Chemistry
Johannes-Kepler-University
Altenbergerstr. 69
4040 Linz
Austria

Petr Jandik
Department of Analytical Chemistry
Johannes-Kepler-University
Altenbergerstr. 69
4040 Linz
Austria

Günther Bonn
Department of Analytical Chemistry
Johannes-Kepler-University
Altenbergerstr. 69
4040 Linz
Austria

Library of Congress Cataloging-in-Publication Data

Jandik, Petr.
   Capillary electrophoresis of small molecules and ions / Petr Jandik and Günther Bonn.
      p.   cm.
   Includes index.
   ISBN 1-56081-533-7.—ISBN 3-527-89533-7 (Germany)
   1. Capillary electrophoresis. I. Bonn, Günther. II. Title.
   QD79.E44J36 1993
   541.3'72—dc20
                                                                                                   93-18350
                                                                                                              CIP

© 1993 VCH Publishers, Inc.
This work is subject to copyright.
All rights reserved, whether the whole or part of the material is concerned, specifically those of translation, reprinting, re-use of illustrations, broadcasting, reproduction by photocopying machine or similar means, and storage in data banks.
Registered names, trademarks, etc., used in this book, even when not specifically marked as such, are not to be considered unprotected by law.

Printed in the United States of America
ISBN 1-56081-533-7

Printing History:
10 9 8 7 6 5 4 3 2

Published jointly by

VCH Publishers, Inc.
220 East 23rd Street
New York, New York 10010

VCH Verlagsgesellschaft mbH
P.O. Box 10 11 61
D-6490 Weinheim
Federal Republic of Germany

VCH Publishers (UK) Ltd.
8 Wellington Court
Cambridge CB1 1HZ
United Kingdom

Contents

1. **Introduction** 1

2. **Fundamentals of Capillary Electrophoresis** 5
   2.1. Elements of Systems for Capillary Electrophoresis 5
   2.2. Basic Theory 11
   2.3. Capillaries and Electroosmotic Flow 21
   2.4. Separation Modes in Capillary Electrophoresis 37
   2.5. Efficiency and Resolution in CE Separations under Optimal Conditions 58
   2.6. The Problem of Joule Heat Generation 64
   2.7. Peak Spreading and Loss of Resolution under Nonoptimal Conditions in CE 67

3. **Instrumentation for Capillary Electrophoresis** 79
   3.1. Sample Introduction 79
   3.2. Selection and Handling of Capillaries 107
   3.3. Detection in Capillary Electrophoresis 118
   3.4. Instrumentation and System Performance 177

4. Selected Applications of Counterelectroosmotic Capillary Electrophoresis 211
    4.1. Counterelectroosmotic CE of Carbohydrates 212
    4.2. Counterelectroosmotic CE of Flavonoids 244
    4.3. Counterelectroosmotic CE of Sulfonamides 247
    4.4. Counterelectroosmotic CE of Biomass Degradation Products 249
    4.5. Counterelectroosmotic CE of Tannins 250
    4.6. Counterelectroosmotic CE of Chiral Compounds 251

5. Application Examples of Coelectroosmotic Capillary Electrophoresis 257
    5.1. Adjusting Selectivity of Anion Separations 260
    5.2. Electromigrative Preconcentration of Anions 268
    5.3. Samples with Disparate Levels of Anions 277
    5.4. Modulation of Selectivity for CE Separations of Cations 281

Index 291

Preface

With capillary electrophoresis, we are currently experiencing the commercial introduction of a third major instrumental separation technique within a relatively short period of 30 years (gas chromatography ca. 1960, HPLC ca. 1970). A large number of analytical chemists, many of them with an extensive experience in GC and HPLC, are making their first experiments with capillary electrophoresis. In a similar way, many of the researchers involved with testing and development of the early prototypes of capillary electropherographs were also highly experienced in the two predecessor techniques. They all probably agree that capillary electrophoresis is an appealingly simple analytical technique with many features that seem familiar to users of chromatographic instrumentation.

However, capillary electrophoresis generally requires of a chemist a more active role than HPLC. Most important advances in the latter technique, especially in the past 10 years, have been made by engineers, frequently forcing chemists into a secondary role of an evaluator or user who develops "applications."

In capillary electrophoresis, on the other hand, considerable progress has been achieved by redesigning chemistries of electrolytes and capillaries. To improve the sensitivity of a given CE method, for example, the analytical chemist does not always have to wait for advances in the design of detectors. By a consequent application of fundamental chemical principles in the optimization of sample introduction procedures, the sensitivity of detection can be often improved a hundred or thousandfold. To obtain a new selectivity of separation, most HPLC users have to purchase a new column chemistry or await the development of new separation columns. The selectivity of CE separations, at least at this stage of development, is governed by the composition of carrier electrolytes. A variety of CE separation modes has been created by finding appropriate compositions of carrier electrolytes.

This book attempts to provide all the necessary information not only for being successful in following the established procedures, but also for the development of new methods and instruments. The small molecules and ions are clearly the main focus of the text, since the governing principles of a new instrumental technique are best explained and more easily understood with simple rather than complex chemistries. In addition to a detailed discussion of principles in Parts 2 and 3, the book also includes a selection of practical applications of capillary electrophoresis in Parts 4 and 5. Large portions of the application-oriented chapters are dedicated to the description of various strategies for adaptation of CE methods to the analysis of difficult samples. Instead of a broad review format, the authors have chosen to discuss primarily the applications and method development stemming from their own laboratories. Results of other workers are presented only if they were found to be an important developmental milestone or a useful point of reference.

In the whole book, the authors have attempted to provide a critical interpretation, rather than a mere listing of relevant information. In the two parts dedicated to applications, the method descriptions are provided mostly in sufficient depth to enable the reader to reproduce the methods of interest, without having to search for the original references in the voluminous literature on the subject of capillary electrophoresis.

August 1992
P. Jandik
Milford, Massachusetts

G. Bonn
Linz, Austria

Acknowledgement

We have been helped by many people in preparing this book. Special thanks are extended to Professor Richard M. Cassidy, University of Saskatchewan, who has read the entire manuscript and offered many valuable suggestions and to Martin Fuchs, M.E., of Waters Chromatography Division of Millipore, who provided essential help in preparation of Chapter 3. Professor Purnendu K. Dasgupta, Texas Tech, is acknowledged for his valuable comments on the section dealing with conductivity detection. Both authors also wish to express their gratitude for important help from members of their research groups. We especially wish to thank William R. Jones, B.S, Waters Chromatography Division, Dr. Andrea Weston, University of Rhode Island, Dr. Peter Oefner, Johannes-Kepler-University and Magister Christian Huber, Leopold-Franzens-University, for numerous helpful discussions which helped to shape the ideas and concepts described in this book.

CHAPTER

1

Historical Introduction

Contemporary capillary electrophoresis represents a change of direction in the advance of modern separation techniques. Only a few years ago, most of us expected an analogous development to capillary gas chromatography (GC) to occur in column liquid chromatography. The advantages of miniaturization were demonstrated convincingly for GC, and a rapid implementation of similar design changes for high-performance liquid chromatography (HPLC) appeared to be imminent. The electrokinetic separation techniques appeared as rather unlikely candidates for a popular method for separations of liquid samples, especially in view of only limited acceptance of isotachophoresis. The sudden popularity of another electrokinetic technique making possible for the first time separations of liquid samples in narrow-bore capillaries was thus a considerable surprise to many analytical chemists. The rapid development of capillary electrophoresis was only possible because of existing fundaments discovered by a large number of researchers over the course of the past 200 years.

The summary in Table 1.1 shows the most important milestones in the development of capillary electrophoresis. For lack of space only a few, in the opinion of the authors, of the most important references can be listed here. A more complete description of historical facts can be found in the References 22-24.

Table 1.1. HISTORICAL MILESTONES OF THE DEVELOPMENT OF CAPILLARY ELECTROPHORESIS

Year	Author's Name	Contribution
1791	Faraday	Laws of electrolysis
1877	Helmholtz[1]	Solvent layer close to an inside wall and the surface of a charged particle acquires an opposite charge
1856	Hittorf[2]	Definition of transport numbers for ions
1897	Nernst[3]	Properties of small ions
1897	Kohlrausch[4]	Kohlrausch function describing the order of electrophoretic migration of ions and their relative concentrations
1930	Tiselius[5]	Thesis—the moving boundary method of studying the electrophoresis of proteins
1923	Kendall and Crittenden[6]	Rare earth metal separation by "ion migration method." First isotachophoresis
1939	Svenson[7]	Development of zone and displacement electrophoresis.
1945	Longsworth[8]	
1948	Tiselius	Nobel Prize for development of the moving boundary method.
1950	Haglund and Tiselius[9]	Electrophoresis tube filled with glass beads, powder material
1955	Smithies[10]	Gel electrophoresis
1958	Hjerten[11]	Electrophoretic analyses in free solution
1967	Martin and Everaerts[12]	Displacement electrophoresis in glass tubes with hydroxyethylcellulose
1967	Hjerten[13]	Elimination of the electroosmosis effects by coating of tubes
1969	Giddings[14]	Nondiffusional model of concentration distribution in free zone electrophoresis
1969	Virtanen[15]	Glass capillaries 0.2–0.5 mm I.D. were used
1970	Everaerts and Hoving-Keulemans[16]	Capillary isotachophoresis, ITP equipment
1970	Arlinger and Routs[17]	Developments in detectors—UV
1972	Verheggen[18] et al.	Conductivity detector
1979	Mikkers et al.[19]	High voltage in narrow-bore Teflon tubes
1981	Jorgenson[20]	Use of 75 μm open tubular glass capillaries; high performance capillary electrophoresis (HPCE)
1984	Terabe et al.[21]	Combination of electrophoretic and chromatographic modes. Micellar electrokinetic chromatography (MECC, MEKC)

References

1. H. V. Helmholtz, *Wiedemanns Ann.* **7**, 337 (1877).
2. J. Hittorf, *Pogg. Ann.* **98**, 1 (1856).
3. W. Nernst, *Z. Electrochem.* **3**, 308 (1897).
4. F. Kohlrausch, *Wiedemanns Ann.* **62**, 209 (1897).
5. A. Tiselius, *Nova Acta Regiae Goc. Sci. Ups. Ser. iV* **4**, 1 (1930).
6. J. Kendall and E. D. Crittenden, *Proc. Natl. Acad. Sci.* **9**, 75 (1923).
7. H. Svensson, *Kolloid. Z.* **87**, 180 (1939).
8. L. G. Longsworth, *J. Am. Chem. Soc.* **67**, 1109 (1945).
9. H. Haglund and A. Tiselius, *Acta Chem. Scand.* **4**, 957 (1950).
10. O. Smithies, *Biochem. J.* **61**, 629 (1955).
11. S. Hjerten, *Arkiv Kemi* **13**, 151 (1958).
12. A. J. P. Martin and F. M. Everaerts, *Anal. Chim. Acta* **38**, 233 (1967).
13. S. Hjerten, *Chromatogr. Rev.* **9**, 122 (1967).
14. J. C. Giddings, *Sep. Sci.* **4**, 181 (1969).
15. R. Virtanen and P. Kivalo, *Suomen Kemistilehti* **B**, 182 (1969).
16. F. M. Everaerts and W. Horing-Keulemans, *Sci. Tools* **1**, 25 (1970).
17. L. Arlinger and R. J. Routs, *Sci. Tools* **17**, 21 (1970).
18. Th. P. E. M. Verheggen, E. C. van Ballegooijen, C. H. Massen, and F. M. Everaerts, *J. Chromatogr.* **64**, 185 (1972).
19. F. E. P. Mikkers, F. M. Everaerts, and Th. P. E. M. Verheggen, *J. Chromatogr.* **169**, 11 (1979).
20. J. W. Jorgenson and K. D. Lukacs, *Anal. Chem.* **53**, 1298 (1981).
21. S. Terabe, K. Otsuka, K. Ichikawa, A. Tsuchiya, and T. Ando, *Anal. Chem.* **56**, 111 (1984).
22. S. Hjerten, *Chromatogr. Rev.* **9**, 122 (1967).
23. O. Vesterberg, *J. Chromatogr.* **480**, 3 (1989).
24. F. Foret and P. Bocek, *Advances in Electrophoresis*, ed. A. Chrambach, M. J. Dunn, and B. R. Radola, VCH, Weinheim (1989).

CHAPTER

2

Fundamentals of Capillary Electrophoresis

2.1 Elements of Systems for Capillary Electrophoresis

The typical capillary electrophoresis (CE) apparatus represents a radical departure from the instruments used in other, more conventional chromatographic and electrokinetic separation techniques such as gas or liquid chromatography, slab electrophoresis, and isotachophoresis. Separation efficiency is greatly improved (see Section 2.5), and, in its present form, CE instrumentation is considerably less complex than, for example, any current chromatographic system. As a bare minimum, any CE system has to contain the modules illustrated in Figure 2.1 and as characterized in the following.

2.1.1 Fused Silica Capillary

Capillary electrophoresis became a viable analytical technique only after the commercial introduction of narrow-diameter (10–100 μm), polyimide-coated fused silica capillaries (see Section 2.3). The narrow capillary diameter facilitates the dissipation of heat generated by the electrical resistance of the electrolyte inside the capillary. Additionally, the fused silica material of the capillary wall is a better heat conductor than any other material used to make narrow-bore capillary tubing. The polyimide coating alleviates the considerable fragility of uncoated fused silica tubing.

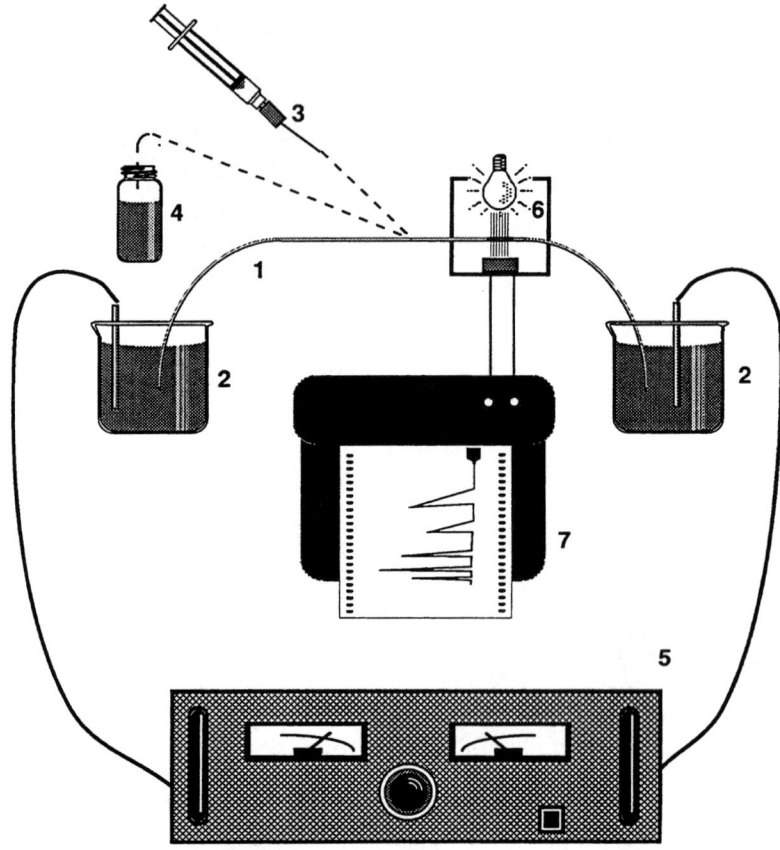

Figure 2.1 Basic components of a capillary electrophoretic system. 1, Fused silica capillary; 2, electrolyte vessels with electrodes; 3, syringe-to-capillary adaptor (replaced in commercial instruments by pressure- or vacuum-driven rinse); 4, sample vial raised to a level necessary for sample introduction by hydrostatic pressure; 5, regulated high-voltage power supply; 6, detector; 7, data acquisition device (recorder, integrator, or computer).

2.1.2 Two Electrolyte Vessels

During separations, the fused silica capillary is immersed on both sides in a suitable electrolyte. Prior to a separation, an electrolyte is introduced into the capillary with the help of a syringe–to–capillary adaptor (Figure 2.2), which is also used to precondition new capillaries by rinsing with 0.1 N NaOH and pure water.

Together with other conditions, the actual composition of a CE electrolyte, also called a carrier electrolyte, defines a capillary electrophoretic method. Unlike in isotachophoresis, both electrolyte vessels in CE systems

FUNDAMENTALS OF CAPILLARY ELECTROPHORESIS

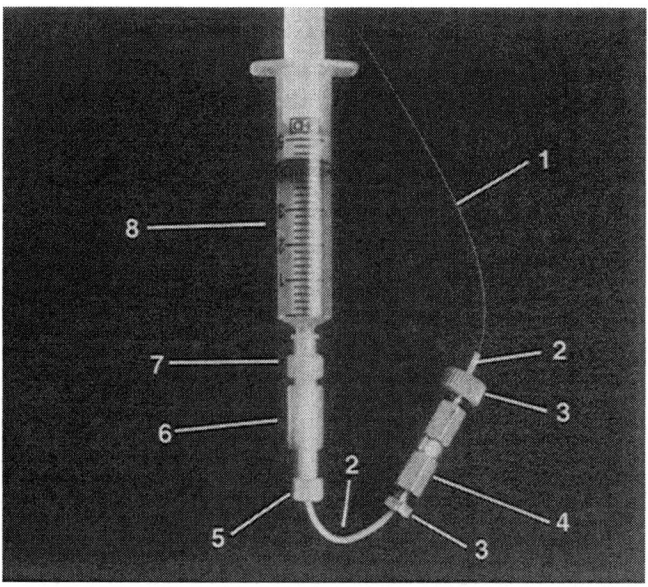

Figure 2.2 Manual priming and rinsing of capillaries. 1, fused silica capillary; 2, narrow-diameter Teflon tubing; 3, stainless steel fitting screw; 4, SS union matched to the SS screw 3; 5, plastic screw with a plastic "gripper" or ferrule matching plastic union 6; 7, luer lok–to–plastic-union adapter; 8, disposable syringe, volume ca. 1–5 ml.

contain an identical electrolyte. A frequently occurring problem with carrier electrolytes in "home-made" instruments, such as the one in Figure 2.1, is air bubbles trapped inside the capillaries. The presence of air bubbles leads to fluctuating baselines in the electropherograms (recordings of electrophoretic separations) and in extreme cases to an interruption of the electrical field required for CE separation. The problem can be successfully controlled either by degassing of electrolytes by one of the three common methods known to HPLC users (helium sparging, microfiltration in vacuum, or sonication) or by application of the syringe-to-capillary adaptor (Figure 2.2) filled with an aliquot of carrier electrolyte. Another important requirement is to position the filling levels of two carrier electrolyte containers at exactly the same height in order to minimize any flow by hydrostatic imbalance.

2.1.3 Sampling Position

The simplest method of sample introduction consists of placing one end of the capillary in a sample container positioned at a certain, always constant height over the level defined by the liquid surfaces in the two electrolyte vessels. Reproducible sampling can be expected only with a rigorously exact positioning of sample containers and for samples having comparable viscosity. Typical injection times lie between 10 and 30 seconds. In most manual instruments, it is very difficult to sample reproducibly with sampling intervals less than 10 seconds. A detailed discussion of sampling techniques is offered in Chapter 3.1.

2.1.4 Power Supply

The voltage required to drive CE separations is provided by a regulated direct current (DC) power supply connected by a pair of insulated cables to the two electrodes, usually made of platinum or other chemically inert metal, immersed in the carrier electrolyte.

Practical considerations limit the maximum value of voltage to 30 kV. Higher voltages could lead to corona discharges through the capillary and elsewhere in the instrument.[1]

The maximum value of the current, 500 to 1000 μA, is dictated mainly by safety considerations.[2] Such home-made instruments, which are not equipped by properly designed circuit interruption contacts and relays, have to be considered unsafe. An operator or uninformed outsider can be exposed to dangerous electric currents by touching the parts under high voltage. The commercial instruments, on the other hand, contain elaborate safety features, making it impossible for even an uninformed person or an absent-minded operator to come in contact with parts of the instrument under high voltage.

Another limitation is imposed by the large levels of heat generated by currents over 500 μA. Even the relatively good thermal conductivity of silica does not allow dissipation of the heat generated at such excessively high currents. Baseline oscillations with most detection techniques and even the boiling of electrolytes inside the capillaries make the use of high currents impractical.

A good CE power supply must have a provision for reversal of polarity. Most power supplies are set up to operate in the constant-voltage mode; however, a constant-current operation may be useful for many analytical problems (for example, excessive concentration ratios; see Section 3.4.2). Specialized applications[3] may require an alternate current (AC) component superimposed on the DC driving potential. This can be accomplished either by using two different types of power supplies or by one integral DC/AC power source. Also discussed in the literature is the programmed change of the applied voltage in the course of a CE separation, requiring computer control of a power supply (see Section 2.2).

The reader is referred to the manufacturer's literature[4] for a more in-depth description of power supply modules. Such information is vital for the design of safe and efficient CE systems, but quite unnecessary for the routine use of commercial CE instruments.

2.1.5 Detector

The simplest and most common detection method in CE is UV detection. A comprehensive discussion of all reported CE detection techniques is offered in Section 3.3. The following provides a brief introduction to the subject of detection in CE.

Several instrument suppliers[5-8] offer accessories for converting an HPLC detector for use with fused silica capillaries. To enable the passage of light, a short length of the capillary (0.5–1 cm) must be stripped of the polyimide layer. This is usually accomplished by a cigarette lighter, although more specialized tools for polyimide stripping have now become available (see Section 3.2). The exposed segment of fused silica is then inserted into the light path of a photometric detector and properly aligned by means of the HPLC–CE conversion module, such as the one shown in Figure 2.3.

In most cases, it is also advisable to remove small residues of burned polyimide from the exposed portion of fused silica capillary prior to its placement in the detector. This may be done by wiping with a soft tissue towel prewetted in methanol.

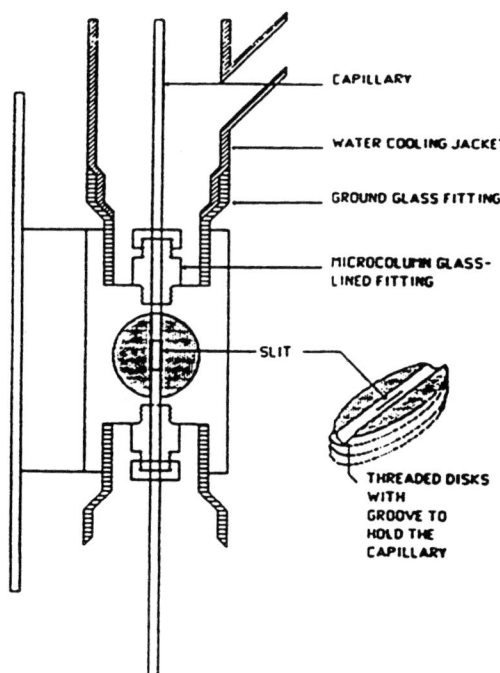

Figure 2.3 Module converting an HPLC UV detector for use in capillary electrophoresis. (Reproduced with permission.[6])

Similar "on-column" detection is currently included in all commercial instruments. It avoids challenging problems in the design of dead volume free cells on a capillary scale. Both variable- and fixed-wavelength photometric detectors are used in commercial CE instruments. The latter type more than compensates for the inconvenience of discrete changes of wavelength by interesting possibilities of measurement at 185 nm and by a generally higher energy output of the mercury lamp in comparison with the deuterium light sources utilized in continuously variable, monochromator-equipped photometric detectors. Since many of the peaks in modern capillary electrophoresis have a width under 3 seconds, the time constant of the detector must be 0.3 seconds or smaller.

2.1.6 Recorder

In principle, all types of devices may be utilized to record electropherograms. (The term *chromatogram* should not be used for CE recordings.) The responsiveness of a recorder is not much of a concern with simple devices such as chart recorders or integrators achieving 75% of full-scale deflection in less than 1 second. However, faster forms of CE such as coelectroosmotic capillary electrophoresis (Chapter 5) require chart speeds as high as 10 or 20 cm/min, resulting in unmanageable quantities of paper.

Most integrators acquire data at frequencies derived from the power main—50–60 Hz. Such a data acquisition rate is adequate even for the fastest of CE separations. Additionally, many of the integrators allow repetitive reprocessing of electropherograms, an essential capability for quantitative analysis at a high daily sample throughput.

Ideally, the CE data acquisition is carried out with the help of a personal computer (PC) having a hard disk drive of at least 20 megabyte and a diskette drive. Most PCs have adjustable data acquisition rates between 1 and 50–60 Hz. No less than 10 Hz should be chosen for an electropherogram with peak widths under 3–5 seconds (Section 3.4.4). The storage capacity of a hard drive can be maximized by a regular transfer of data files to a diskette and by a delayed start of recording.

For example: Most separations in capillary ion analysis occur within 60 to 90 seconds, with the first peak reaching the detector only after approximately 2 minutes. A delayed acquisition start at 2 minutes can thus reduce the data quantity by two-thirds.

PC recorders allow not only very convenient data management, but are also useful for their ability to perform certain calculation routines automatically, such as, for example, internal standard calibration. As discussed in Section 3.4.5, internal standard calibration improves the precision of migration times and peak areas. It largely eliminates well-known sources of imprecision in CE stemming from temperature and current changes between analytical runs.

FUNDAMENTALS OF CAPILLARY ELECTROPHORESIS

Table 2.1. GLOSSARY OF TERMS

HPLC	CE
Eluent, mobile phase	Carrier electrolyte
Pump	High-voltage power supply
Flow rate	Electroosmotic flow
Injection	Sample introduction (hydrostatic, electromigration)
Column	Fused silica capillary
Retention time	Migration time
Chromatogram	Electropherogram

2.2 Basic Theory

At the beginning of an analytical run, one of the two ends of the CE capillary holds a segment of an unresolved mixture of analytes (Figure 2.4a), contained in a very small volume (typically 10 to 100 nanoliters) of either the original sample solvent (hydrostatic sample introduction) or in the carrier electrolyte (electromigrative sample introduction). A desirable outcome of

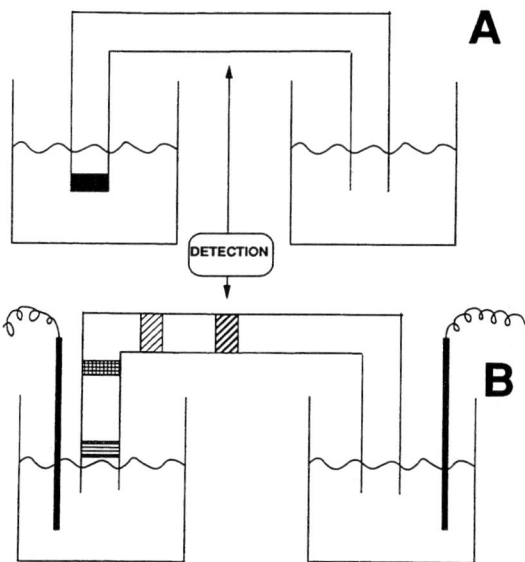

Figure 2.4 Two initial stages of CE separation. A. Following a sample introduction, the capillary is transferred back into the electrolyte vial for application of a separation potential. B. The separation potential has been applied for a time period coinciding with the migration time of the first of the several analyte zones.

CE separations is the migration of separated zones of all analytes toward the detector (Figure 2.4b).

Similarly as in HPLC, the time interval between the beginning of the separation and the passage of an analyte zone through the point of detection (retention time in HPLC, migration time in CE) is used for identification of sample components. The information contained in observed migration times (t_m) can be made more universally comparable by a conversion to observed migration rates or velocities of migration (v_{obs}) [Eq. (2.1)]. The inclusion of the term *observed* reminds us that the velocity is due not only to electrophoretic migration, but that it may contain other contributions as well.

$$v_{obs} = L_d/t_m \, [\text{cm/sec}], \tag{2.1}$$

where L_d [cm] is the capillary length between the points of sample introduction and detection. The calculation of migration rates allows a comparison of migration behavior between capillaries of different length. The L_d dimension should thus be included in all descriptions of CE conditions along with L_t [cm], the total capillary length. To compare migration data not only for varying L_d, but also at different separation potentials P [V] applied to different L_t, it is necessary to calculate the observed electrophoretic mobilities (μ_{obs}):

$$\mu_{obs} = v_{obs}/E = L_d L_t / P \, t_m \, [\text{cm}^2 \, \text{V}^{-1} \, \text{sec}^{-1}] \tag{2.2}$$

P in volts divided by L_t defines the applied field strength E [V/cm]. The values of observed electrophoretic mobilities are generally in the range of 10^{-4} cm^2 V^{-1} sec^{-1}, translating to migration times in 10^2 sec or several minutes in 10^1 cm capillaries at 10^4 V [$10^{-4} = 10 \times 10 \times 10^{-4} \times 10^{-2}$, from Eq. (2.2)].

Entering the next higher level of inquiry, we shall now consider the several most frequently occurring contributions to the observed electrophoretic mobility. As shown in Figure 2.5, the observed or apparent mobility μ_{obs} is usually a sum of at least two contributions.

All mobility contributions have not only a magnitude, but also a direction, and so it is correct to treat them as vectors.

The conditions depicted in Figure 2.5 describe the most frequently reported type of capillary electrophoresis. The analyte of interest is transported toward a detector mainly by the electroosmotic flow. A discussion of the origins of electroosmotic flow is given in Section 2.3. The inherent ionic mobility is utilized only to provide a selectivity by slowing down different ions to a different degree. Separations of anions in untreated fused silica capillaries can be named as an example. The anions show a tendency to migrate toward the anode (to the left in Figure 2.5), while the electroosmotic flow is directed toward the cathode (to the right side of Figure 2.5). Controlled utilization of additional mobility contributions is a more recent ad-

FUNDAMENTALS OF CAPILLARY ELECTROPHORESIS

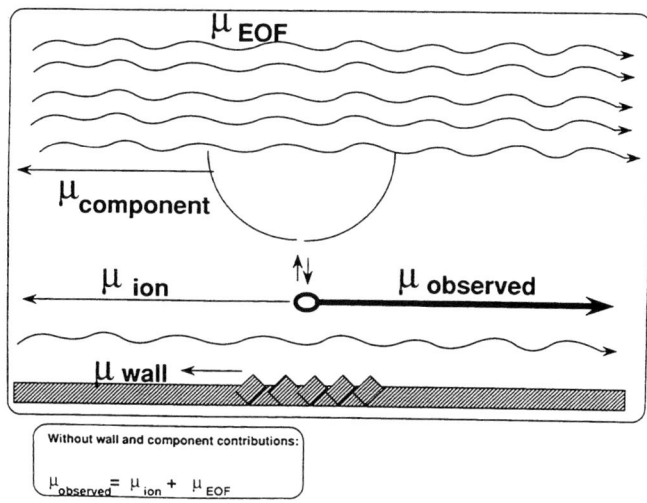

Figure 2.5 Possible contributions to the directions and magnitude of the resulting apparent or observed analyte mobility μ_{obs} in the applied electric field inside a capillary: μ_{eof} is the mobility contribution of the electroosmotic flow; μ_{ion} describes the inherent ionic mobilty of the ionic analyte; μ_{wall} introduces the possibility of analyte–capillary-wall interactions; and $\mu_{component}$ is the mobility of an electrolyte component, in solution or as a suspended particle, capable of interacting with the analyte ion.

dition and will be discussed in the Section 2.4, dealing with the separation modes in capillary electrophoresis.

As explained more thoroughly in Section 2.3, electroosmotic flow originates from interactions of carrier electrolyte components with charges on the capillary wall in the applied electric field. The polarity of the capillary wall charges can be made either negative or positive, resulting in two possible directions of the electroosmotic flow. The electroosmotic flow rate v_{eof} and mobility μ_{eof} can be calculated from the migration times of easily detectable uncharged compounds under conditions shown in Figure 2.6 and with the help of Equation (2.2).

Figure 2.7 shows that the electroosmotic flow rate increases in going from Teflon to Pyrex and with increasing pH.

The second contribution to the observed electrophoretic mobility, the ionic mobility μ_{ion}, can be calculated by one of the following methods:

In CE systems with the known μ_{eof}, the ionic mobility is obtained from the migration time of an ionic analyte t_m [sec]:

$$\mu_{ion} = L_d L_t / P t_{m(ion)} - \mu_{eof} \qquad (2.3)$$

A useful source of numerical values for net ionic mobilities μ_{ion} is in the tables of limiting ionic equivalent conductances λ_{equiv} [cm² equivalent⁻¹

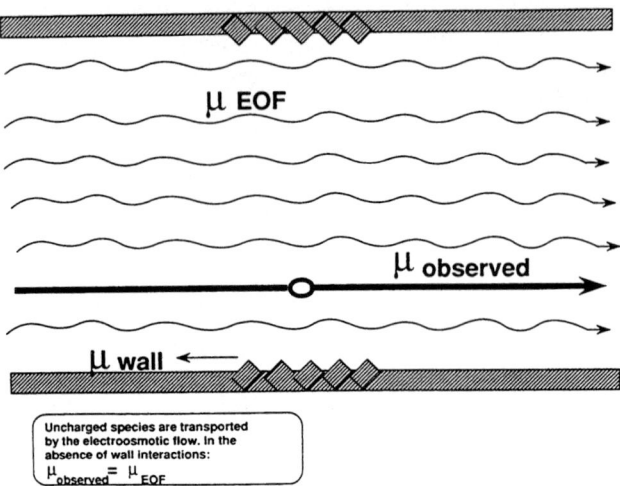

Figure 2.6 Sample components that are not ionized are transported by the electroosmotic flow. Their mobility is thus identical with that of the electroosmotic flow.

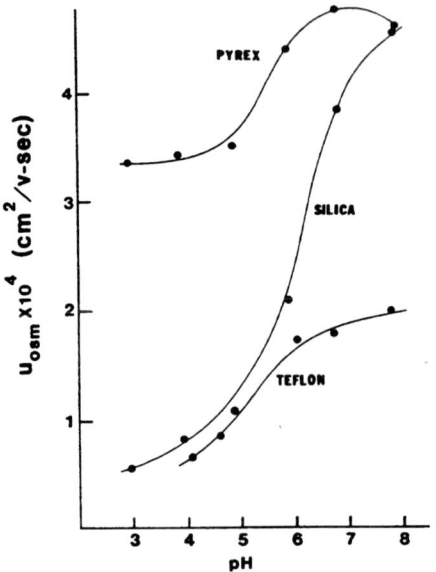

Figure 2.7 Electroosmotic mobilities—mobilities of uncharged species transported by electroosmotic flow—in capillaries made of three different materials with changing pH. The ionic strength was kept constant at $I = 0.06$ M. (Reproduced with permission.[9])

ohm^{-1}],[10] which can be converted to mobilities with the help of Faraday constant F [9.6487×10^4 A sec equivalent^{-1}]:

$$\mu_{ion} = \lambda_{equiv}/F \tag{2.4}$$

For those ions that are not included in the tables of conductances, λ_{equiv} can be estimated from a simple conductometric measurement. The only requirement for such determinations is the availability of a conductometer and a pure salt with a known counterion. Prior to the actual measurement, the cell constant of the instrument has to be verified and adjusted. Also critical is the control of temperature to within ± 1°C. However, many commercial conductometers are equipped with temperature sensors and automatically convert a conductivity value from a given temperature to 20 or 25°C. The equivalent conductances of an unknown ion then results from the following equation[10]:

$$\lambda_{\text{unknown ion}} = (10^{-3}\, GK/M) - \lambda_{\text{equiv of known counterion}} \tag{2.5}$$

where G is the conductivity readout in μS [microsiemens or ohm^{-6}], K is the cell constant [cm^{-1}], and M is the concentration in equivalents per liter. While the equivalent conductances calculated from this equation are useful in predicting the migration behavior in CE, it should be remembered that the values listed in most tables are the limiting equivalent conductances. Several λ_{equiv} values are obtained at the lowest measurable concentrations and extrapolated to infinite dilution. Now that we have learned how to measure the value of electroosmotic flow rate or mobility and how to calculate the migration times from the ionic equivalent conductances and electrophoretic mobilities, let us consider the practical value of all possible combinations of the two most important mobility vectors (μ_{ion} and μ_{eof}) in a capillary electrophoretic system. The terms *slow* and *fast ions* are used in the following text to describe the relative magnitudes of the μ_{ion} and μ_{eof} vectors. With a very slow electroosmootic flow, as observed, for example, in Teflon capillaries at pH 3 to 5 (Figure 2.7), most anions would be considered "fast" according to the present frame of reference, being able to overcome the electroosmotic flow conflicting with their inherent migrational direction. If, on the other hand, the electroosmotic flow is increased, as, for example, under alkaline conditions in fused silica capillaries (Figure 2.7), most anions would be described as slow, not being able to overcome the electroosmotic flow oriented against the direction of their own migration vector. The upper half of Figure 2.8, Cases A–E, shows the same orientation of the electroosmotic flow vector with respect to the electric field as in Figure 2.6. The injection side is positive, and the negatively charged capillary wall generates an electroosmotic flow toward the detection side. Fast anions migrate very slowly under such conditions, with a retention time frequently longer than 10 or 20 minutes (Case B). The large majority of inorganic anions are fast, exceeding the mobility of the electroosmotic flow under most conditions and in conse-

16 CAPILLARY ELECTROPHORESIS OF SMALL MOLECULES AND IONS

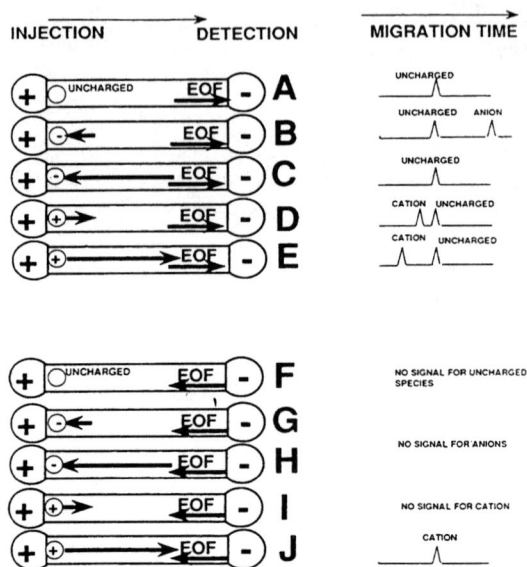

Figure 2.8 Combinations of mobility vectors in a CE system with the positive polarity at the sample introduction side and negative polarity at the detector side. The right side of the diagram indicates approximate positions of peaks for neutral and charged species. The sample solvent is usually swept through the detector at the migration time of the neutral species. Without additional mobility contributions, neutral species remain unresolved under the conditions described in this and the following figure.

quence migrate away from the detection side toward the sampling end of a capillary (Case C). The configuration with the positive injection side is generally more suitable for cations than for anions. Even the slowest of cations move with reasonable migration times through the detector cell (Case D), and the majority of highly mobile inorganic cations can be separated and detected with total run times not longer than 3–5 min (Case E). As already indicated, the direction of electroosmotic flow can be reversed by certain electrolyte additives or by a chemical modification of the capillary wall. Under the conditions of Cases F and G in Figure 2-8, such flow reversal appears to be counterproductive for three out of four combinations. Only the most highly mobile cations are capable of migrating toward the point of detection and producing a signal (Case J).

Contrary to the preceding diagram, the configurations depicted in Figure 2.9 can be described as highly suitable for the separations of fast anions. An analytical signal is always obtained for highly mobile anionic species. It is possible to choose between short (Case E) and long (Case J) run times for such species, opening the possibility to analyze successfully a wide range of different samples. To analyze, for example, a trace concentration of a fast

FUNDAMENTALS OF CAPILLARY ELECTROPHORESIS

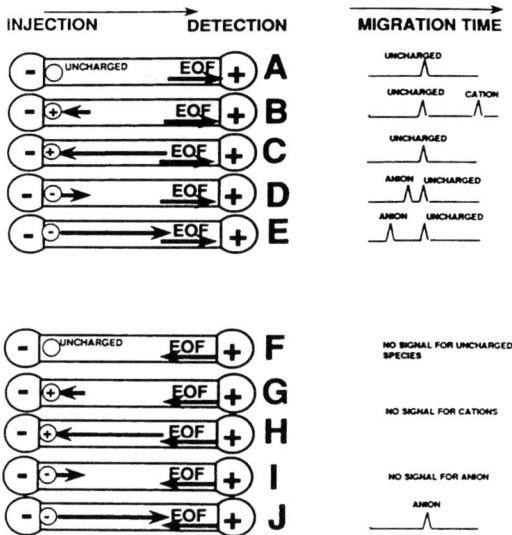

Figure 2.9 Different combinations of mobility vectors in a CE system with the negative polarity at the sample introduction side and a positive detector side. The right half of the figure indicates approximate positions of peaks for neutral and charged species.

anion in the presence of an excess of a slower anion, the proper choice would be Case E. Note that under these conditions the cations would also be well separated from the analytical signal. The fast cations would move in an entirely different direction—toward the cathode Case C—and the slow cations would reach the detector only after a very long time, Case B. Slowly migrating anions can also be separated relatively well, but the options are somewhat more limited; notice that in Figure 2.9 no analytical signal is generated for a slow cation, which is swept toward the anode due to the larger value of the μ_{cof} vector. The utility of the anion separations under these conditions is demonstrated in the electropherogram of thirty anions shown in Figure 2.10.

A possibility exists to calculate ionic mobilities from hydrated ionic radii or the Stokes radii of hydrated species, r_i:

$$\mu_{ion} = (10^7 z_i e)/(6\pi\eta r_i) \tag{2.6}$$

where z_i is the valence number of an ion, determining by its sign one of the two possible directions of the electrophoretic mobility vector, e is the charge of an electron (1.602×10^{-19} Coulombs), and η is the dynamic viscosity of the electrolyte medium in Poise. The numerical factors stem from the conversion of joules to ergs in the equation, giving the force acting on a charged particle in an electric field ($f = 10^7 Fze$), and from the Stokes law, describing

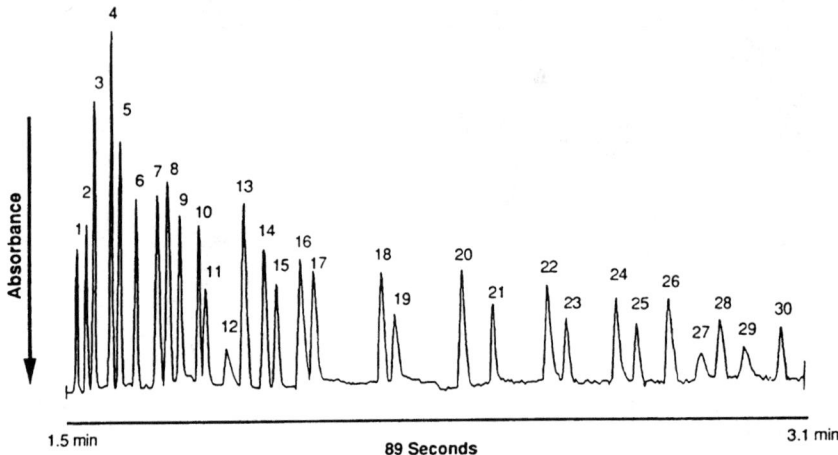

Figure 2.10 Electropherogram of thirty anions. Carrier electrolyte: 5 mM chromate, 0.5 mM electroosmotic flow modifier (Waters OFM Anion-BT), pH 8.0, fused silica capillary I.D. 50 μm, voltage 30 kV (negative), electromigrative injection 15 sec at 1 kV, indirect UV detection at 254 nm. Peak identities: 1, thiosulfate 4 ppm; 2, bromide 4 ppm; 3, chloride 4 ppm; 4, sulfate 2 ppm; 5, nitrite 4 ppm; 6, nitrate 4 ppm; 7, molybdate 10 ppm; 8, azide 4 ppm; 9, tungstate 10 ppm; 10, monofluorophosphate 4 ppm; 11, chlorate 4 ppm; 12, citrate 2 ppm; 13, fluoride 1 ppm; 14, formate 2 ppm; 15, phosphate 4 ppm; 16, phosphite 4 ppm; 17, chlorite 4 ppm; 18, galactarate 5 ppm; 19, carbonate 4 ppm; 20, ethanesulfonate 4 ppm; 21, acetate 4 ppm; 22, propionate 5 ppm; 23, propanesulfonate 4 ppm; 24, butyrate 5 ppm; 25, butanesulfonate 4 ppm; 26, valerate 5 ppm; 27, benzoate 4 ppm; 28, L-glutarate 5 ppm; 29, pentanesulfonate 4 ppm; 30, D-gluconate 5 ppm. (Reproduced with permission.[12])

the friction opposing the movement of a spherical particle in a medium of certain viscosity ($f = 6\pi\eta r$).[13] Since the limiting ionic equivalent conductances are always evaluated under constant conditions (dilute aqueous solutions, constant temperature), it is possible to show that the tabulated values of λ_{equiv} are just "thinly disguised" values of aqueous charge–to–radius ratios:

$$\lambda_{equiv} = (10^7 \times 1.6 \times 10^{-19}\, 96{,}487)/(6 \times 3.14 \times 10^{-2}) \times z_i/r_i \quad (2.7)$$
$$= 8.19 \times 10^{-7} \times z_i/r_i$$

Equation (2.7) is thus a more fundamental version of Equation (2.4). Using Equation (2.6) for expressing the term μ_{ion} and Equation (2.2) with its measurable parameters L_d, L_t, P, and t_{eof} for describing the term μ_{eof} in Equation (2.3), it is possible to arrive at a new relationship connecting apparent ionic mobilities μ_{obs} and hydrated ionic radii:

$$\mu_{obs} = (10^7 z_i e)/(6\pi\eta r_i) + L_d L_t/Pt_{m(eof)} = \mu_{ion} + \mu_{eof} \quad (2.8)$$

FUNDAMENTALS OF CAPILLARY ELECTROPHORESIS

The values of μ_{obs} or observed mobilities can be easily calculated from the migration times in Figure 2.10 by means of Equation (2.2). The hydrated ionic radii r_i are available from the literature.[10-13] It is of considerable interest to evaluate the correlation between the observed (apparent) mobility data from the electropherogram of thirty anions on one hand and the ionic radii on the other. Similar correlations had been attempted previously for peptides and proteins without success.[14-17] The interrelationship between the electrophoretic behavior of low-molecular-weight anions and their ionic radii can be expected to be more straightforward than that of more complex biomolecules. But even in the case of simpler ions a certain measure of judgment has to be applied. First, the dissociation behavior of weakly acidic anions can change the observed mobilities over a broad range. Second, the ionic radii are calculated from the limiting equivalent conductances assuming that the nominal charge of a species equals the effective charge in solution. While this may be correct for infinite dilution, in the carrier electrolyte used in Figure 2.10, the effective charges are likely to be different from the nominal charges. For these two reasons (dissociation behavior and unknown effective charges), we shall attempt a correlation of only a selected seventeen ions from the thirty-anion separation. The relevant data for those anions are summarized in Table 2.2.

Table 2.2. OBSERVED MOBILITIES (μ_{obs}) AND HYDRATED IONIC RADII (r_i) FOR SOME OF THE ANIONS FROM FIGURE 2.10

Anion	μ_{obs}[a]	$10^8 \, r_i$[b]
Bromide	0.0010690	1.0505
Chloride	0.0010569	1.0750
Iodide	0.0010340	1.0679
Nitrite	0.0010250	1.1423
Nitrate	0.0010010	1.1488
Azide	0.0009683	1.1886
Chlorate	0.0009269	1.2656
Fluoride	0.0008888	1.4806
Formate	0.0008710	1.5021
Chlorite	0.0008293	1.5773
Bicarbonate	0.0007669	1.8430
Ethanesulfonate	0.0007252	2.0711
Acetate	0.0007046	2.0054
Propionate	0.0006744	2.2910
Propanesulfonate	0.0006641	2.2110
Butyrate	0.0006396	2.5150
Benzoate	0.0006018	2.5317

[a] μ_{obs} was obtained from Equation (2.2) ($L_d = 52$ cm, $L_t = 60$ cm, and $P = 30,000$ V) using migration times observed in Figure 2.10.
[b] Ionic equivalent conductances from Reference 10 were converted to hydrated ionic radii using Equation (2.7).

The best fit for the data in Table 2.2 is given by the equation:

$$\mu_{obs} = 7.78 \times 10^{-12}/r_i + 3.3 \times 10^{-4} = \mu_{ion} + \mu_{cof} \qquad (2.9)$$

A theoretical calculation yields a value of 8.45×10^{-12} for the coefficient $(10^7 e)/(6\pi\eta)$ of Equations (2.6) and (2.9), in good agreement with the value found by curve fitting (Figure 2.11). The magnitude of the constant in the best-fit equation translates to $t_{cof} = 5.25$ min [$60 \times 52/30000 \times t_{cof}$ according to Eq. (2.2)], again in rather good agreement with the observed migration times of the water peak under the conditions of the separation of the thirty-anion mixture.

The fact that it is possible to correlate the migration data directly to a simple physical parameter like ionic radius distinguishes capillary electrophoresis of low-molecular-weight ions from all other separation methodologies carried out in the liquid phase. Unlike CE migration data, the retention times in liquid chromatography are a composite result of multiple parameters such as polarity, dipole moment, ionic strength, ion pairing, molecular size, pore size, surface area of stationary phase, etc. The theoretical explanation and prediction of retention behavior is possible either with semiempirical approximations (i.e., eluotropic series for normal- and reverse-phase liquid chromatography) or not at all (elution sequence in ion exchange). Thus capillary electrophoresis is not only much simpler experimentally, but it also has a much more straightforward theoretical foundation. This observation, together with much shorter analysis times, increased separation efficiency, and greater flexibility, adds up to a considerable practical advantage over all other separation techniques.

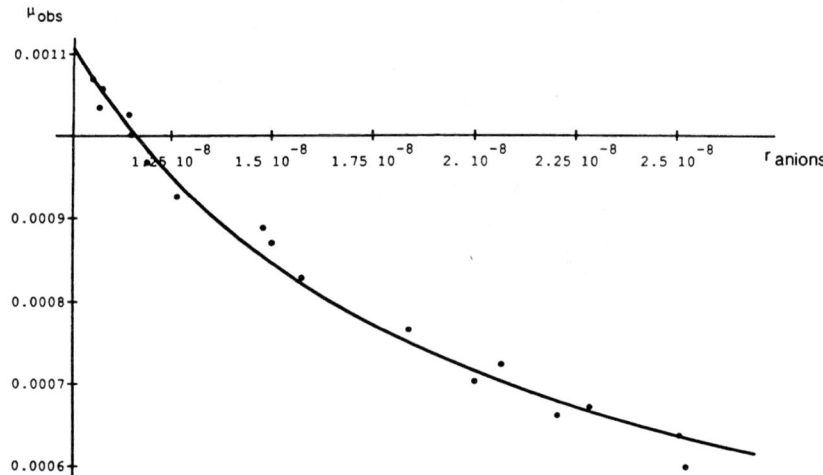

Figure 2.11 Curve fit of the migration data from the thirty anion separation by the theoretical Equation (2.6) relating actual mobilities and hydrated ionic radii.

2.3 Capillaries and Electroosmotic Flow

In the previous section we discussed the usefulness of electroosmotic flow in optimization of CE methods aimed at shortening analytical run times. We avoided any thorough explanation of the origins of the electroosmotic flow, saving it for the discussion that follows. A good quantitative and qualitative (and intuitive) understanding of the electroosmotic flow is essential for anyone working with modern capillary electrophoretic techniques.

In the first years following the inception of capillary electrophoresis, that is, between approximately 1967 and 1984, the potential role of electroosmotic flow in CE was not understood. Many reports published during that period discuss electroosmosis as an undesirable phenomenon interfering with the proper functioning of the CE apparatus. This was at least partly influenced by the fact that most early CE researchers were experienced users of isotachophoresis. In isotachophoresis electroosmosis hardly ever fulfills a useful purpose and therefore has to be minimized for optimal results. In tune with the prevailing opinion, the initial commercial suppliers of CE advertised chemically modified capillaries for their ability to eliminate electroosmotic flow. Electroosmosis still seems to inspire suspicion and negative reactions. Electroosmotic flow is deemed to cause irreproducible results by being "inherently unstable" or by interfering with the sample introduction, etc. Such attitudes may at least partly originate from an incorrect assumption that electroosmosis is a poorly understood phenomenon, unique to CE and related techniques.

The opposite is true. Electroosmotic flow is part of well-understood electrokinetic phenomena involving interactions between the charged walls of a cylindrical container filled with liquid. Let us consider the three best-known electrokinetic experiments depicted in Figure 2.12.

If a pressure drop is generated from one end of the cylindrical container to another, forcing the liquid inside it to flow, it is possible to measure the so-called *streaming potential* (Figure 2.12a). In the beginning of commercial air transportation, the streaming potential was responsible for sparking from the hoses used for pumping jet fuel. This potentially dangerous situation was eliminated by measures aimed at decreasing the charge on the refuelling hose wall, leading to lower values of the streaming potential and decreased incidence of sparking. Initially conductive additives were used. A more recent approach[18] involves chemical modifications of the hose walls. The relevant fact for our discussion of capillary electrophoresis is that a method for measuring the streaming potential had to be developed along with a calculation[19,20] linking the measured value of the streaming potential S [V] to the charge on the wall, which became to be known as the zeta potential ζ [V]:

$$S = 8.0 \times 10^{-5} \zeta\Delta/C\Lambda \tag{2.10}$$

where C is the electrolyte concentration in equivalents per liter, Δ the pressure drop causing the flow [atm], and Λ is the equivalent conductivity in cm^2

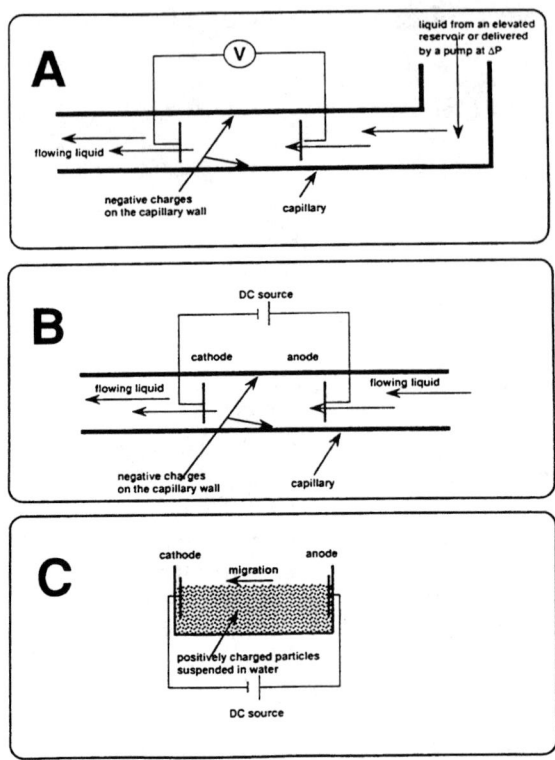

Figure 2.12 (A) Measurement of streaming potential; (B) electroosmosis; (C) electrophoresis.

Siemens equiv^{-1}. The above-mentioned safety problem with aviation fuel, along with other observed phenomena, led to measurements of streaming potentials and indirectly to the evaluation of zeta potentials on a large variety of materials.

In a slightly different variation of the previous experiment, it is possible to eliminate the pressure drop between the two ends of the cylindrical container (hose, pipe, tube, or capillary) and replace it by a voltage drop imposed, for example, by a pair of well-placed inert electrodes (Figure 2.12b). The outcome of such an experiment would be a flow of the fluid through the tube in a direction determined by the sign of ζ and at a flow rate that is proportional to the magnitude of the zeta potential. This phenomenon remained a curiosity with only limited practical use in the years preceding the commercial availability of today's capillary electrophoresis equipment.

Considerably more attention was given originally to the experimental arrangement in Figure 2.12c, which has become a mainstay of an independent discipline—colloid science. The rate of the movement of the particles in the electric field depends also on a zeta potential, with one important differentiation. The parameter ζ is a property of the surface of a single bead and not, as in Figure 2.12b, of the wall of the container. It can be calculated,[21] for example, from the velocity of a bead u_e [cm/sec] in a volume of water between two electrodes L[cm] apart and at a known potential P[mV]:

$$\zeta = 12.9\, u_e P/L \tag{2.11}$$

The controlled use of particles with known zeta potentials has become rather common in many industrial situations (paints, textile, electrically enhanced filtration and sedimentation, etc.). Correspondingly, there has been a strong need for zeta potential data for a large number of materials. This information has been generated and published in relevant journals (*J. Colloid Interface Sci., Colloid Polymer Sci.,* and *J. Appl. Polym. Sci.*) and in several books.[22-24]

The large body of data on the measured and calculated zeta potentials may be utilized directly for predicting the magnitude and direction of the electroosmotic flow v_{eof} in capillary electrophoresis according to the Smoluchowski equation[25]:

$$v_{eof} = -(\varepsilon\zeta/\eta)E \tag{2.12}$$

where ε is the dielectric constant of the electrolyte, ζ the zeta potential [V], η the viscosity [Poise], and E the applied electrical field [V/cm].

Figure 2.13 shows schematically how the polarity of the wall charges determines the direction of the electroosmotic flow. In the immediate vicinity of a capillary wall carrying negative charges, we observe two types of species: first, the oriented dipoles of water molecules, and second, the relatively larger hydronium cations. Since a certain portion of wall charges is covered only relatively "lightly" by the dipoles oriented with their positive sides toward them, it is possible that thermal and other disturbances create a ongoing movement of dipolar molecules and hydronium ions to and from the wall charges. This leaves a number of negative charges without any positive counterparts in their immediate vicinity, and thus a negative value of the zeta potential is achieved. The excess negative charges without a cation immediately next to them are counterbalanced from a larger distance by a corresponding number of cations. We are able to recognize two regions of the fluid profile next to the solid interface: first, the relatively rigid and stagnant inner Helmholtz layer closer to the wall, and second, the more diffuse and less organized outer Helmolholtz layer a little farther from the solid liquid interface. If both Helmholtz layers are situated in an electrical field with an orientation parallel to the solid surface, it is possible to induce a movement of less tightly attached hydronium cations in the outer Helmholtz layer toward the cathode. These hydronium ions are interlocked within the water structure, and so their movement is possible only if the bulk of the

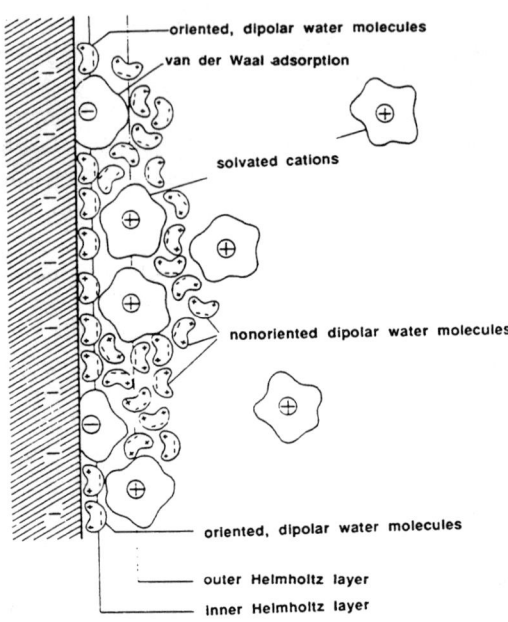

Figure 2.13 Interactions of the wall charges with ions and molecules in the liquid phase. The origins of the electroosmotic flow. (Adapted from Reference 26.)

water is moving along, resulting in electroosmotic flow. As Figure 2.14 shows, this kind of flow exhibits a different hydrodynamic profile than the one usually observed if a liquid flow is generated by a pressure differential between the inlet and outlet of a capillary. As we shall see (Section 2.5), the special electroosmotic flow profile is one of the sources of increased separation efficiency in capillary electrophoresis compared to liquid column chromatography.

As illustrated in Figure 2.7, different capillary materials generate different values of the zeta potential, corresponding to a different direction and magnitude of the electroosmotic flow. During the pioneering stages of the method, researchers evaluated several capillary materials—fused silica, borosilicate glass, fluoroethylenepropylene, and tetrafluoroethylene, to name just a few. While the results were encouraging with most of these materials, current users of CE prefer fused silica capillaries (see Figure 2.15), almost to the exclusion of any alternatives. The explanation for this lies in the ease of use and considerable elegance of modern fused silica capillaries.

It could be argued that the rapid expansion of CE was made possible by the commercial availability of the polyimide-coated, narrow-bore capillaries. The two most important suppliers of fused silica capillaries are Polymicro Technologies, Inc., in Phoenix, Arizona, and Scientific Glass Engineering, Inc., in Austin, Texas. As we shall see in Section 2.5, the narrow

FUNDAMENTALS OF CAPILLARY ELECTROPHORESIS

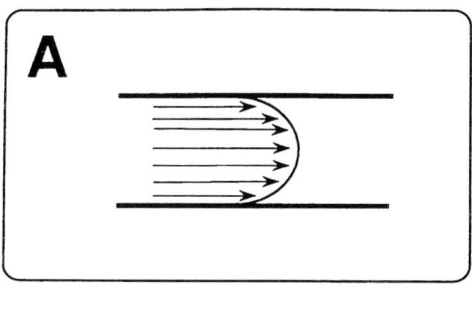

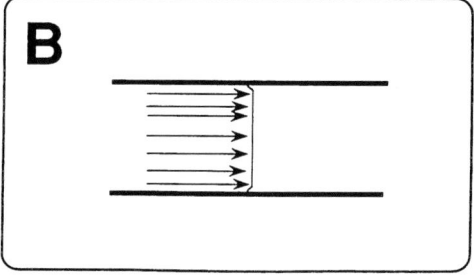

Figure 2.14 Hydrodynamic profiles of (A) flow created by a pressure difference and (B) electroosmotic flow.

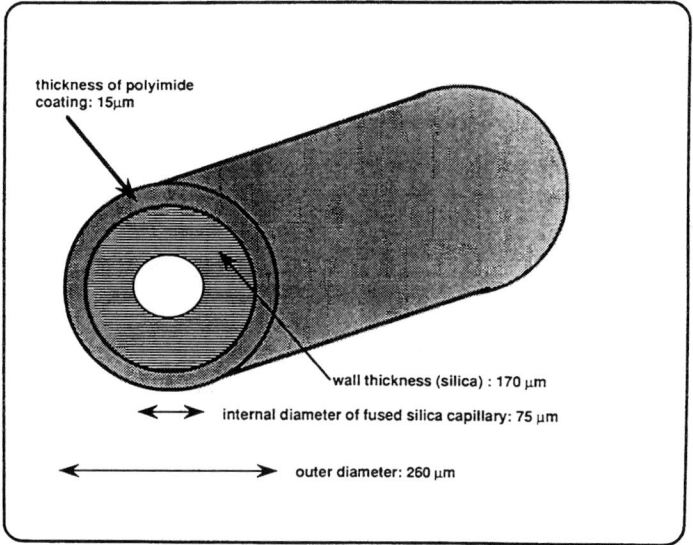

Figure 2.15 Cross section of a polyimide-coated fused silica capillary.

diameter is a necessary precondition for an efficient Joule heat dissipation through the wall of a capillary. At too-large diameters (above ca. 120 μm) it becomes practically impossible to impose the electrical field necessary for the movement of ionic species in the direction of the corresponding electrode. Because of an insufficient heat dissipation, the electrolyte inside the capillaries overheats and begins to boil, leading at first to strong fluctuations of the electrical current and finally to its interruption due to the numerous bubbles present in the advanced stages of evaporation.

The polyimide coating renders the fused silica material, which by itself is very brittle and fragments easily, unusually flexible. The same capillary, which without the protective coating breaks easily, can be wound into narrow loops due to the flexibility of the polyimide material.

As discussed in more detail in Section 3.2, a short length of the protective polyimide coating has to be removed to enable optical detection in all current CE systems. Technically, fused silica is made by melting the purest forms of naturally occurring silicon dioxide to temperatures reaching approximately 1100°C. The silicon atoms at the surface of solid forms of silica show a strong tendency to adjust to a more relaxed tetrahedral configuration with the help of oxygen, which at room temperature is usually accomplished by binding with the hydroxyl groups stemming from water. The hydroxide groups bound to silicon atoms are termed *silanols*. Figure 2.16 shows several different types of silanol groups described in the literature[28].

At temperatures used for manufacturing fused silica capillaries (i.e., 1100°C and higher), the silanols condense to water, allowing the two neighboring silicon atoms to create a relatively strained siloxane bond:

$$\begin{array}{c} H \\ / \\ O \\ | \\ Si \end{array} + \begin{array}{c} H \\ / \\ O \\ | \\ Si \end{array} = Si \begin{array}{c} O \\ / \ \backslash \end{array} Si + H_2O \qquad (2.13)$$

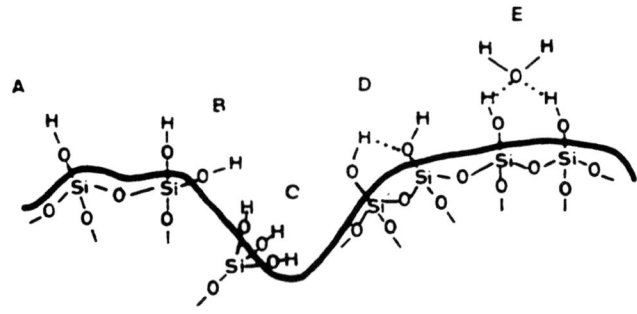

Figure 2.16 Various types of silanol groups at the surface of fused silica.

FUNDAMENTALS OF CAPILLARY ELECTROPHORESIS

One of the suppliers[27] of fused silica capillaries specifies less than 1200 ppm hydroxyl (silanol) content in the bulk of their fused silica material. After cooling down from temperatures at which the capillaries are being drawn from the bulk material and exposed to the moisture in the atmosphere, the surface of the silicon dioxide begins a slow process of relaxation of the strained siloxane groups by accepting hydroxide groups from water, gradually developing a coverage of silanol groups, as shown in Figure 2.16.

Another important property of silanol groups at the surface is their acidity. The pK value of the first dissociative equilibrium of orthosilicic acid is:

$$Si(OH)_4 = Si(OH)_3^- + H_3O^+, \qquad pK = 9.9 \tag{2.14}$$

The pK value of the surface silanol groups obtained from infrared data[29] is ~7.7. The difference from the value for the orthosilicic acid in solution is explainable by the $(d-p)\pi$ interactions along the chain of Si–O bonds in the silica structure, resulting in partial positive charge at the silicon atom of the dissociated silanol group.[28] Due to the differences in the chemical environment of the single silanol groups indicated in Figure 2.16, it is probably correct to think of pK for silanol groups as being in a certain range rather than at a single defined value. In pure water and in other neutral–to–alkaline solutions, fused silica exhibits a negative value of the zeta potential. The absolute value of the potential is dependent on the number of dissociated silanol groups and will increase with pH. At ~ pH 2 the dissociation of all surface silanol groups is completely suppressed, and the value of the zeta potential approaches zero. Correspondingly, the electroosmotic flow at pH 2 should also be minimal or approaching zero. Below pH 2, the silica wall exhibits a positive charge according to the reaction:

$$\equiv Si\text{-}OH + HX = \equiv SiOH_2^+ + X^- \tag{2.15}$$

The positive value of the zeta potential and the validity of Equation (2.15) could be verified by direct electrophoretic measurements at pH 1.2.[28]

It is thus clear that, depending on the previous use of a capillary and its storage conditions (humidity, temperature, duration), the number and dissociation of silanol groups can vary over a wide range, giving large changes of the zeta potential. The changes of the zeta potential affect the reproducibility of migration times under CE conditions. The gradual conversion of capillary wall surface siloxane groups to silanols [Eq. (2.13)] is a lengthy process. In a dry state, it is measured in weeks rather than days. Without proper conditioning, the conversion to silanol groups may be too slow, even if the capillary walls are exposed to a liquid electrolyte. In order to achieve a uniform state of fused silica capillaries at the beginning of their use in CE, many workers have adopted an alkaline pretreatment of the inner capillary wall, for example, by forcing a 10–100 μl volume of 0.1 M sodium hydroxide through the capillary. Without such alkaline pretreatment, the fused silica capillaries were shown to exhibit a considerable hysteresis effect within the μ_{eof} versus pH coordinates (see Figure 2.17).

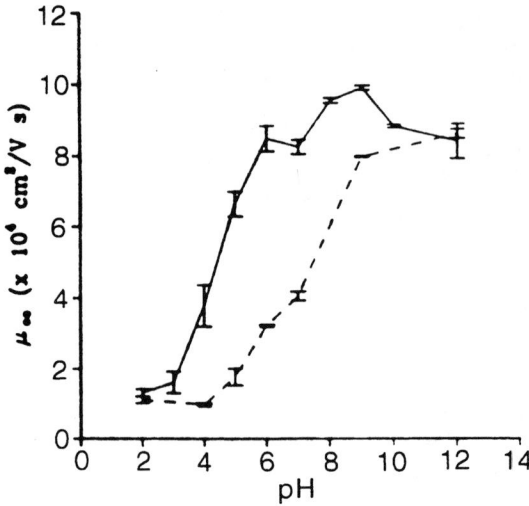

Figure 2.17 Hysteresis of electrophoretic mobilities (μ_{eof}) with pH in untreated capillaries. (Reproduced with permission.[30])

For example, at pH 6, the electrophoretic mobility of the same capillary differs by a factor of four depending on the previous history of unconditioned capillary (ca. 2×10^{-4} μ_{eof} going from acidic to alkaline solutions vs. ca. $8 \times 10^{-4} \mu_{eof}$ in the reverse direction). Significantly, the same report also demonstrates how the hysteresis can be avoided by pretreatment with an alkaline solution. The pH range of 4 to 6 is identified as requiring special attention for successful prevention of pH hysteresis.

Practical experience shows that after the initial alkaline conditioning, well-defined and unchanging storage protocols are sufficient in the majority of the cases for a good constancy of electrophoretic mobilities and a satisfactory reproducibility of migration times in capillary electrophoresis in fused silica capillaries. As determined by the acidic properties of the silanol groups, the electroosmotic flow in fused silica capillaries is usually directed toward the negative electrode - cathode, the predominating species in the secondary Helmholtz layer being the hydronium cations. As explained in the discussion of Figure 2.8, such a direction of the flow is conducive to a rapid analysis of cationic analytes (see also the separation in Figure 2.18) but quite counterproductive for the analysis of anionic analytes requiring electroosmotic flow toward the anode.

The reversal of the sign of the zeta potential in fused silica capillaries after the addition of a long-chain alkyl ammonium salts was first reported by Everaerts and co-workers in 1983.[31] The authors evaluated the effectiveness of the suppression of electroosmotic flow by several different additives with the objective to find new ways to improve results in isotachophoresis. The

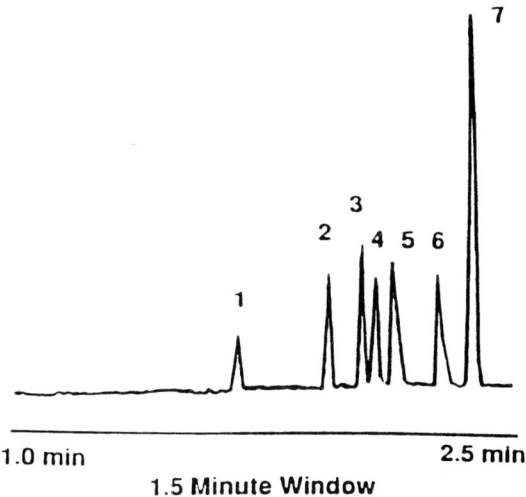

Figure 2.18 Alkali metal and alkaline earth cations separated using the natural direction of electroosmotic flow. Fused silica capillary 50 × 65 cm × 75 μm, + 20 kV. Indirect detection at 214 nm. Carrier electrolyte: 5 mM UV CAT 1 (Waters), 0.021 mM citrate. Peak identities: (1) potassium, (2) barium, (3) strontium, (4) sodium, (5) calcium, (6) magnesium, (7) lithium.

measurements of streaming potentials for the calculation of ζ were carried out not only in fused silica but also in Teflon capillaries. The evaluated additives along with the values of calculated Zeta Potentials are listed in Table 2.3.

Of the additives listed in Table 2.3, only CTAB reverses the polarity of the zeta potential on the walls of both Teflon and fused silica capillaries, causing the electroosmotic flow to move from the cathode toward the anode.

Table 2.3. VALUES OF ζ POTENTIAL OBTAINED BY EVERAERTS AND CO-WORKERS IN 1983[a]

Additive	Concentration	ζ (mV)	
		PTFE	Fused Silica
None		−41	−25
Polyvinylpyrrolidone (PVA)	0.1%	−3	−4
Triton X-100	0.1%	−4	−9
Poly(vinylalcohol) (PVA)	0.05%	−1	−2
Hydroxyethylcellulose (HEC)	0.2%	−1	−1
Hydroxypropylmethylcellulose (HPMC)	0.2%	−2	−2
Cetyltrimethylammonium bromide (CTAB)	0.1 mM	+40	+18

[a] In 0.01 M histidine chloride, pH 6.

In 1987, Tsuda et al.[32] investigated the effect of CTAB on the walls of a fluoroethylene capillary and were the first to document the utility of electroosmotic flow reversal for the separation of anionic analytes.

In 1989, Zare and co-workers[33] investigated another additive having a better water solubility than CTAB, tetradecyltrimethylammonium bromide, or TTAB. They found that in fused silica capillaries with their generally higher electroosmotic flow rates than in fluoropolymeric capillaries, a complete analysis of mixtures of fast and slow anions was possible only after reversing both electric field and electroosmotic flow (Figure 2.19).

Under "standard CE conditions" (Case B, Figure 2.8), only the slow anions are detected; correspondingly, the two fastest anions, formate and acetate, are missing in the upper electropherogram. The polarity reversal—electropherogram (b)—without flow reversal creates conditions under which only the fastest of the six carboxylates can be detected (Cases I and J, Figure 2.9). The optimum conditions for the separation of both fast and slow anions are achieved by reversing not only the polarity but also the direction of the electroosmotic flow electropherogram (c) and Cases D and E in Figure 2.9.

In the year preceding the report by Zare's group, Altria and Simpson[34] published results of a comprehensive investigation of electroosmotic flow modification in fused silica capillaries with the help of CTAB. They were

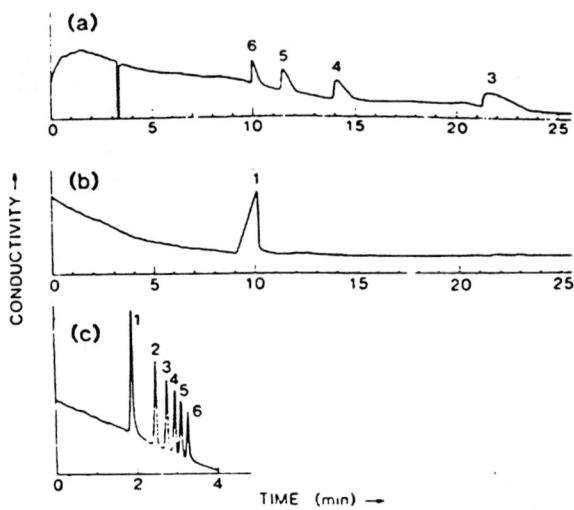

Figure 2.19 Separations of six carboxylates. (a) "Standard CE conditions": injection side positive at +20 kV, electrolyte 10 mM MES/His pH 5.9. (b) As (a) but with negative injection side −20 kV. (c) Polarity as in (b) with 0.5 mM TTAB added to the electrolyte. Fused silica capillary: L_t 42 cm, L_d 40 cm × 75 μm. Conductivity detection. Peak identities: (1) formate, (2) acetate, (3) propionate, (4) butanoate, (5) pentanoate, (6) hexanoate. All sample components at 5×10^{-4} M. (Reproduced with permission.[33])

able to achieve electroosmotic mobilities ranging from $+18 \times 10^{-4}$ to -12×10^{-4} cm^2 V^{-1} sec^{-1}, concluding that "the combination of negative potential and solutions of CTAB may be useful for anion analysis."

We have thus seen that, in agreement with the principles postulated in Section 2.2, the CE analysis of anions requires a different system polarity and a different direction of electroosmotic flow than cation analysis. Both sets of analytical conditions are achievable with standard fused silica capillaries. The origins of the electroosmotic flow in a capillary having a negative value of the zeta potential (i.e., by dissociation of silanol groups on fused silica or by the adsorption of negative ions on the surface as in the case of fluoropolymeres) are illustrated in Figure 2.13.

Our understanding of charge reversal on the fused silica capillary walls is largely based on the fundamental work by D. W. Fuerstenau and co-workers[19,20] published between 1953 and 1964. Employing a scheme similar to that in Figure 2.12a, Fuerstenau's group measured streaming potentials in alkylammonium salt solutions flowing through a bed (plug) consisting of "Brazilian quartz beads." They discovered that the zeta potential on their quartz beads—calculated from Equation (2.10)—could be reversed in alkyltrimethylammonium salt solutions with alkyl chain lengths between C10 and C18 (Figure 2.20).

In solutions of ammonium acetate and decyltrimethylammonium acetate, the charge on the quartz wall remains negative even at the highest concentrations of the additive. The reversal of the charge also depends on the concentration of a suitable additive. Below $\sim 10^{-5}$ moles per liter the sign of the

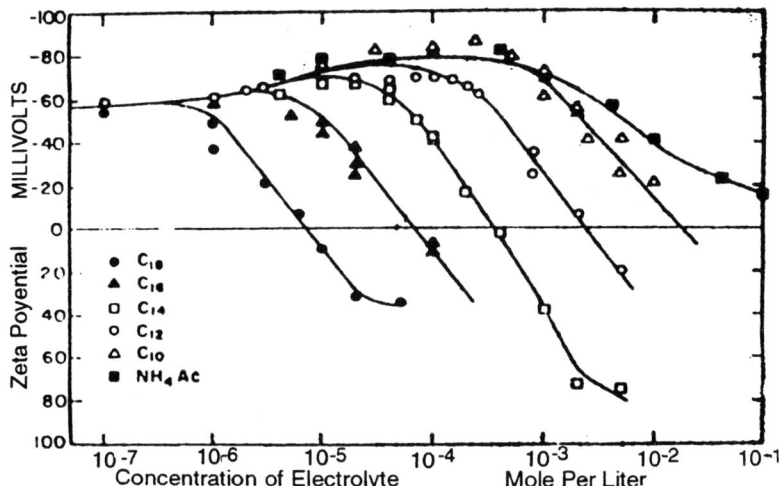

Figure 2.20 Effect of hydrocarbon chain length on zeta potential of quartz in solutions of alkylammonium acetates and in solutions of ammonium acetate. (Reproduced with permission.[20])

zeta potential remains negative with all ammonium salts investigated. To explain the charge reversal on the quartz/liquid interface, Fuerstenau postulates a concentration-dependent formation of so-called hemimicelles, as shown in Figure 2.21.

The creation of micelles is observed in aqueous solutions, for example, for anionic surfactants, if their concentration exceeds a certain level, termed the *critical micelle concentration*. The bulk concentration necessary for hemimicelle formation was defined by Fuerstenau as the point in Figure 2.20 where the ζ–log C curve of an additive separates from the curve for ammonium acetate. The values of such concentrations were in a similar ratio to the critical micellar concentrations for four of the investigated compounds, opening the possibility of estimating the required hemimicellar concentrations from a large number of published critical micellar concentrations.[35]

From Figure 2.21 it is obvious that once the hemimicelles are formed, the charge of the wall potential is reversed. Furstenau's work thus provides a fundamental insight into the mechanism of reversal of polarities of the zeta potentials and of the change of direction of the electroosmotic flow in capillaries, especially since the values of required hemimicellar concentrations in Figure 2.20 are in good agreement with the observed conditions for the reversal of electroosmotic flow in capillary electrophoresis.[12,33,34]

The recent increased use of additives in capillary electrophoresis indicates a possible analogy to the development in the early days of modern liquid chromatography. Just as the first expansion of CE was based on fused silica capillaries, the first examples of HPLC separations were achieved on silica particles due to their excellent mechanical properties. To extend the

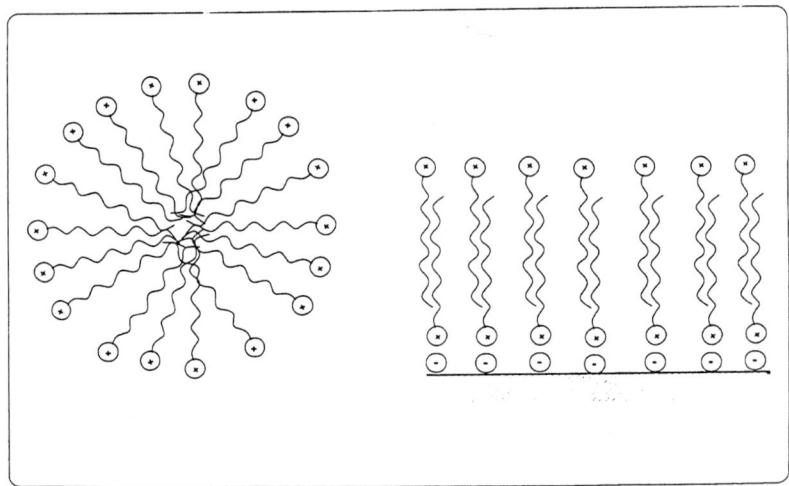

Figure 2.21 Micelles and hemimicelles.

scope of the method, researchers soon began modifying the silica particles, first by additives to the mobile phase and later by either applying various types of polymeric coatings to the silica surface or by covalently attaching functional groups to surface silanols. If the present frequent use of additives in CE can be taken as a basis for extrapolation, widespread application of chemically modified capillaries can be expected in the next stage of development. However, one should also keep in mind the major differences between separation principles in HPLC on one hand and in CE on the other. Certainly the main point of differentiation is that in HPLC the chemistry of "naked" dynamically and permanently coated silicas undergoes a direct interaction with the separated species as the main part of the separation mode. One example is ion exchange separations, where ionized solutes interact with ion exchange groups bound through so-called "spacers" (usually trialkylsilyl groups) to the silica surface. Another frequent example is reversed-phase separations, based on the different degree of lipophilic interaction of organic molecules with alkyl chains bound to the silica stationary phase. In complete contrast to HPLC, in CE any interaction of the analytes with the silica surface of the capillary wall is considered undesirable. This is true not only for the "naked" unmodified capillaries, but also for dynamically coated (hemimicelles) and permanently coated capillaries as well. In CE, chemical modifications of capillaries serve the sole purpose of influencing the electroosmotic flow. The only exception from this rule is the modification of capillaries to prevent reversible or even irreversible adsorption of some large molecules (i.e., proteins, peptides, naturally occurring polymeric phenols, etc.) on the capillary wall. The analogy between HPLC and CE chemistries is thus somewhat less general than postulated at the beginning of this paragraph. The similarity lies mostly in the need, existing in both HPLC and CE, to make useful modifications of silica, originally achieved by dynamic coating, more permanent by covalently bonding molecules or polymers to the silica surface. Today the researchers engaged in the development of chemistries for permanent modification of fused silica capillaries have an advantage over early workers in HPLC. Unlike 15 or 20 years ago, there is now a large body of experience and knowledge of reactions of silanol groups.[28] However, the narrow geometries of fused silica capillaries require considerable changes of the apparatus used to carry out the chemical reactions on the surface of silica (see Figure 2.22).

A dry (moisture-free) solution of dimethyloctadecylchlorosilane in N, N-dimethylformamid is placed in a polypropylene centrifuge tube (A). A capillary (B) is inserted through a disposable pipette tip (C). At the bottom of the tip is placed a small ball of cotton (D), overlayed by calcium chloride (E) to keep atmospheric moisture from entering the reagent. Another layer of cotton (F) keeps the calcium chloride in place. The capillary end inside the tube containing the reagent is held ~30 cm higher than the other end. At the beginning of the procedure, the reagent is sucked through the dried capillary by means of a syringe attached with a length of narrow-diameter poly-

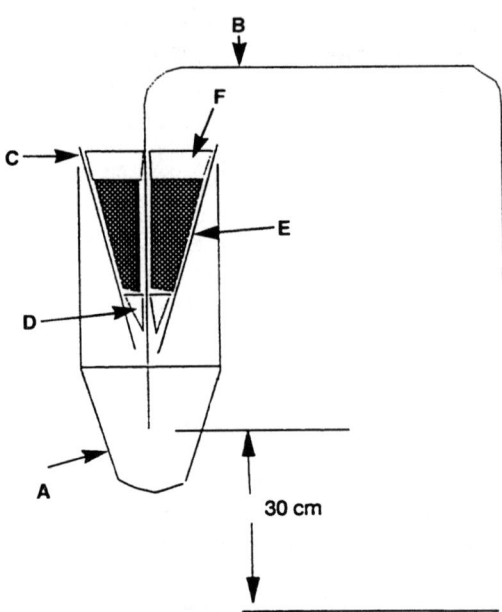

Figure 2.22 Apparatus for performing chemical reactions on the inner surface of capillaries. See text for explanation of symbols. (Reproduced with permission.[36])

ethylene tubing. To carry out the chemical reaction, the reagent was allowed to run through the capillary by gravity. Figure 2.23 illustrates several possible chemistries of unmodified and permanently coated capillary walls.

Chemical modifications of capillaries can be achieved not only by a "point-by-point" reaction of single silanol groups with suitable silanes, but also by at least two different polymerization procedures, as shown in Figure 2.24.

The surface polymerization approach is based on a polymerization reaction at the capillary wall relying on the microscopic roughness of fused silica and on the rigidity of the polymeric coating to keep it in place. The anchored polymeric layer appears to be an optimum solution of the problem of both stability and completeness of chemical modification. This approach has been shown by Hjerten,[37] with a bifunctional reagent (γ-methacryloxypropyltrimethoxysilane). The first functional group (silane) can be covalently attached to the interior wall, and the second is capable of being polymerized (methacrylate).

Extending the discussed analogy between the past development of HPLC and the expected direction of development in CE a little further, it is also possible to expect an increased use of polymeric materials for some time in the future. The polymeric packings in HPLC were introduced at a relatively

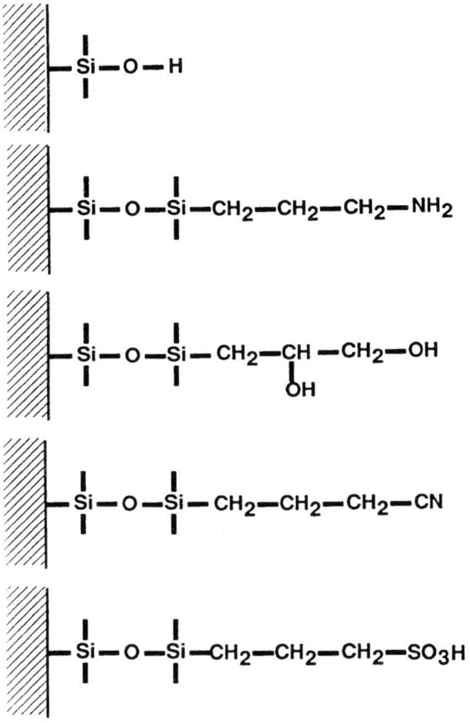

```
|—Si—O—H

|—Si—O—Si—CH₂—CH₂—CH₂—NH₂

|—Si—O—Si—CH₂—CH—CH₂—OH
              |
              OH

|—Si—O—Si—CH₂—CH₂—CH₂—CN

|—Si—O—Si—CH₂—CH₂—CH₂—SO₃H
```

Figure 2.23 Various chemistries for controlling electroosmotic flow in capillaries.

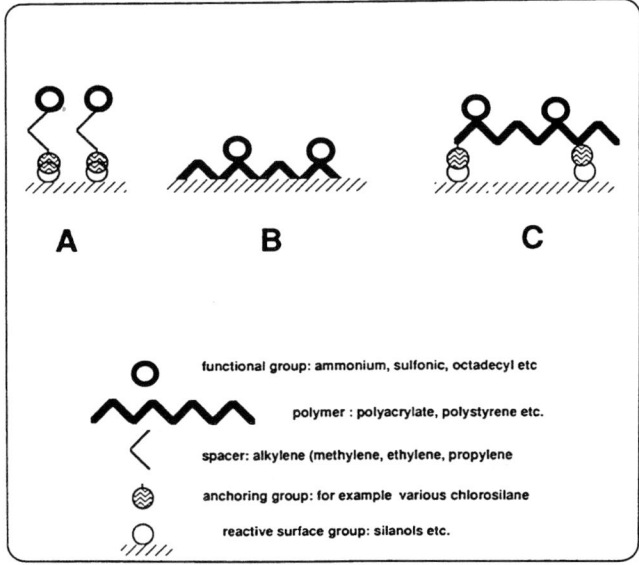

Figure 2.24 Three different approaches to chemical modification of capillary walls.

Table 2.4. COMMERCIAL SOURCES OF CE ADDITIVES AND MODIFIED CAPILLARIES[a]

Type	Chemistry	Function	Supplier
Additive	Zwitterion	Prevents adsorption to capillary wall	Waters[b]
Additive	Alkylammonium	Reverses EOF	Waters[b]
Capillary	C18	Prevents adsorption	Bio Rad[c]
Capillary	Anion exch.	Reverses EOF	Sarasep[d]

[a] Gel-filled capillaries are not included in the table.
[b] Waters Chromatography Division of Millipore, 34 Maple St., Milford, MA 01757.
[c] Bio-Rad Lab., 1414 Harbour Way South, Richmond, CA 94804.
[d] Sarasep, 1600 Wyatt Dr., Suite 10, Santa Clara, CA 95054.

late stage, when it became obvious that certain, usually alkaline, mobile phases, could not be employed with silica-based packing materials. The recommended range for silica-based chromatographic packings is between pH 2 and 8.5. In a similar fashion, it will probably become desirable to find new capillary materials for use with carrier electrolytes having pH higher than 12. Whenever this is going to occur, the researchers involved in the search for new capillaries will have available a large amount of experimental data, allowing a reliable prediction of electroosmotic properties of materials in question. Examples of such data are included in Figure 2.7 and Table 2.3 of this chapter.

According to some recent reports, the electroosmotic mobility can be controlled not only by chemical means (i.e., electrolyte additives or chemical modification of capillaries), but also by applying an external electric field, perpendicular to the direction of the field generated to effect the electrophoretic separation. Utilizing the experimental setup in Figure 2.25, Lee and co-workers[38,39] demonstrated changes of magnitude as well as directional changes of the electroosmotic flow in fused silica capillaries. At the time of

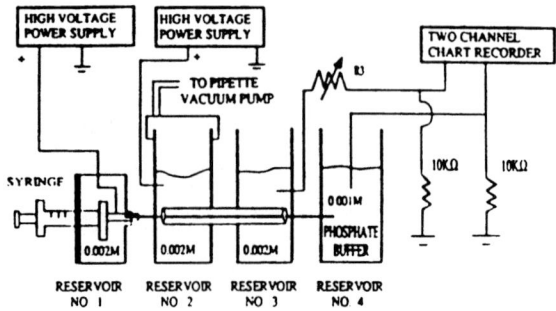

Figure 2.25 Apparatus for examining the effect of external field on the electroosmosis. (Reproduced with permission.[38])

publication of this book a practical application of this approach has yet to be demonstrated.

Another group of researchers[40] discussed a similarity between the apparatus for control of electroosmosis and certain semiconductor devices—"metal–oxide–semiconductor field-effect transistor, or MOSFET." A new designation, "metal–insulator–electrolyte–electrokinetic field-effect device, or MIEEKFED," was proposed for the capillary electrophoresis with electric control of electroosmotic flow.

2.4 Separation Modes in Capillary Electrophoresis

The preceding two sections discussed the underlying principles of the transport of ions in capillaries filled with a suitable electrolyte and placed in an oriented electric field. Main contributions to the movement of ions under such conditions are, first, the electrophoretic mobility of the analyte ion, depending on the charge/ionic radius ratio and, second, the electroosmotic flow depending on the polarity and magnitude of the charge on the capillary walls. The relevant combinations of electroosmotic and electrophoretic vectors are summarized in Figures 2.8 and 2.9.

Since electrophoresis and electroosmosis are the main sources of selectivity in CE, it is useful to define two of the main separation modes with the help of the two possible configurations of the electrophoretic and electroosmotic vectors. Depending on whether or not both kinds of mobility vectors point in the same direction, we can thus distinguish between a **coelectroosmotic** (same direction of electrophoresis and electroosmosis) and **counterelectroosmotic** CE. The third main separation mode is then the molecular sieving in gel filled capillaries. The capillary gel electrophoresis is performed without any contribution by electroosmosis.

For many ions, the mere combination of electrophoresis and electroosmosis does not offer a sufficient selectivity for a satisfactory resolution from other peaks in the sample mixture. The analysts are frequently forced to refine CE separations by utilizing analyte interactions with suitable components in the electrolytes or as in the case of separation of particles, by inducing interaction between the particle to be separated and the capillary wall. An overview of several possible contributing mobilities to the resulting observed mobility was given in Figure 2.5. Questions that arise are: How are the different combinations of mobility vectors defining a CE separation mode accomplished under practical conditions? What is the interrelationship among the reported separation modes? Is it possible to combine more than one mode for a difficult separation? In order to obtain some answers to these questions, let us consider all types of combinations of electroosmotic and electrophoretic vectors reported in the literature.

Table 2.5 contains a list of most common descriptive terms for CE separations. This listing is by no means complete, as we shall see in the following

Table 2.5. GLOSSARY OF TERMS FOR SEPARATION MODES

Name	Abbreviation	Description
Capillary electrophoresis	CE	General name for all electrophoretic separations in capillaries. Also called high-performance capillary electrophoresis (HPCE)
Capillary zone electrophoresis	CZE	Originally synonymous with CE. Used mostly for counterelectroosmotic separations. Another version of the name is high-voltage capillary zone electrophoresis (HVCZE)
Capillary ion analysis	CIA	Coelectroosmotic CE with indirect UV detection
Capillary gel electrophoresis	CGE	CE separations in gel-filled capillaries
Countermigration capillary electrophoresis	CMCE	Molecular sieving of macromolecules under CZE conditions
Micellar electrokinetic capillary chromatography	MECC	CZE separations combined with interaction of analytes with micelles. Another term is micellar electrokinetic chromatography (MEKC)
Ion-exchange electrokinetic chromatography	IEEKC	Counterelectroosmotic CE enhanced by addition of polymeric ion exchangers to carrier electrolytes

discussion. Currently, there are not enough widely recognized terms to provide unequivocal names for all available CE separation modes.

2.4.1 Electrophoretic Molecular Sieving in Narrow-Bore Capillaries

This sieving can be viewed as a link between the traditional slab gel electrophoresis and the more recent version of electrophoresis performed in capillaries. Such separations are used exclusively for biologically relevant large-molecular-weight compounds and are thus somewhat outside of the scope of this book. A brief discussion of this mode is offered solely for the sake of completeness.

2.4.1.1 Capillary Gel Electrophoresis

Analytes carrying an electric charge, usually large biomolecules (proteins, peptides, fragments of nucleic acids, etc.) move through a gel-filled capillary in an applied electric field. The rate of migration of a charged macromolecule through the gel can be characterized by coefficients of diffusion.[41]

Larger molecules experience more resistance or "friction" during their migration through suitable (optimum degree of cross-linking) gels, resulting in longer migration times in comparison with smaller molecules, as shown in Figure 2.26. The electroosmotic flow is considered undesirable for this

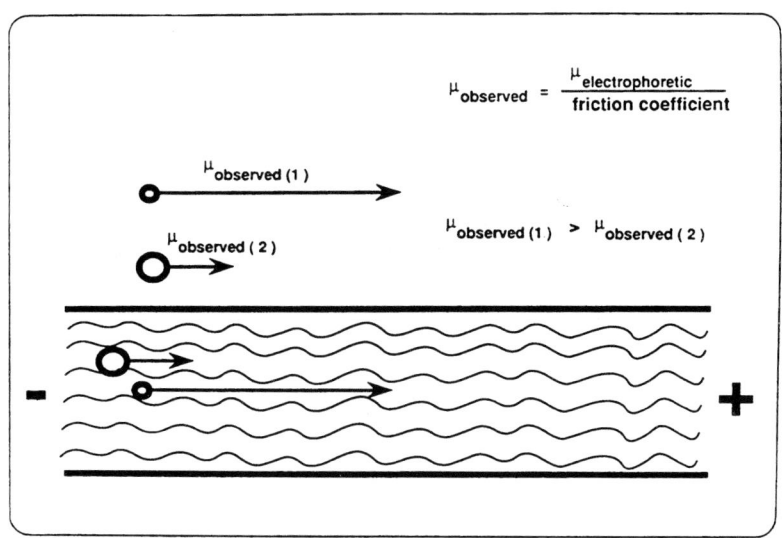

Figure 2.26 Gel electrophoresis in capillaries.

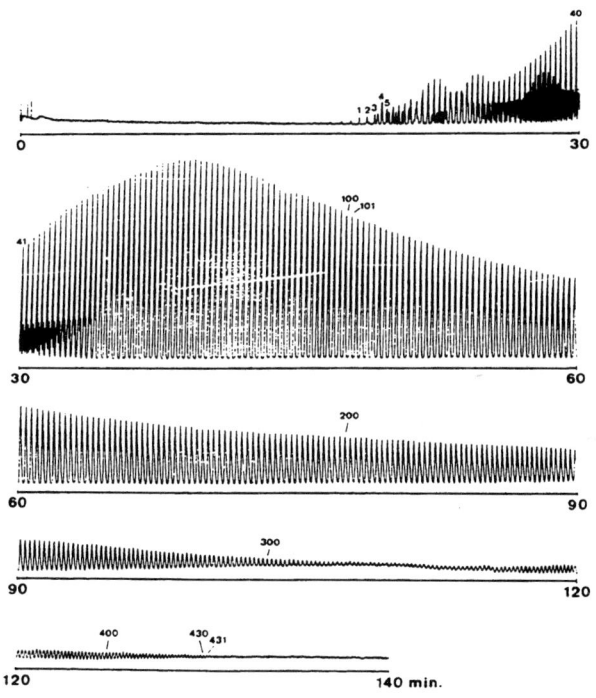

Figure 2.27 CGE separation of poly(uridine 5′-phosphate). The numbers in the electropherogram indicate the number of nucleotide units. Polyacrylamide gel capillary 45 cm × 100 μm. Buffer: 0.1 M tris, 0.025 M boric acid, 7 M urea. Separation conditions: 300 V/cm. Detection: UV at 260 nm. Electrokinetic sampling at 5 kV and 2 sec. (Reproduced with permission.[46])

separation mode. The co- and counterelectroosmotic designations cannot be applied.

Gel electrophoretic separations in capillaries were first achieved and further advanced by Karger and co-workers[42,43] as well as by Schomburg's group.[44] At its most recent stage of development, capillary gel electrophoresis (CGE) is discussed as an alternative to classical slab gel separations used in DNA sequencing. The gel-filled capillaries for such applications have become commercially available.[45] The high resolution of such systems is shown in Figure 2.27.

2.4.1.2 Capillary Electrophoresis in Sieving Carrier Electrolytes

This represents an attempt to overcome the problems usually encountered in CGE. It is relatively difficult to make gel-filled capillaries reproducibly. Additionally, the current gel-filled capillaries do not last as long as other types of CE capillaries or HPLC columns used for similar separations. Sieving CE, also termed counter-migration capillary electrophoresis (CMCE), uses solutions of polymers capable of interacting with the analytes. The methodology was first described by Zhu and co-workers.[47] CMCE is usually carried out under the conditions illustrated in Figure 2.28.

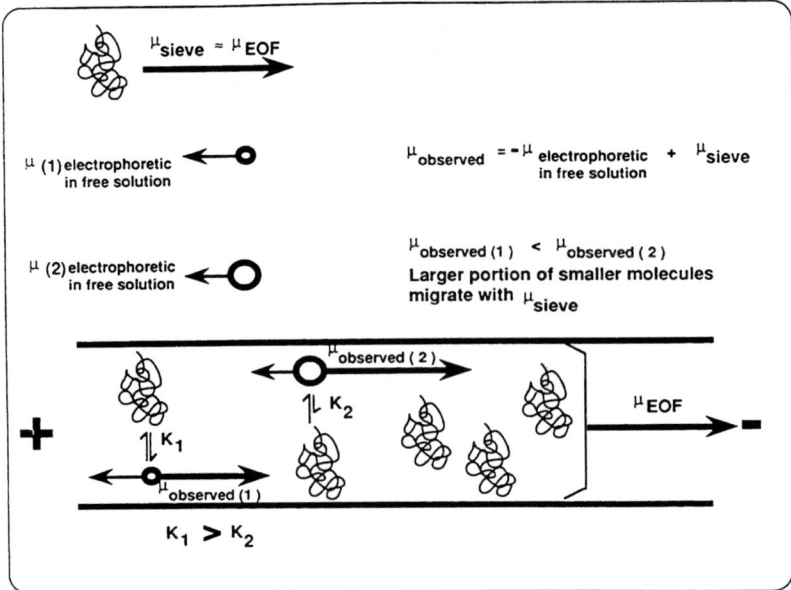

Figure 2.28 Molecular sieving in dissolved polymers under the conditions of capillary electrophoresis.

A major difference to CGE, discussed in the preceding paragraph, is in the sequence of migrating polymeric analytes. In CMCE the high molecular analytes reach the detector first, followed by lower molecular sample components. This is attributed to a deeper penetration of smaller macromolecules into the loose webbing of the dissolved additive molecules that are moving in the opposite direction with the electroosmotic flow. In consequence, the inherent migration of smaller molecules is reversed to a larger degree than that of bigger molecules. This attraction enhances the natural electrophoretic differentiation, which is also due mostly to differences in molecular size. Additionally, as found by Bonn and co-workers,[48] enhanced interactions between analytes and polymeric additives can lead to improvements in overall efficiency and selectivity of sieving separations. (See Figure 2.29.)

In this separation, HEC serves as a sieving agent. The addition of ethidium bromide is made to improve interactions of DNA fragments with HEC. Spectroscopic and hydrodynamic studies[49] of complexes of ethidium bromide and DNA suggest that the phenanthidium ring of ethidium bromide intercalates between two G–C base pairs, derotating the DNA duplex. Additionally, cationic charges on nitrogen atoms interact with phosphate

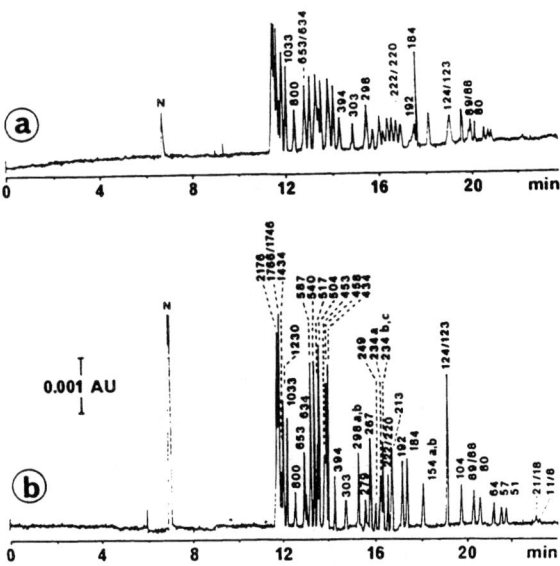

Figure 2.29 Effect of ethidium bromide on the CMCE separation of a complex mixture of DNA standards. (a) CMCE without ethidium bromide. The electrolyte components are: 10 mM TRIS borate, 0.5% hydroxyethylcellulose (HEC), 25 mM NaCl, and 0.1 mM EDTA at pH 8.7. (b) CMCE with 1.27 μM ethidium bromide in the electrolyte having otherwise the same compositions as previously. Capillary: 72 × 52 cm, 50 μm fused silica. Voltage: +15 kV. Temperature 35°C. Vacuum sampling 8 sec. UV detection at 260 nm. (Reproduced with permission.[49])

groups. The overall resulting change of DNA conformation enhances its interaction with the hydrophilic cellulose derivative. Strictly speaking, molecular sieving by CMCE represents already one of the auxilary separation modes explained in section 2.4.4. It is only discussed here because of its close relationship to CGE.

2.4.2 Counterelectroosmotic Capillary Electrophoresis

This is a separation mode in which the mobility of at least one analyte is in the opposite direction to that of the electroosmotic flow. In most published reports to date, the term CZE is used to describe that type of electrophoresis. In its most frequent version CZE is carried out with the positive polarity of the injection side. In such a case the cations migrate the fastest, followed by a solvent peak marking the electroosmotic velocity. Depending on the relative electrophoretic and electroosmotic velocities, it is also possible to obtain peaks for some of the slowest anions in the mixture. Counterelectroosmotic CE is thus capable of simultaneous analysis of species carrying opposite charges. The first use of counterelectroosmotic separations in capillaries of a relatively large diameter was reported by Hjerten in 1967.[50] The first counterelectroosmotic separations in narrow-bore fused silica capillaries were published by Jorgensen and co-workers in 1981.[51]

In its simplest form, counterelectroosmotic capillary electrophoresis relies only on two types of electrokinetic vectors to achieve a separation, first, the electrophoretic mobilities of analytes, and, second, on the mobility con-

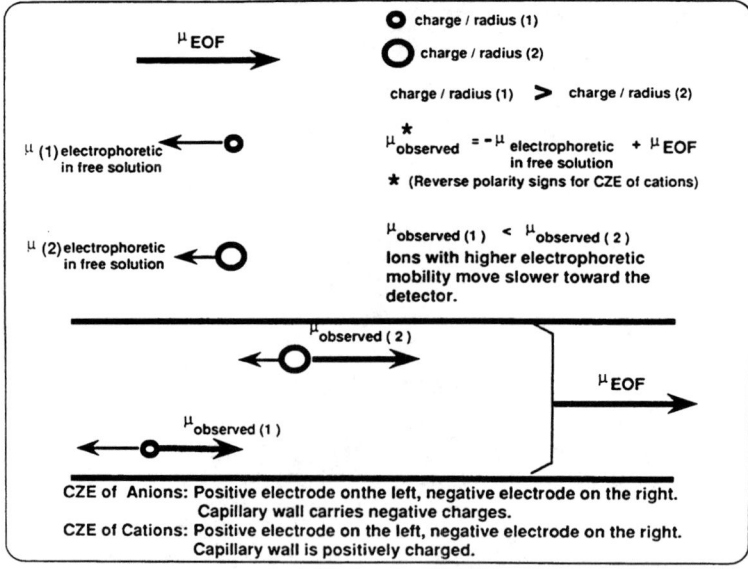

Figure 2.30 Mobility vectors in counterelectroosmotic capillary eletrophoresis.

tribution of the electroosmotic flow. With only two types of mobilities effecting the separation, the scope of this approach is necessarily limited to mixtures of ions having their respective mobilities within a limited range. The ionic components outside this range are mainly those with vectors opposing the direction of the electroosmotic flow and exceeding the magnitude of the electroosmotic flow. This limitation is clearly not a problem for polymers and large biomolecules with absolute values of electrophoretic mobilities seldom exceeding the magnitude of the electroosmotic flow. However, the same limitation is a very critical one for low-molecular-weight ions that are the main focus of this book. (See cases C and I in Figures 2.8 and 2.9.) Until only a few years ago, counterelectroosmotic CE was the only practiced version of capillary electrophoresis. Consequently, in a number of published reports low-molecular-weight ions have been separated under conditions where their electroosmotic mobility was opposed by the direction of electroosmotic flow, resulting in migration times longer than in HPLC separations of the same compounds (see cases B and J in Figures 2.8 and 2.9). An example of this approach is the electropherogram in Figure 2.31. However,

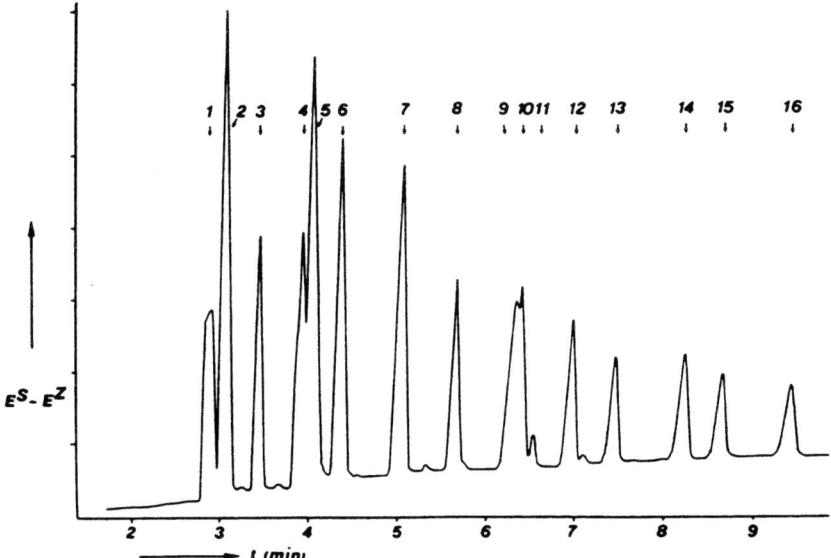

Figure 2.31 Counterelectroosmotic CE (CZE) separation of anions. Carrier electrolyte: 0.01 M 2-(N-morpholino)ethanesulfonic acid/histidine, 0.1% hydroxyethylcellulose, pH 6.05, capillary 0.2 mm, current: 20 µA, sample volume: 0.7 µl. Peak identities: 1, chloride; 2, sulfate; 3, chlorate; 4, malonate; 5, chromate; 6, pyrazole-3,5-dicarboxylate; 7, adipate; 8, acetate; 9, propionate; 10, b-chloropropionate; 11, unidentified; 12, benzoate; 13, naphthalene-2-monosulfonate; 14, glutamate; 15, enanthate; 16, benzyl-DL-aspartate. Total amount of sample: 17.5×10^{-12} mole of each constituent. E_S-E_Z, difference in electrical field strength between the carrier electrolyte and the sample zone. (Reproduced with permission.[52])

in this particular example, the total run time is reduced and the range of analyzable ions is extended by the use of hydroxyethylcellulose as an additive suppressing the electroosmotic flow. The previous chapter offers an extensive discussion of various approaches to manipulation of electroosmosis in capillaries.

2.4.3 Coelectroosmotic Capillary Electrophoresis

This is an approach optimizing CE separations for a group of ions of the same type of charge. The systematic approach for analytes with like charges was first introduced under the name electrophoretic capillary ion analysis.[53] The anions are analyzed under conditions that are distinctly different from optimum conditions for cations. The main advantage of this approach is that both "fast" and "slow" ions can be separated in a single electropherogram within run times that are much shorter than those of any other conventional separation techniques—for example, ion chromatography or HPLC.

By choosing the optimized coelectroosmotic conditions, the analyst deliberately excludes the oppositely charged and neutral species from the electropherogram. This results not only in much shorter run times, but also, among other things, in decreased interferences by various sample matrices. The time savings achieved by coelectroosmotic CE stem from the addition of the mobility vectors. The relatively longer run times in the counterelec-

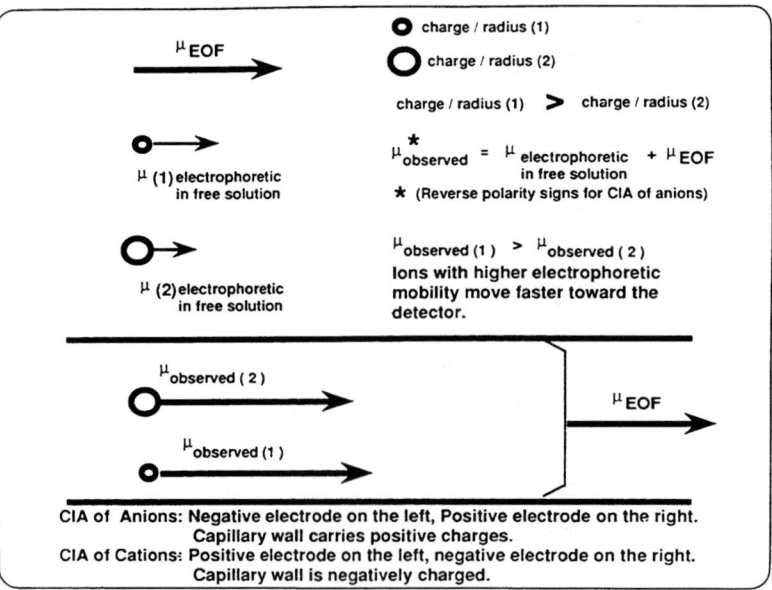

Figure 2.32 Orientation of mobilities in coelectroosmotic capillary electrophoresis.

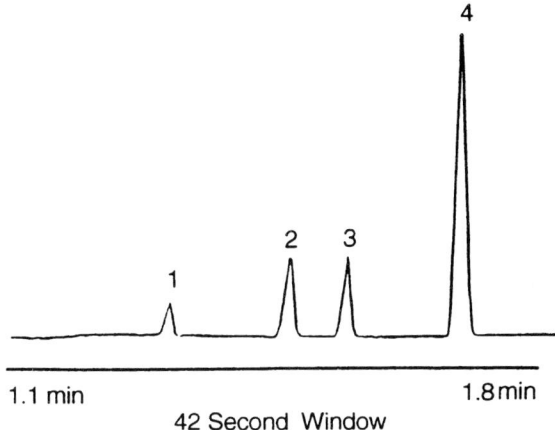

Figure 2.33 Separation of monovalent cations by coelectroosmotic capillary electrophoresis. Conditions: +25 kV, 5 mM morpholineethanesulfonate at pH 8.5, capillary 60 × 52 cm × 75 μm, indirect detection at 214 nm. Peak identities: 1, potassium; 2, ammonium; 3, sodium; 4, lithium. (Reproduced with permission.[54])

troosmotic mode are caused by the subtraction of the electrophoresis and electroosmosis vectors. The decreased matrix interference is due to the oppositely directed vectors of cationic species and zero electrophoretic mobility of neutral compounds. As can be deduced from cases D and E in Figures 2.8 and 2.9, the neutral and oppositely charged species are made to migrate well after the last peak of interest in coelectroosmotic CE. Unlike in column chromatography, in CE it is not necessary to wait until all sample components are out of the system. A standard procedure in CE consists of interrupting the separation after the last peak of interest had been detected. After rinsing and filling the capillary with a new volume of carrier electrolyte, which in most cases is done automatically, the system is ready for another sample. An example of coelectroosmotic CE applied to a complex mixture of anions is given in Figure 2.10. Note that the polarity of the injection side is negative and the electroosmotic flow is directed toward the positive electrode positioned behind the detector.

On the other hand, the hardware configuration for cations is similar to that in conventional CZE (i.e., positive polarity at the injection side), but the vectors of cationic analytes and electroosmotic mobility have the same direction (as in Figure 2.8 F and G; see also Figure 2.33).

2.4.4 Auxiliary Separation Modes

So far we have limited our discussions mostly to the combinations of inherent analyte electrophoresis and electroosmosis induced by capillary walls.

The only exception was the use of molecular sieving on dissolved polymers performed as counterelectroosmotic CE. The interactions of analytes with the dissolved polymer represent an auxiliary separation mode in the frame of reference of the following discussion. Additional interactions contributing to the separation mechanism are outlined in Figure 2.5. Even if we discuss the analyte–electrolyte component interactions as an independent separation principle, we must not forget that they almost always occur in combination with electroosmotic flow and thus for the sake of classification become an additional separation parameter either in counter- or coelectroosmotic CE. Only for a very small number of applications in which the electroosmosis is suppressed to zero does the component interaction with analyte become an independent and exclusive separation principle. Probably the most frequent interaction between analytes and electrolyte components involves the role of hydronium and hydroxide ions in the acid–base equilibria. The electrolyte pH is an important parameter with all CE separation modes and, for that and other reasons (i.e., the role of pH in electroosmosis), it is not practical to discuss pH adjustments as an independent separation principle.

The role of pH will be discussed in detail in the sections dedicated to different separation modes.

2.4.4.1 Ion Exchange Interactions

Such interactions are between analyte ions and polymeric ions in solution. So far there has been only one published example[55] of the use of polymeric ion exchangers for enhancement of ionic separations. Just as capillary gel electrophoresis can be seen as a link between preceding methodologies and CE, the use of ion exchangers dissolved in carrier electrolytes can be considered as a link to the more conventional separation techniques for ions, such as ion chromatography. The interactions operative in this auxiliary separation mode are illustrated in Figure 2.34. The authors[55] have chosen to call this method "ion-exchange electrokinetic chromatography" (IEEC), stressing the analogy with the conventional column chromatographic techniques.

Their approach is ideally suited for separations of ions with close to identical electrophoretic mobilities. The two electropherograms in Figure 2.35 demonstrate the difference between the crude separation by the number of anionic charges in capillary zone electrophoresis without any enhancement and the counterelectroosmotic CE separation enhanced by ion exchange in solution.

2.4.4.2 Complexing Interactions

Such interactions between the cations or anions from a sample mixture on one hand and a suitable ligand on the other can be utilized either to decrease

FUNDAMENTALS OF CAPILLARY ELECTROPHORESIS

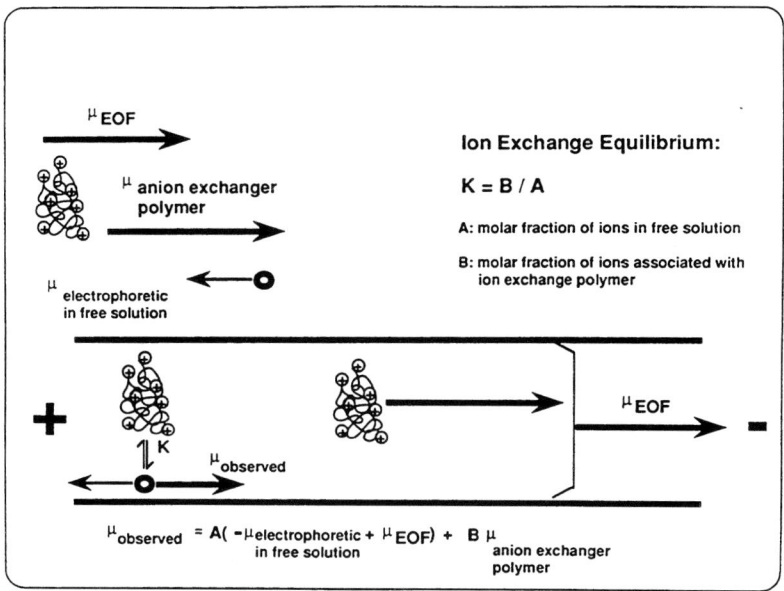

Figure 2.34 Effect of ion exchange on polymers in solution on the observed mobilities of analyte ions in capillary electrophoresis.

the mobility of a free ion or to induce selectivity for two or more ions with similar mobilities. The complexed species, being larger and possessing a lesser charge, migrates usually much more slowly than the fully dissociated ion. Since the stability constants for metals, even from the same period or group of the periodic table, vary in a broad range, it is possible to modulate the migration of a large number of ions. This allows successful CE separations of ionic mixtures with very close values of electrophoretic mobilities that would otherwise be very hard to separate had only inherent and electroosmotic mobilities been available.

The first example of a CE separation utilizing complexation was published in 1983.[56] Janak and co-workers used cadmium added to the carrier electrolyte to slow down the migration of the fast anions nitrate, sulfate, and chloride to make possible their separation under counterelectroosmotic CE conditions with the positive polarity at the injection side—converting case C to case B in Figure 2.8. In 1990, Bocek and coauthors introduced an application of organic ligands for the separation of cations with similar rates of electromigration.[57] The separation of bismuth and copper with the help of lactic acid in the carrier electrolyte published by Hjerten in 1967[50] represents the earliest example of the use of an organic ligand under the conditions of capillary electrophoresis.

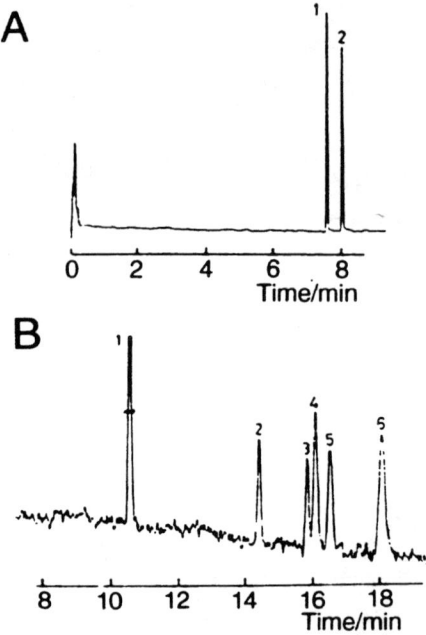

Figure 2.35 Counterelectroosmotic (CZE) and counterelectroosmotic/ion exchange (IEEC) separations of a mixture of naphthalene sulfonates. (A) *CZE electropherogram*: Peak identities: (1) five naphthalenedisulfonates, (2) two naphthalenesulfonates. Carrier electrolyte: 50 mM phosphate, pH 7.0, with 0.1% hydroxypropylcellulose for suppression of electroosmosis. Capillary: 75 × 46 cm × 50 μm fused silica. Separation voltage: +20 kV. (B) *IEEC electropherogram*: Peak identities: (1) 1- and 2-naphthalenesulfonates; (2) 2,6-; (3) 2,7-; (4) 1,6-; (5) 1,5-; (6) 1,7-naphthalenedisulfonates. Carrier electrolyte: 2% of (diethylamino)ethyldextran in the elctrolyte for unenhanced CZE. Capillary: 75 × 50 cm × 50 μm. Separation voltage +20 kV. The UV detection for both separations was carried out at 210 nm. The sampling was by hydrostatic method. (Reproduced with permission.[55])

A more recent version of such approach utilized to improve a coelectroosmotic CE separation of lanthanides is shown in Figure 2.37.

Other examples of the use of complexation equilibria include CE separations of amino acid enantiomers by ligand exchange with copper(II)–aspartame complexes in carrier electrolyte[57] and separations of carbohydrates complexed with borate.[58,59] The latter examples along with another report describing the use of borate for separations of catecholamines[60] represent a borderline between this and the following separation mode.

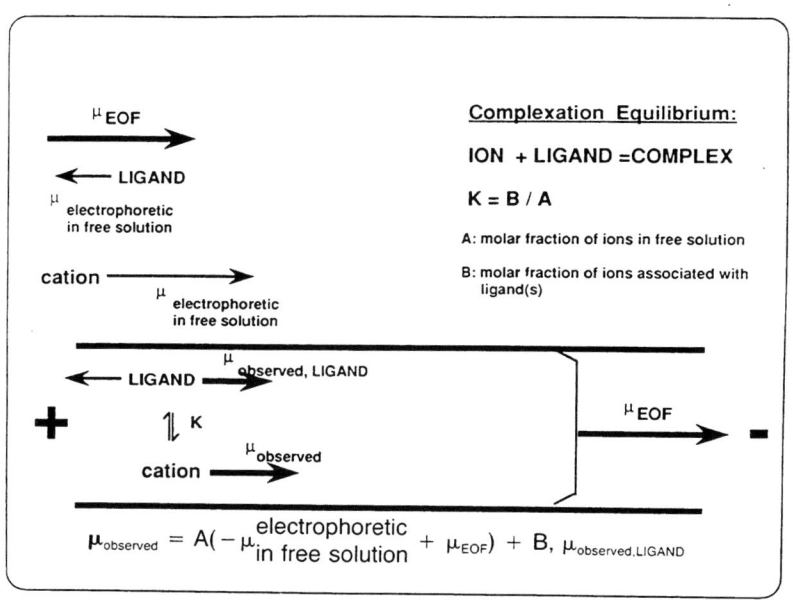

Figure 2.36 Influence of complexation on the observed electrophoretic mobility.

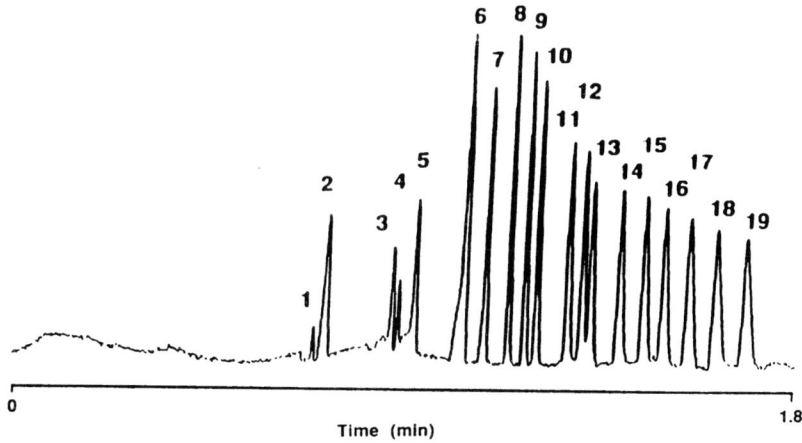

Figure 2.37 Simultaneous separation of alkali metals, alkaline earths, and lanthanides. Carrier electrolyte: 10 mM Waters UVCat-1, 4.0 mM hydroxyisobutyric acid, pH adjusted to 4.4 by acetic acid, capillary: 36.5 cm × 75 μm fused silica, voltage +30 kV, hydrostatic sampling: 20 sec/10 cm, indirect UV detection at 214 nm. Peak identities: 1, rubidium 2 ppm; 2, potassium 5 ppm; 3, calcium 2 ppm; 4, sodium 1 ppm; 5, magnesium 1 ppm; 6, lithium 1 ppm; 7, lanthanum 5 ppm; 8, cerium 5 ppm; 9, praseodymium 5 ppm; 10, neodymium 5 ppm; 11, samarium 5 ppm; 12, europium 5 ppm; 13, gadolinium 5 ppm; 14, terbium 5 ppm; 15, dysprosium 5 ppm; 16, holmium 5 ppm; 17, erbium 5 ppm; 18, thulium 5 ppm; and 19, ytterbium 5 ppm. (Reproduced with permission.[53])

2.4.4.3 CE Separations of Neutral Molecules Made Possible by Charge Association

A possible approach to separation of uncharged molecules in CE is to convert them to charged species either by association with a suitable ion or by another appropriate chemical reaction (see Figure 2.38). Two examples mentioned in the preceding paragraph involving a complexation reaction between borate and carbohydrate or between borate and catecholamine have one thing in common. The originally uncharged species—catecholamines and carbohydrates—could not be separated by CE. Their reaction with borate converts them to anions with different charge to ionic radius ratios, and consequently into species relatively easily separated by electrokinetic techniques.

The first example of charge generation was by solvophobic association, and was demonstrated by Walbroehl and Jorgenson in 1986[61] (see Figure 2.39). The authors separated a mixture of organic molecules with the larger, less polar analytes migrating faster due to their stronger interaction with the tetrahexylammonium ion. Formamide, with its minimal interaction with the alkyl ammonium cation, served as a marker compound indicating the rate of the electroosmotic flow. Since both analyte mobility and electroosmotic mobility were in the same direction, the overall conditions can be described as those of coelectroosmotic CE. The charge association then represents an

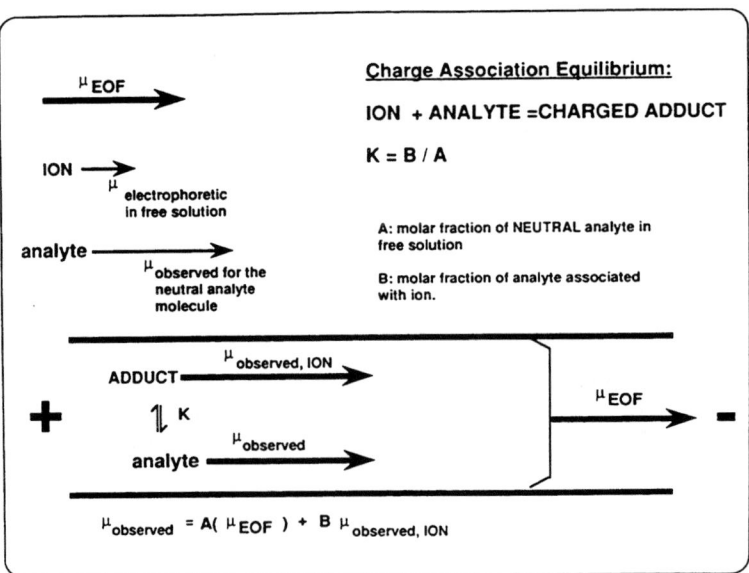

Figure 2.38 Electromigration of an originally uncharged molecule after charge association.

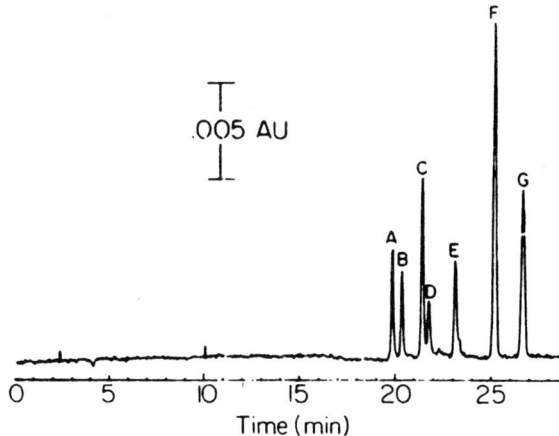

Figure 2.39 Coelectroosmotic separation of uncharged organic compounds after charge association. (A) benzo[ghi]perylene, (B) perylene, (C) pyrene, (D) 9-methylanthracene, (E) naphthalene, (F) mesityl oxide, and (G) formamide. Naphthalene concentration in the standard with 5×10^{-4} M. All other standard components were at 1×10^{-4} M. The carrier electrolyte composition: 0.025 M tetrahexylammonium perchlorate in acetonitrile/water (50/50 v/v). Capillary: 100 cm × 75 μm fused silica. Separation voltage: +20 kV. Sampling was by electromigration at +2.5 kV for 20 sec. UV detection was carried out on the column at 229 nm. (Reproduced with permission.[61])

auxiliary mechanism enabling the coelectroosmotic separation mode. (See Figure 2.39.)

2.4.4.3 Micellar Interactions

The use of micelles in carrier electrolytes was first described by Terabe and co-workers in 1984.[62] He introduced the name micellar electrokinetic chromatography (MEKC) for this method. In the meantime, the technique has become even more popular under a slightly different name—micellar electrokinetic capillary chromatography (MECC). According to the recent literature, [62a]MECC can be considered as a separation technique occupying some intermediate position between capillary electrophoresis and liquid chromatography due to its special separation mechanism, which is described in Figure 2.40. At first glance, MECC appears to be very similar to CE because it uses the same instrumentation as CE and requires only the addition of a charged surfactant to the carrier electrolyte. The main difference is in the fact that the analytes are distributed between an aqueous electrophoretic phase and micelles that are formed owing to the application of the surfactant in concentrations larger than the critical micellar concentration. On the other hand, MECC differs from chromatography insofar as that micellar phase is not a stationary but a moving phase (usually it is called a "pseudostationary phase").

The separation principle of MECC given by Terabe et al.[62b] is illustrated in Figure 2.40. In this example, the analysis of neutral molecules, by the more usual separation mode employing an anionic surfactant and fused silica capillaries, is considered.

When a high voltage is applied across a capillary, an aqueous solution is forced to flow toward the cathode by electroosmosis (Section 2.3), while anionic micelles in the carrier electrolyte are considerably retarded by the electrophoretic migration to the anode, thus acting as a dynamic stationary phase. If the electroosmotic flow is higher than the electrophoretic mobilities of the micelles, they will move in the same direction as the bulk of the buffer, but with a slower velocity. Analyte molecules will be distributed between the aqueous and the micellar phase. Therefore, depending upon their partition coefficients, they will migrate with different mobilities that are intermediate between μ_{EOF} and $\mu_{observed,\ micelle}$. Up to this point we have utilized capillary electrophoretic terminology. In the following, we reproduce a chromatographic terminology used by Terabe et al. in their discussion of MECC. Very polar molecules, which do not interact with micelles, will elute with the retention time t_0, while highly hydrophobic compounds, being completely included in the micelles, will elute at the same retention time as mi-

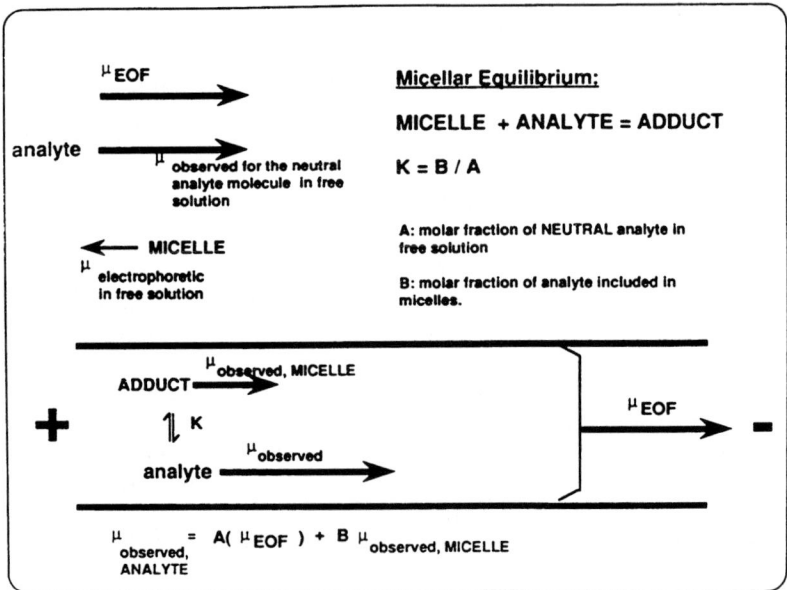

Figure 2.40 Separation mechanism based on interactions of analytes with micelles moved by electroosmotic flow.

celles, t_{MC}. For all the other analytes, a partition process is vital for the separation, and they must elute with t_R between t_0 and t_{MC}. For neutral species, a measure of partition of a solute between the aqueous and micellar phase, the capacity factor (k') is derived from the retention data as follows[62b]

$$k' = \frac{t_R - t_0}{t_0(1 - t_R/t_{MC})} = \frac{(1/v_R - 1/v_0) v_0}{(1 - v_{MC}/v_R)} \qquad (2.16)$$

where v_0, v_R, and v_{MC} are the electroosmotic flow velocity of the eluent and the elution velocity of the solute and micelles, respectively. The same expression can be obtained for anionic solutes, taking into account the proper electrophoretic velocity.

It is interesting to note that, in its current version, MECC is always carried out in a counterelectroosmotic mode; that is, the mobilities of analytes and micelles are directed against the electroosmotic flow. This leads to relatively long analytical run times in comparison with coelectroosmotic capillary electrophoresis. However, the use of MECC in a coelectroosmotic mode appears to be relatively difficult. The usual procedures for the reversal

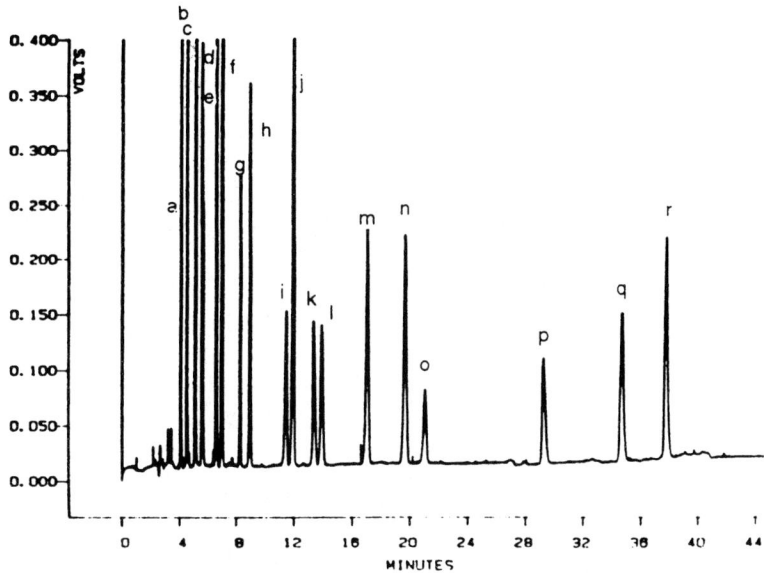

Figure 2.41 MECC forensic drug screen. Conditions: capillary 25 cm × 50 μm, voltage 20 kV, temperature 40°C, carrier electrolyte 85 mM SDS, 8.5 mM borate, 15% acetonitrile, pH 8.5. UV detection at 210 nm. Each sample component at 250 μg/ml. Peak identities: (a) psilocybin, (b) morphine, (c) phenobarbital, (d) psilocin, (e) codeine, (f) methaqualone, (g) LSD, (h) heroin, (i) amphetamine, (m) lorazepam, (n) diazapam, (o) fentanyl, (p) PCP, (q) cannabidiol, (r) Δ^9-THC. (Reproduced with permission.[65])

of electroosmosis fail due to interactions between, for example, the anionic surfactants and the positive charges of the capillary wall, leading to much higher required surfactant concentrations in the electrolytes and to a reversal of the wall charges back to the original negative polarity. The original goal of MECC was to make CE separations of uncharged or weakly ionic compounds possible. However, in the past few years the scope of the method was expanded to the analysis of ionic species, including a large variety of low-molecular-weight ions.[63] The partition between the aqueous and micellar phases can increase the selectivity (as the additional equilibrium influences the effective charge/mass ratio) and allows the separation of ions with very similar mobilities. Another example of practical MECC applications is the separations of chiral compounds.[64] The high stereospecific resolution can be attained by exploiting the same strategies as in liquid chromatography, for example, the formation of diastereomers or stereoselectivity obtained via a chiral stationary (micellar) phase. Figures 4.30 and 4.31 show electropherograms of racemic mixtures of amino acids obtained under CE and MECC conditions. This comparison manifests a certain advantage of the latter technique in terms of selectivity and efficiency.

Thus, even at the present stage of development, MECC allows a number of previously difficult separations. One of such previously difficult separations of high practical utility is the screening method for forensic drugs shown in Figure 2.41.

2.4.4.4 Formation of Inclusion Complexes

The formation of inclusion complexes is observed in molecules having a well defined cavity defined by their structure. The two best known examples are the crown ethers and cyclodextrines.

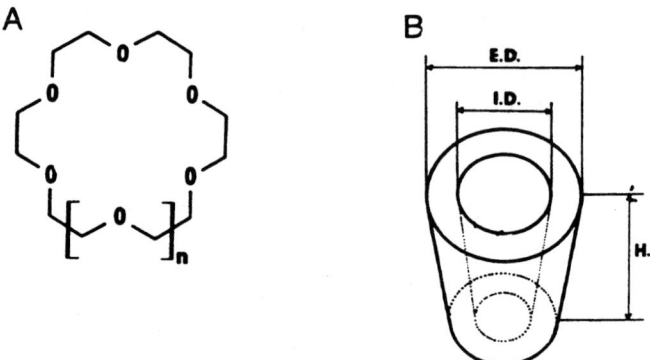

Figure 2.42 Structures of (A) crown ethers and (B) cyclodextrines. Courtesy J. Snepek.

FUNDAMENTALS OF CAPILLARY ELECTROPHORESIS

The dimension of the cavity in crown ethers makes those compounds ideally suitable for selective interactions with alkali cations. The cyclodextrines make possible inclusion complexation of a relatively large number of charged and uncharged organic ions. It seems that the original uncharged cyclodextrines are suitable only for the CE analysis of charged analytes. If nonionic compounds have to be separated, the cyclodextrines should be modified by an ionizable substituent.

Inclusion complexes provide additional selectivity for CE separations of cations, anions, and nonionic compounds (see Figure 2.43). Enhanced cation selectivity can be achieved, for example, by an addition of a suitable crown ether to carrier electrolytes.[66] Applications of host–guest complexation by means of cyclodextrines were demonstrated for separations of racemic mixtures of amines. The first reported application of cyclodextrines is due to Terabe and co-workers,[67] who were able to separate substitution isomers of benzene by inclusion complexation.

Inclusion complexation adds selectivity to CE separations by providing different values of complexing equilibria between the host molecule on one hand and the guest molecules or analyte ions on the other hand. The resolving power is considerably greater for ionic species having an inherent mobility than for nonionic analytes.[68] However, as demonstrated by Terabe and co-workers, even in the latter case, the resolving power is large enough to separate, for example, aromatic substitution isomers. (See Figure 2.44.)

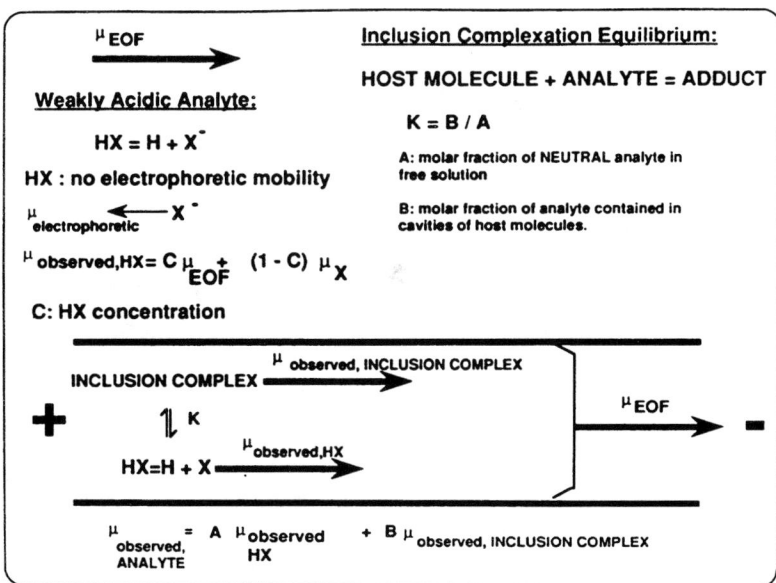

Figure 2.43 Functional principle of separations involving inclusion complexation.

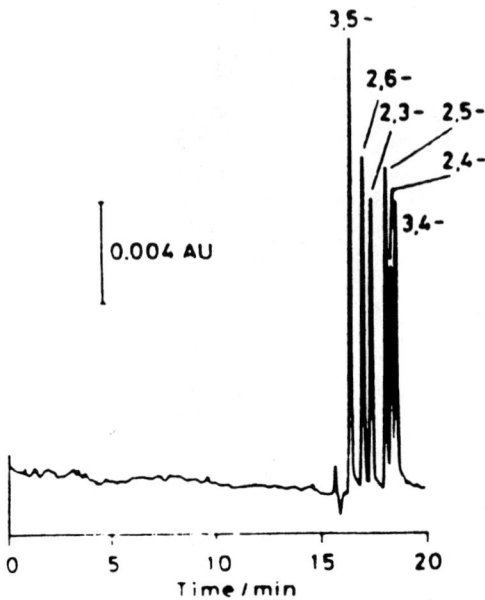

Figure 2.44 Separation of xylidine isomers by inclusion complexations with 2-o-carboxymethyl-β-cyclodextrin (β-CMCD). Carrier electrolyte: 0.025 M β-CMCD in 0.1 M phosphate, pH 7.0. Fused silica capillary: 65 cm × 50 cm × 50 μm. Separation voltage: +12 kV. UV detection at 210 nm. (Reproduced with permission.[67])

Interestingly, the inclusion complexation mode was also shown to be useful for the separation of different isomers of cyclodextrines.[69] In this case, the uncharged cyclodextrines associate with a carrier electrolyte component carrying a negative charge (aromatic carboxylate). The degree of such association is sufficiently specific to enable separations of α, β, and γ cyclodextrines not only from each other but also from the water peak occurring with indirect UV detection.

2.4.5 Separations of Particles by Capillary Electrophoresis

Capillary electrophoresis is able to close an important gap in the sizing methodologies for microscopic particles. The conventional instrumental techniques can provide adequate resolution down to about a 1 μm particle size. The CE technique was found to be useful not only in the 1–100 μm range, but also for smaller particle sizes in the nanometer range. It is important to note that CE application to particles is the only known example where the interaction of analytes with the capillary walls as a separation principle (not as an interference as in the case of proteins and peptides) is not only assumed but also supported by the experimental data.

FUNDAMENTALS OF CAPILLARY ELECTROPHORESIS

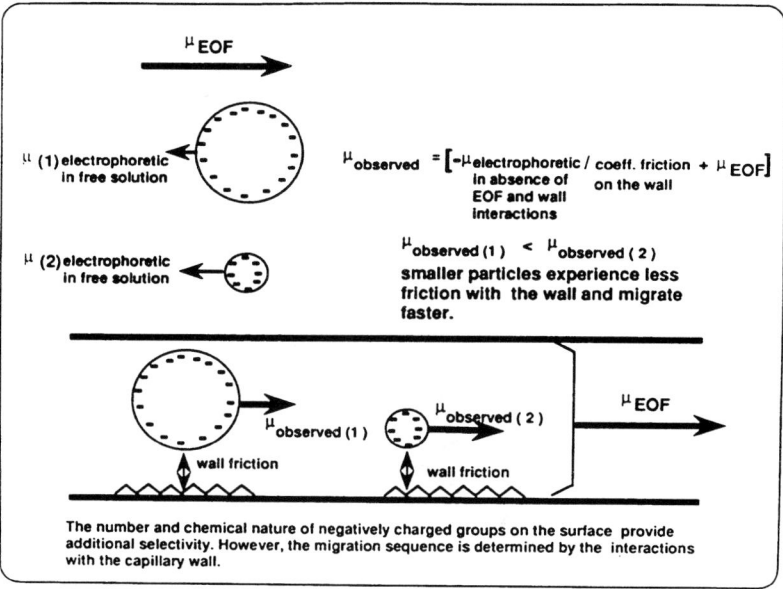

Figure 2.45 Particle interaction with the capillary wall as a separation principle.

VanOrnam and McIntire[70] reported the first successful CE separation of nanometer-range particles. They observed that a suitable ratio of particle diameters to internal diameter of fused silica capillary was one of the necessary requirements for successful separations. Particles larger than about 700 nm could not be transported through a 50 μm ID capillary. However, the same capillary diameter was sufficient for 39 nm particles. Zhu and Chen,[71] who used fluorinated ethylene–propylene copolymer capillaries to separate different types of red blood cells, made a similar observation. No peaks were obtained for red blood cells, which are usually about 6–9 μm in size, in capillaries having internal diameters smaller than 300 μm. Satisfactory separations were achieved in capillaries of internal diameter between 450 and 500 μm.

According to Jones and Ballou,[72] and also in view of the results of Zhu and Chen,[71] CE is capable of separating small particles not only according to their size but also according to their different chemical properties. The electropherogram in Figure 2.46 gives an example of such a type of separation.

Summary: The overall classification of separation modes is facilitated by recognizing molecular sieving in gel filled capillaries (CGE), counterelectroosmotic CE (CZE), and coelectroosmotic CE (CIA) as the three main separation modes. All the other contributions to separation can then be classified as auxiliary.

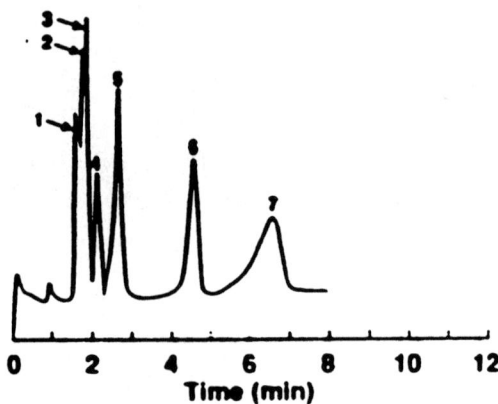

Figure 2.46 Electropherogram of a seven-component mixture of latex particles. Sizes and surface chemistries of particles in the seven peak zones: (1) 30 nm, sulfate groups; (2) 79 nm, sulfate groups; (3) 70 nm, carboxylates; (4) 100 nm, carboxylates; (5) 200 nm, carboxylates; (6) 500 nm, carboxylates; and (7) 1.16 μm, carboxylates. Electrolyte was 1 mM NaOH and 5 mM Na_2HPO_4. Fused silica capillary: 55 × 40 cm × 75 μm. Separation voltage: +30 kV. The concentration of sample components was between 10^8 and 10^{14} particles per ml. The sampling was carried out hydrostatically. (Reproduced with permission.[72])

Molecular sieving in solution (CMCE), complexation reactions, micellar interactions, charge association, etc. have so far been employed only in conjunction with coelectroosmotic or counterelectroosmotic separation modes. The new nomenclature highlights the importance of all contributions to the observed mobilities of analytes. This is hoped to inspire new approaches in the optimization of modern CE separations.

2.5 Efficiency and Resolution in CE, Separations under Optimal Conditions

The efficiency of CE separations is evaluated by calculation of theoretical plates. The concept of theoretical plates was originally developed for distillation and countercurrent extraction. Later, it was successfully applied to evaluations of separation efficiency in gas and liquid chromatographic techniques. The gradual evolution of the term *plate* from its real meaning in distillation and extraction to the abstract term *theoretical plate* in column chromatographic techniques is a subject of entry-level instrumental analytical chemistry textbooks.[73] In an article published in 1969,[74] Giddings justified the expansion of the usage of the concept of theoretical plates to measurements of efficiency in electrophoresis and undertook to predict the maximum attainable separation efficiency for that technique.

The fundamental concept making the application of theoretical plates possible to a broad range of separation techniques is the "rate of generation of variance," $d\sigma^2/dL$. The term *variance*, σ^2, originates from the theory of probability. The parameter σ describes the standard deviation of the Gaussian distribution as related to the width of Gaussian distribution curves[75] at half height. L is the length of a separation device in centimeters. In our case, it is the length of the capillary from the sample introduction to the point of detection, L_d.

The goal of chromatographic and electrophoretic techniques is to keep separated zones as narrow as possible. Therefore, the systems with a lower rate of variance generation (analyte zone spreading, peak broadening) are described as more efficient. The two drawings in Figure 2.47 represent two separations of identical selectivity. The selectivity is measured as the distance of the highest points of peaks. The separation in A is usually described as incomplete. Large portions of the analyte zones are still overlapping. The same selectivity as in Figure 2.47A yields the complete separation in Figure 2.47B. There the peaks are much narrower, more efficiently separated, than in Figure 2.47A.

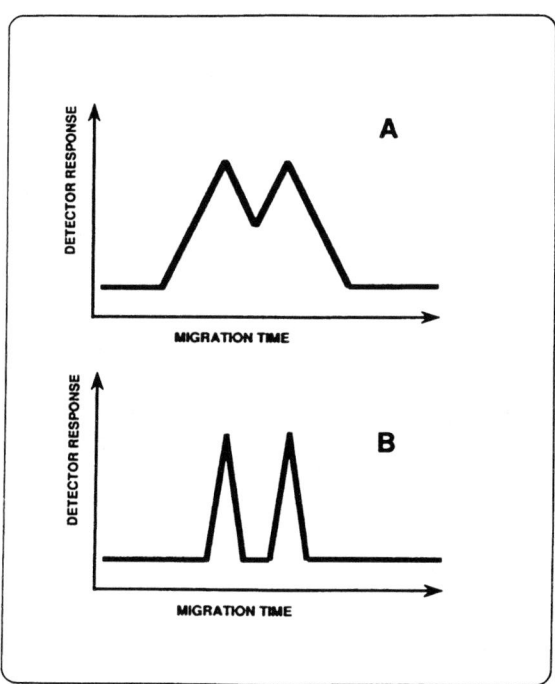

Figure 2.47 The influence of peak efficiency in a difficult separation.

The following text, reproducing the original discussion by Giddings, is based on the assumption of optimal conditions in electrophoresis. Under such conditions the only cause of peak broadening is molecular diffusion. According to Einstein, a "variance" or width of a zone of a dissolved substance in the liquid phase in time t after its introduction as a thin segment having a negligible value of σ^2 is dependent on the coefficient of molecular diffusion D of the substance in question:

$$\sigma^2 = 2Dt \tag{2.17}$$

The rate of generation of variance $d\sigma^2/dL$, the height of the theoretical plate H, and the number of theoretical plates N in electrophoresis can then be given by the following relationships:

$$\frac{d\sigma^2}{dL} = 2D \frac{dt}{dL} \tag{2.18}$$

or, since $dL/dt = v$:

$$H = \frac{d\sigma^2}{dL} = 2D/v \tag{2.19}$$

and the dimensionless number of theoretical plates N is then:

$$N = L_d/H = L_d\, v/2D \tag{2.20}$$

where v is the rate of migration [cm/sec]. Equation (2.20) specifies that the separation efficiency improves to the same degree as the rate of electrophoretic migration v becomes more significant than the rate of diffusion. The magnitude of the rate of diffusion is determined by the diffusion coefficient D. In HPLC v corresponds to the linear flow rate [cm/sec], but it is important to note that Equations (2.19) and (2.20) do not apply to column liquid chromatography. In that technique the spreading of analyte zones is due to additional contributions, next to which the molecular diffusion is relatively small in magnitude. We should also recall that in LC the peak spreading can be characterized only by semiempirical relationships (see vanDeemter and equation in Figure 2.48).[76]

Having developed the fundamental theoretical relationship for the migration rate and the theoretical plates in CE, we can also derive an expression connecting the separation efficiency to thermodynamic parameters. If we replace D by RT/ζ in Equation (2.20), with ζ the friction constant per mole for a given medium ($\zeta = 1/\mu$), R the gas constant, and T the absolute temperature, we get:

$$N = L_d\, v\, \zeta/2RT = L_d\, v/2\mu RT \tag{2.21}$$

Expressing the mobility μ with the help of Equation (2.2) (or the friction coefficient with $\zeta = E/v$), we obtain:

$$N = L_d\, E/2\, RT \tag{2.22}$$

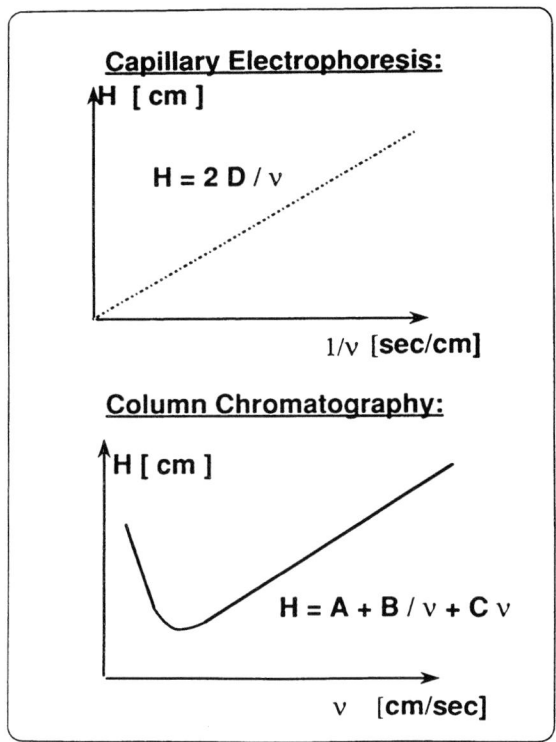

Figure 2.48 Peak spreading as described by Equation (2.19) and the vanDeemter equation.

The product $L_d E$ represents the difference of the electrochemical potential ($-\Delta\mu_{el.\ chem.}$) between the sampling end of the capillary and the point of detection:

$$L_d E = -\Delta\mu_{el.\ chem.} = z_i FP \quad (2.23)$$

with z_i signifying the number of charges of an ion, F the Faraday constant [96,487 Coulomb], and P the electric potential in volts. Equation (2.20) can now be rewritten in a more fundamental form, allowing predictions of the separation efficiency from the thermodynamic parameters:

$$N = z_i FP/2RT \quad (2.24)$$

In 1969, Giddings calculated 10^5 as a theoretical limit for the number of theoretical plates in electrophoresis. This was based on the assumption that the maximally attainable value of $z_i P$ was $\sim 10^4$. The product $z_i FP$ was $\sim 10^9$ J/mole and $2RT \approx 5000$ J/mole.

Jorgenson and Lukacs[77] discussed another form of Equation (2.20). By substituting $\mu \left(\dfrac{P}{L_d}\right)$ for v in Equation (2.20), they obtain:

$$N = \mu P/2D \qquad (2.25)$$

As in Equation (2.24), N is shown to be directly proportional to the separation potential P [V], but in addition to that, the separation efficiency is represented as a direct function of the observed electroosmotic mobility. As discussed in the preceding two sections, observed electroosmotic mobility results from vectorial addition of all electroosmotic and electrophoretic contributions to the movement of the ionic (or nonionic) analyte in the electrical field. Equation (2.25) thus represents another important prediction: The maximum efficiency of CE is achievable only in those cases where the electroosmosis and electrophoresis vectors carry the same sign (see coelectroosmotic CE, Sections 2.4 and 2.5). Only then is the mobility contribution to the separation efficiency maximized. However, the optimization of CE separations is made more complicated by a divergence of optimal conditions for separation efficiency *(N)* and for resolution *(R)*. According to Giddings[74] and Jorgenson et al.,[77] the resolution is determined by the following equation:

$$R = (N^{1/2}/4)(\Delta v/v) \qquad (2.26)$$

where $\Delta v/v$ is the relative migration velocity of the two analyte zones to be resolved.

$$\Delta v/v = (\mu_{obs.1} - \mu_{obs.2})/\mu_{average} \qquad (2.27)$$

Remembering that $\mu_{obs} = \mu + \mu_{EOF}$, where μ stands for the inherent electrophoretic mobility directly connected to the charge to radius ratio, the above equation can be simplified to:

$$\Delta v/v = (\mu_1 - \mu_2)/(\mu_{average} + \mu_{EOF}) \qquad (2.28)$$

From Equation (2.28) it follows that an increase of μ and μ_{EOF} will decrease the value of relative velocity for any given peak pair. With decreased relative velocity, the resolution *(R)* is also going to be reduced. Fortunately, this is only true for those cases where the inherent ionic mobility and the electroosmotic mobility are the only contributions to the mobility observed. Such conditions were the only ones to be considered and discussed in the early stages of CE. Correspondingly, the correct conclusion at that time was to utilize methodologies in which the term μ_{obs} could be minimized. A good example of such conditions is CZE or counterelectroosmotic CE (Sections 2.4 and 4), with opposing directions of electrophoresis and electroosmosis yielding only very small values for μ_{obs}.

In the intervening years, multiple ways have been described (see Section 2.4) to modify observed analyte mobilities over a broad range. It has become increasingly possible to adjust selectivity by additional, auxiliary separation modes. The selectivity can now be modified so much that a decrease of relative velocities does not have to mean a loss of resolution. The advantage of the use of auxiliary separation modes under coelectroosmotic conditions lies in considerably reduced analytical run times and an increased ability to analyze complicated mixtures having a large number of components. (See Figure 2.49.)

The increased availability of auxiliary separation modes makes it possible to consider a change in the general strategy for optimization in capillary electrophoresis. Counterelectroosmotic electrophoresis, according to the definition in the preceding section, is a method of choice in all those cases, where close migration of analytes cannot be improved by auxiliary separation modes (i.e., complexation, dissociation equilibria, etc.). In all other cases with different enough values of μ_{obs}, the conditions of coelectroosmotic electrophoresis can be expected to give superior results. The former is the case with many biological macromolecules, and the latter can be applied to the majority of low-molecular-weight ionic compounds.

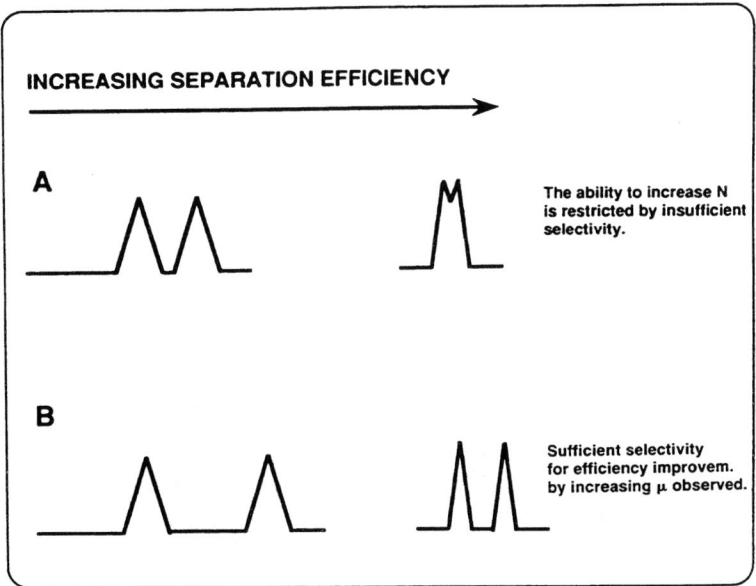

Figure 2.49 Improvement of separation efficiency occurring simultaneously with a deteriorating resolution. (A) Maximizing μ_{obs} and N for a separation with a limited selectivity. (B) Adjusting selectivity to make possible good resolution under conditions where N is also maximized.

2.6 The Problem of Joule Heat Generation

Anyone familiar with the separation efficiencies attainable in HPLC (ca. 20,000 theoretical plates) and not familiar with the experimental realities of CE as discussed in the following paragraphs will have difficulty understanding the slow rate of introduction of CE for routine use. Although the paper defining the theoretical limit for electrophoresis as 10^5 plates was published in 1969, the first commercial instruments began to appear in 1989. Why did it take so long for the method to become widely available to average users? One answer can be found in the high rate of Joule heat generation in electrophoresis. Without adequate heat dissipation, the heating power due to electric currents at high separation potentials will rapidly evaporate any liquid medium for electrokinetic separations. The first practical solution of the heat dissipation problem is due to Hjerten,[78] who in 1967 utilized rotating capillaries of 3 mm inner diameter. The more recent solution to Joule heating evolved from the usage of submillimeter inner diameter tubing. The following discussion, adapted from the original report by Knox,[79] provides an explanation of the necessity to reduce capillary diameters in CE for optimum heat dissipation.

To move dissolved ions electrokinetically over 10–100 cm distances within 10–30 min requires separation potentials in the kilovolt range. The Joule heat generation Q [W/cm^3] in an electrolyte solution is a function of the applied field strength E [V/cm], molar conductivity of electrolyte Λ [cm^2 mol^{-1} Ω^{-1}], and concentration c [mol/1000 cm^3]:

$$Q = E^2 \Lambda c \tag{2.29}$$

For a typical case in CE where E = 500 V/cm (25 kV/50 cm), Λ = 15 cm^2 mol^{-1} Ω^{-1}, and c = 10 mM, Knox[79] calculates a heat generation of Q = 375 W/cm^3. This represents 1570 cal/ml, a heating power sufficient to vaporize any electrolyte many times over. Under real conditions, however, the Joule heat can always be expected to dissipate from the electrolyte through the capillary wall into the medium in which the capillary is placed (i.e., air or thermostatting fluid). As illustrated in Figure 2.50, such heat dissipation can be indicated by the temperature at four different points across the profile of a capillary. T_1 is the temperature at the midpoint of a capillary, T_2 is the temperature at the inside capillary wall, T_3 can be measured at the outside wall of the capillary, and T_4 is the ambient temperature. Knox[79] evaluated three parameters termed *temperature excess*, θ [K]. The temperature excess (gradient) within the electrolyte in the direction from the center toward the wall is $\theta_{core} = T_2 - T_1$. The temperature gradient across the capillary wall is defined as $\theta_{wall} = T_3 - T_2$, and finally $\theta_{ambient} = T_4 - T_3$ is the temperature excess causing the final heat exchange step between the capillary wall and the environment.

The temperature profile within the capillary is represented as a parabolic curve. Another report,[80] based on a more detailed treatment, postulates a

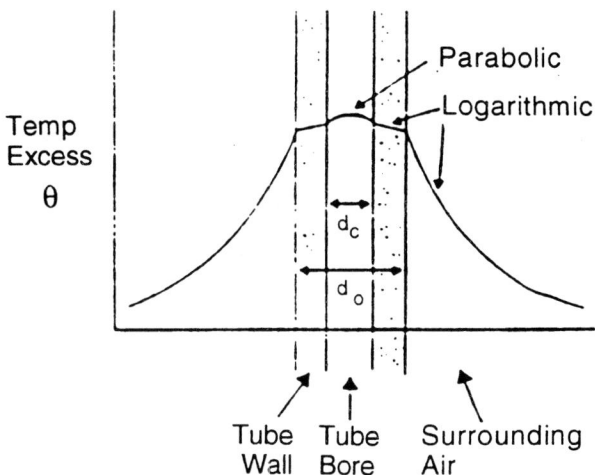

Figure 2.50 Semiquantitative representation of a temperature profile across a tube containing electrolyte heated by passage of an electric current. (Reproduced with permission.[79])

Bessel function as an appropriate description of that temperature profile and provides a discussion of temperature changes in a capillary having polyimide coating on the outside wall. Since the limiting heat transfer step is from the outside capillary wall into the environment and the condensed phases all have thermal conductivities of a similar order of magnitude and sufficiently different from gases (see Table 2.6), it is possible to discuss the heat release problem with the simplified diagram shown.

The temperature gradients can be defined by the following relationships:

$$\theta_{core} = Qd_c^2/16\kappa = (E^2 \Lambda c)(d_c^2/16\kappa) \tag{2.30}$$

giving $\theta_{core} = 0.59$ K, for $E = 500$ V/cm (25 kV/50 cm), $d_c = 100$ μm, and $\kappa = 0.4$ W m^{-1} K^{-1}.

$$\theta_{wall} = (Q\,d_c^2/8\,\kappa_{wall})\ln(d_o/d_c) \tag{2.31}$$

giving $\theta_{wall} = 0.11$ K with the same separation field as in Equation (2.31) and with $d_o/d_c = 2.0$, $\kappa_{wall} = 1.0$ W m^{-1} K^{-1}.

Table 2.6. TYPICAL THERMAL CONDUCTIVITIES, κ [W m^{-1} K^{-1}], FROM REFERENCE 79.

Air	0.025
Water	0.60
Methanol	0.20
Quartz	1.40
Borosilicate glass	1.0

As we can see, the temperature gradients inside the capillary electrolyte and within the capillary wall are only a fraction of a degree under standard CE conditions. This of course is true only if the third heat transfer step dissipates the Joule heat fast enough into the environment to prevent temperature increases in the walls and in the carrier electrolytes. The third heat transfer step is indicated by the value of $\theta_{ambient}$. Knox generated the plot in Figure 2.51, allowing the prediction of $\theta_{ambient}$ at several different levels of heating power and for different diameters of borosilicate glass capillary.

As calculated, the heating power provided under regular CE conditions is ~ 300 W/cm³. The need for miniaturization of tubes for electrophoresis is readily apparent from the diagram. Internal diameters bigger than ~100 μm lead to temperature gradients in excess of 50 to 100 degrees following the application of separation voltages. Such temperature gradients can be expected to cause a rapid degassing, boiling, and even evaporation of carrier electrolytes. All three phenomena lead to interruptions of electrical current, making CE separations impossible. It is only with capillaries having internal diameters smaller than 75 μm that the temperature gradients become manageable either by forced air convection—that is, fans blowing air at high rates, or by immersion of capillaries in thermostatted fluids.

According to a recent report,[81] the rates of heat dissipation can be optimized not only by the choice of narrow capillaries but also by the geometry

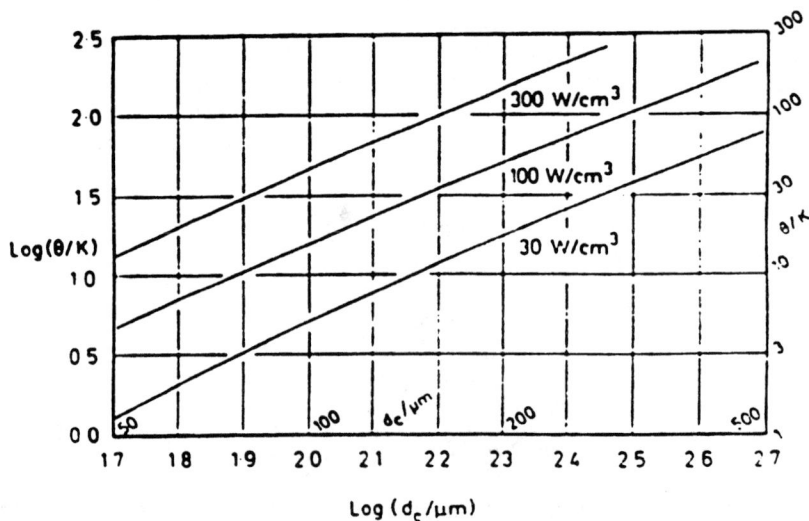

Figure 2.51 Temperature gradient between the capillary wall and ambient air assuming natural convection only. θ/K indicates that temperature gradients are given in degrees Kelvin. (Reproduced with permission.[79])

of the profile of CE tubes. In capillaries with a rectangular profile, the temperature gradients between the electrolyte at the center of the capillary and the outside wall were found to be only a fraction of those found in cylindrical capillaries made of the same material. The same report also describes the advantages of using silicon capillaries in comparison with fused silica or glass. The latter observation appears to be in good agreement with the values of temperature conductivities and the theoretical equations discussed elsewhere in this chapter.

2.7 Peak Spreading and Loss of Resolution under Nonoptimal Conditions in CE

The spreading and deformation of peak zones can be caused by numerous experimental parameters. The optimization of separations is facilitated by the fact that the phenomena causing the peak spreading are recognizable from resulting peak shapes. Thus far, the following causes of deformation and spreading of peaks have been described in the literature: injection artifacts,[82] hydrostatic flow superimposed on the electrokinetic movement of analytes,[82] radial temperature gradients,[79,83] analyte adsorption on the capillary wall,[83] and conductivity and/or pH differences between analyte zones and the bulk of carrier electrolyte.[83,84] In addition to such primary causes of peak deformation, it is also possible to recognize several secondary or composite contributions, for example, the influence of capillary diameter. With large diameters the separation efficiency generally decreases due to larger radial temperature gradients and possibly also to much more pronounced injection artifacts. Not all contributions are equally important. Hydrostatic flow is easily reduced to a minimum by maintaining the electrolyte surfaces at the same level on both sides of a CE system. Low-molecular-weight ions, the main focus of this book, usually exhibit only weak interactions with capillary walls under standard CE conditions. Strong interactions are more frequent with large biomacromolecules. The following discussion of contributions to decreased separation efficiency is limited to those factors that influence peak shapes for low-molecular-weight ionic compounds.

2.7.1 Injection Artifacts

Grushka and MacCormick[82] derived Equation (2.32) relating an "acceptable" increase of plate height E (deterioration of separation efficiency) and length of the sample plug l. The terms l and E are used here as proposed by the authors and are not to be confused with the symbols for current and electric field elsewhere in the text. The sample zone or plug resulting from electromigrative sample introduction can be expected to have an ideally rectangular shape according to the authors. The same is probably true, even if only to a lesser degree, for a sample zone after a hydrostatic sample's introduction.

$$l = (24DEt)^{1/2} \tag{2.32}$$

where D is the diffusion coefficient in cm^2/sec [see also Eq. (2.19)] and t the migration time in seconds. For a small molecule having a diffusion coefficient of approximately 10^{-5}, we can calculate that the maximally admissible length of injection segments is about 0.12 cm. This calculation is based also on an accepted deterioration of separation efficiency by 10% (i.e., $E = 0.01$) and migration time of 600 sec.

The same authors also describe an inadvertent sample introduction into the capillary occurring even before the electromigrative sampling potential or hydrostatic pressure is applied. Three possible mechanisms for such sample introduction extraneous to the intended sample procedure were postulated:

1. Displacement of small sample segments into the capillary caused by turbulence during the insertion of the capillary into the sample solution.
2. Outflow of carrier electrolyte from the capillary and subsequent inflow of a sample volume caused by viscosity, surface tension, and density differences.
3. Diffusion of analytes into the capillary due to incorrect timing of delay or injection periods.

The inadvertent sample injection causes larger than expected lengths of sample segments and in extreme cases leads to a deterioration of separation efficiency, in a seeming contradiction to results of Equations (2.20) and (2.33).

2.7.2 Joule Heat as a Cause of Peak Spreading

In Section 2.6, we discussed the urgent need for narrow diameters of capillaries used in electromigrative separations of ions. We observed that capillary diameters smaller than 100 μm were necessary to prevent boiling or excessive outgassing of carrier electrolytes due to the currents observed at the required values of separation potentials. As we shall see, the temperature gradients, between the center of the capillary and intermediate environment at the capillary outside walls, are a cause of additional concern. If allowed to become sufficiently large, these temperature gradients can lead to excessive peak spreading. If the temperature at the center of a capillary is higher than in the liquid layer close to the capillary wall, it is possible to observe a difference in viscosity between the two points. Equations (2.6) and (2.12) describe the respective dependencies of electrophoretic and electroosmotic mobilities on viscosity. Higher temperature means lower viscosity. A decrease in viscosity leads in turn to an increase of migration velocity

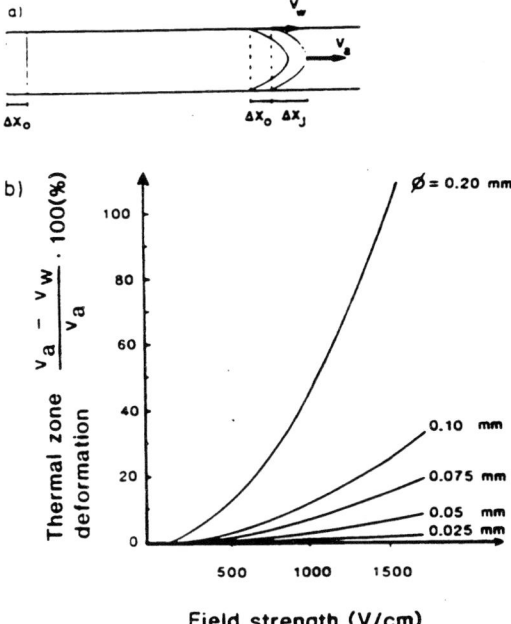

Figure 2.52 Zone distortion by temperature gradients. (a) The longer of the two arrows indicates the greater observed velocity in the center of the capillary. The initial width ΔX_O of the sample plug at time zero grows to $\Delta X_O + \Delta X_J$ during the separation. ΔX_J is the thermal zone broadening at the peak base. (b) Plot of Equation (2.34) for various parametric values of capillary diameter. Other parameters have the following values: $B = 2400$ K, $\kappa = 367$ μS/cm, $\lambda = 5.73 \times 10^{-3}$ J/(sec cm deg) and $T_O = 274$ K. (Reproduced with permission.[83])

in capillary electrophoresis. Figure 2.52a shows the distortion of a peak zone caused by a larger migrational velocity at the center compared with the regions closer to the capillary wall. A quantitative relationship, allowing predictions of the magnitude of peak distortion, $(v_r - v_w)/\omega_w$, for any given value electric current, was derived in 1967 by Hjerten[78]:

$$(v_r - v_w)/v_w = [BI^2(R^2 - r^2)]/(4 \kappa \pi^2 R^4 \lambda T_0^2) \quad (2.33)$$

where v_r and v_w are the respective velocities at the center and at the wall, I the current, B a parameter with a value of 2400 K, R the capillary radius, r the distance from the center of the electrophoretic capillary, κ the electrical conductivity, λ the thermal conductivity, and T_0 the temperature of the environment at the outside capillary wall. For $r = 0$ and $F = I/(\kappa \pi R^2)$, Equation (2.33) simplifies to:

$$(v_a - v_w)/v_w = (B\kappa/\lambda)(FR/2T_0)^2 \quad (2.34)$$

where v_a is the velocity of an analyte at the center of the capillary. Equation (2.34) was used to generate a plot of the dependency of zone distortion $(v_a - v_w)/v_w$ on the field strength in Figure 2.52b.

The diagram in Figure 2.52b can serve as a quick reference for minimizing peak spreading due to temperature gradients. We can see that, for example, at 20 kV and a total capillary length of 60 cm, corresponding to 333 V/cm, the thermal zone deformation is negligible for an internal diameter of 75 μm. For 30 kV and otherwise the same set of parameters (i.e., 500 V/cm), the peak distortion can already approach 5%.

It also obvious that 100 μm I.D. capillaries will exhibit considerable thermal peak distortion for the majority of useful combinations of experimental parameters.

2.7.3 Peak Deformations Caused by Conductivity Differences between Zone of Analytes and the Carrier Electrolyte

Considering the separation of alkylsulfonates shown in Figure 2.53, it is possible to distinguish clearly between three types of peaks. Peaks 1–3 exhibit a type of asymmetry that in an HPLC separation would be called *fronting*. The fourth peak corresponding to butanesulfonate is symmetrical, whereas the last three peaks show a strong tailing, if column chromatographic terminology is to be used again.

As explained by Mikkers, Everaerts, and Verheggen[84] in 1979, such peak asymmetries can be predicted with the help of relative values of electrophoretic mobilities or ionic equivalent conductances [recall that: μ_{ion} = equivalent conductance/Faraday constant, see Eq. (2.3)]. The analytes having higher electrophoretic mobilities than that of the main coion in the carrier electrolyte (i.e., electrolyte ion carrying the same charge as the analyte ions—benzoate anion in Figure 2.53) can be expected to give fronting peaks. Tailing peaks are then due to analytes that are less mobile than the electrolyte coions. Only those sample components with electrophoretic mobilities closely matching that of a given electrolyte coion can be expected to produce fully symmetrical, Gaussian peaks.

A quantitative explanation of peak asymmetry caused by conductivity/mobility differences was developed recently by Hjerten.[83] The conductivity difference between a sample zone and the carrier electrolyte $\Delta\kappa$ is described as:

$$\Delta\kappa = (c_B/\mu_A)(\mu_A - \mu_B)(\mu_R - \mu_B) \qquad (2.35)$$

with subscripts B, A, and R corresponding, respectively, to sample ion, electrolyte coion, and electrolyte counterion, μ standing for inherent electrophoretic mobilities, and c_B the amount of charge in coulombs/ml transported by the analyte ion [c_B = (valency × molar concentration of B × Faraday constant)/1000]. According to the widely recognized convention, the mobil-

FUNDAMENTALS OF CAPILLARY ELECTROPHORESIS 71

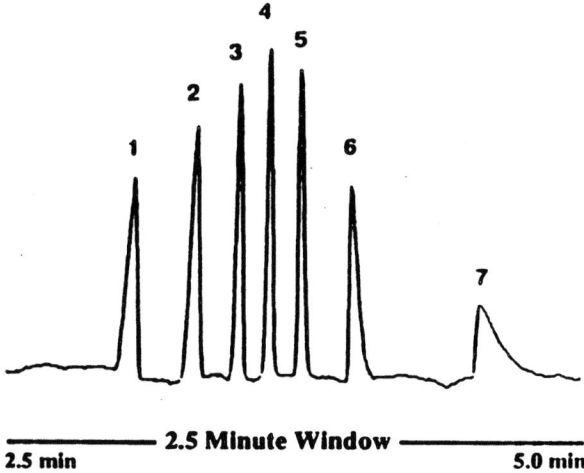

Figure 2.53 Separation of homologous alkylsulfonates. Carrier electrolyte: 5 mM benzoate, pH 6.0, containing 0.5 mM Waters OFM BT reagent for the reversal of the electroosmotic flow. Separation voltage was at −20 kV. Indirect UV detection was carried out at 254 nm. Fused silica capillary dimensions: 60 × 52 cm × 75 μm. Peak identities: 1, methanesulfonate; 2, ethanesulfonate; 3, propanesulfonate; 4, butanesulfonate; 5, pentanesulfonate; 6, hexanesulfonate; and 7, heptanesulfonate. All analytes at 10 ppm. (Reproduced with permission.[85])

ities and concentration of cations have a positive sign, while the same parameters are negative for anions.

To explain the mechanism causing peak distortion, we shall first consider a case where the conductivity of a sample zone is lower than that of a carrier electrolyte solution. This is often due to a relatively higher mobility of carrier electrolyte ions ($\mu_A > \mu_B$). We will also assume that the rate of diffusion is exceeded by the rate of electrophoretic migration: $v^\beta > v_{\text{diff}}$ in Figure 2.54. The upper diagram in Figure 2.54 shows two situations for a sample zone inside the capillary. On the left side, we see a sample zone before the application of separation potential. In the absence of inadvertent injection discussed in Section 2.7.1, such a sample zone will have a rectangular shape and a width of ΔX_0. The movement of ions by diffusion is independent of the application of separation voltage, and diffusion thus occurs even before the actual separation begins. This is indicated by the presence of three different types of analyte ions in the diagram. M_3 are the analyte ions transferred by diffusion into the electrolyte segment to the left of the sample zone, and M_2 are the sample ions unable to escape from the sample zone. M_1 then signifies all the analyte ions diffusing from the sample zone into the carrier electrolyte deeper in the capillary.

On the right side of the diagram, we see the sample zone some time after the application of a separation potential. The three types of sample ions are

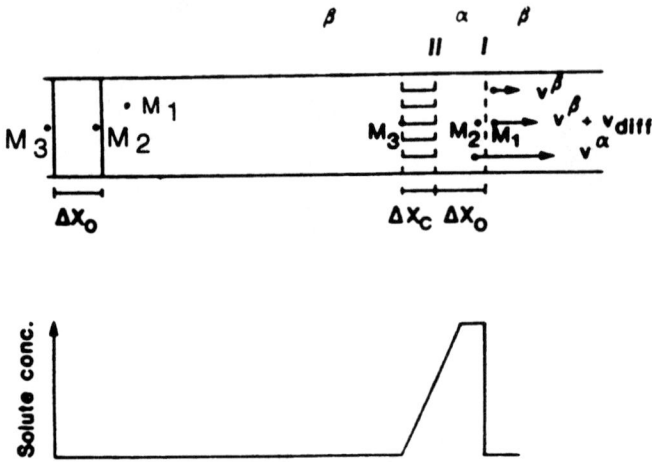

Figure 2.54 Zone broadening caused by a conductivity difference between sample (α) and carier electrolyte (β) zones. v_{diff} represents the rate of diffusion. See discussion in the text for additional explanations. (Reproduced with permission.[83])

now going to be exposed to different values of field strength resulting from conductivity differences between the sample and carrier electrolyte zones. Under the conditions of the present experiment, the sample zone α is less conductive and will thus experience a higher field strength than the rest of the capillary filled with carrier electrolyte β. Ions with identical electrophoretic mobilities will migrate faster when exposed to a higher electric field [see Eq. (2.2)]. The velocity of analyte ions in the sample zone v^α is thus higher than that of analyte ions in the carrier electrolyte v^β. Sample ions M_1 will thus always be overtaken by the M_2 ions migrating at the same velocity as the sample zone. This situation leads to a creation of a sharp and well-defined Boundary I. In the concentration versus migration distance plot shown in the lower part of Figure 2.54, this is indicated by a perpendicular line for Boundary I. The ions M_3 transferred by diffusion from the sample in the time period preceding the separation, as well as from the migrating sample zone α into the electrolyte segment β following the sample zone, are experiencing a lower electric field. Their migration rates will thus also be slower than for analyte ions remaining in the sample zone. However, due to their position relative to the migrating zone, M_3 ions will lag behind by the distance ΔX_O:

$$\Delta X_O = L(\Delta \kappa / \kappa^\beta) \tag{2.36}$$

where $\Delta \kappa$ is the conductivity difference calculated from Equation (2.35) and κ^β is the conductivity of the carrier electrolyte. L then signifies that distance, which an individual ion M_3 traveled in the electrolyte behind the sam-

FUNDAMENTALS OF CAPILLARY ELECTROPHORESIS   73

ple zone. Ions are diffusing not only before the analysis, but also throughout the separation across Boundary II. Different values of L caused by diffusion for individual M_3 ions thus explain the sloping line for Boundary II in the concentration versus migration distance plot shown in the lower part of Figure 2.54.

Finally, we have to recall that the concentration plot in Figure 2.54 shows the dependence on distance only. The electropherograms, on the other hand, are plotted versus migration time. In this frame of reference, Boundary I will be recorded first (ions traveling the longest distance at any given time will be detected first), and the analyte peak is going to have a characteristic "tailing" shape as observed for the low-mobility analyte peaks in Figure 2.53. An analogous explanation of peak fronting can be developed along the same lines, simply by reversing the relative values of v^α and v^β.

In his 1990 report, Hjerten also discusses the case where the rate of diffusion exceeds the rate of electromigration, and shows that the resulting asymmetry will be much less pronounced than that indicated previously. It

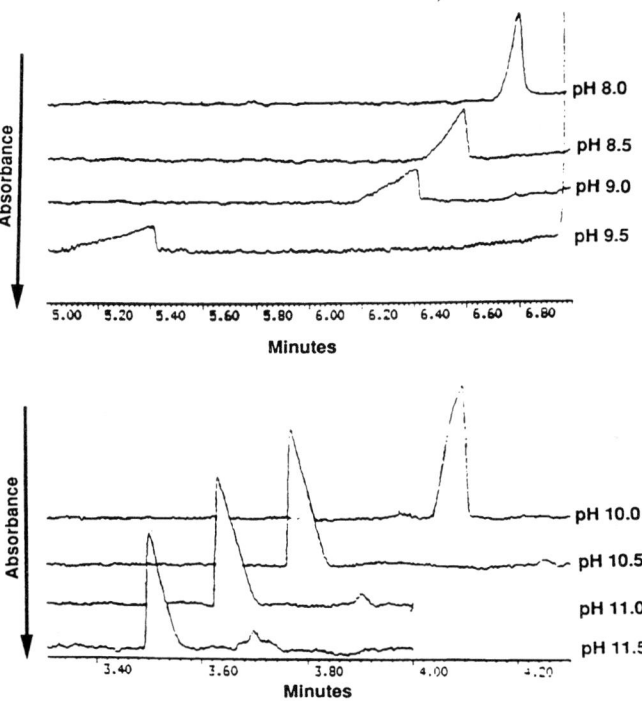

Figure 2.55 Peaks for boric acid at several different pH values of carrier electrolyte (5 mM sodium chromate and 5 mM Waters OFM BT). Capillary: 60 × 52 cm × 75 μm. Separation voltage: −20 kV. Indirect detection at 254 nm. (Reproduced with permission.[86])

is also interesting to note that the pH-caused peak distortions can be explained using a similar framework as for asymmetries stemming from different values of electrophoretic mobilities. The asymmetries due to pH are observed, for example, for weak acids such as boric acid in Figure 2.55.

In this series of electropherograms, we observe not only shifting migration times but also changing types of asymmetries for the borate peak depending on the pH of the electrolyte. Recalling that the pK value for borate lies at 9.24, it is possible to explain the observed gradual shortening of migration times in going from a lower to a higher pH of the electrolyte. The dissociation form of borate, which is dictated by the electrolyte pH, slowly shifts from a low percentage of mobile anion to a higher concentration of that mobile species. As discussed in Section 2.4, the observed mobility of an ion in equilibrium between mobile and less mobile species is equal to the sum of the respective mobility–molar-fraction products. Borate is a slower-migrating anion than chromate, which explains the tailing shapes of peaks between pH 10.5 and 11.5. The fronting at pH values below 10.5 is explainable by a pH influence. As long as the pH of an electrolyte leaves considerable portions of borate in boric acid form, the autodissociation of boric acid will produce a relatively high concentration of highly mobile hydronium ions, increasing the sample zone conductivity over that of the carrier electrolyte. The higher sample zone conductivity is then responsible for the fronting of peaks at lower values of pH.

2.7.4 Peak Resolution under Nonoptimal Conditions in CE

A peak deformation usually has more than one cause. The electropherogram in Figure 2.55 is a good illustration of multiple causes of peak spreading. It shows peak broadening due to diffusion alone (discussed in Section 2.5) as well as three additional effects leading to distortions of analyte peaks. Using a similar approach as Jorgensen,[77] Hjerten[83] arrives at the complex expression in Equation (2.37), summarizing all known parameters affecting resolution under nonoptimal conditions. As stated in Section 2.5, under optimal conditions the only cause of peak broadening in CE is molecular diffusion. Additional causes of peak spreading convert any given CE system from optimal to nonoptimal.

$$R = (\Delta\mu/\mu) \, [(\Delta X_0^2/12L^2) + (2D/\mu FL) + (1/12)(B\kappa/\lambda)^2 \, (RF/2T_0)^4 + AF/L]^{-1/2} \qquad (2.37)$$

From Equation (2.37), it follows that under nonoptimal conditions the resolution increases with decreasing observed mobility (μ), width of the sample plug (ΔX_0), coefficient of diffusion (D), and conductivity of carrier electrolyte (κ). On the other hand, the resolution deteriorates with longer capillaries (L) and higher thermal conductivity of carrier electrolyte. The term A in Equation (2.37) is a proportionality constant between the height of a theoretical plate and field strength. It should be stressed again that this equation, in its full form, becomes relevant only under nonoptimal conditions. The ability to optimize CE separations to near-optimal conditions is within the reach of educated users of the technique. The term ΔX_0 is minimized by reducing sample volume or by utilizing isotachophoretic pretreatment of sample segments prior to separation. The influence of electrical conductivities and thermal effects is largely eliminated by matching closely the equivalent conductivities of the analytes to those of carrier electrolytes. Additionally, thermal distortions are also reduced by choosing smaller capillary diameters. The fact, that large portions of CE method optimization are within the competence of an average operator represents a feature unique among analytical separation methods. With the elimination of artifacts discussed in Sections 2.7.1 to 2.7.3, Equation (2.37) simplifies to a form similar to the ideal resolution equation, which results from the combination of Equations (2.25) and (2.26).

References

1. R. G. Brownlee and S. W. Compton, *American Biotechnology* **6**, 10 (1988).
2. *Handbook of Laboratory Safety*, 2nd Edition, p. 521, Norman E. Steere ed., CRC Press, Inc., Boca Raton, FL, 1971.
3. US Patent 4,909,919.
4. RHR Series High Voltage Power Supply, Spellman Corp., Plainview, NY.
5. Isco Technical Bulletin, ISCO, Inc., Lincoln, NE.
6. Knauer Technical Bulletin, Knauer GmbH, Berlin, Germany.
7. Linear UVIS 200 Detector, Care and Use Manual, Linear Instruments Corp., Reno, NV.
8. SpectroVision, AD-200 Absorbance Detector, SpectroVision, Inc., Chelmsford, MA.
9. K. D. Lukacs and J. W. Jorgenson, *HRC* **8**, 407 (1985).
10. *Handbook of Chemistry and Physics*, 66th Edition, CRC Press, Boca Raton, FL, 1985.
11. D. T. Gjerde and J. S. Fritz, *Ion Chromatography*, 2nd Edition, Huethig Verlag, Heidelberg, 1987.
12. W. R. Jones and P. Jandik, *J. Chromatogr.* **546**, 445 (1991).
13. A. W. Adamson, *A Textbook of Physical Chemistry*, Academic Press, New York, NY, 1973.
14. D. J. Pennino, BioPharm September 1989, p. 41.
15. P. D. Grossman, J. C. Colburn, and H. H. Lauer, *Analytical Biochemistry* **179**, 28 (1989).
16. L. G. Daignault, D. P. Rillema, and D. C. Jackman, *HRC* **14**, 293 (1990).
17. P. D. Grossmann, J. C. Colburn, H. H. Lauer, R. G. Nielsen, R. M. Riggin, G. S. Sittampalam, and E. C. Rickard, *Anal. Chem.* **61**, 1186 (1989).

18. B. J. Herren, S. G. Shafer, J. Van Alstine, J. M. Harris, and R. S. Snyder, *J. Colloid Interface Sci.* **115**, 46 (1987).
19. D. W. Fuerstenau, *J. Phys. Chem.* **60**, 981 (1956).
20. P. Somasundaran, T. W. Wealy, and D. W. Fuerstenau, *J. Phys. Chem.* **68**, 3562 (1964).
21. P. McFayden, *American Laboratory* **19**, 64 (1987).
22. R. J. Hunter, *Zeta Potential in Colloid Science*, Academic Press, New York, 1981.
23. A. W. Adamson, *A Textbook of Physical Chemistry*, Academic Press, New York, 1973.
24. D. J. Shaw, *Electrophoresis*, Academic Press, New York, 1969.
25. M. v. Smoluchowski, *Physik. Z.* **6**, 530 (1905).
26. C. H. Hamann and W. Vielstich, *Electrochemie I*, 2nd Edition, VCH, Weinheim, Germany, 1985.
27. Technical Bulletin, Polymicro Technologies, Inc. Phoenix, AZ, 1991.
28. K. K. Unger, Porous Silica, *J. Chromatogr. Library*, Vol. 16, Elsevier Co. Amsterdam, 1979.
29. M. L. Hair and W. Hertl, *J. Phys. Chem.* **74**, 91 (1970).
30. W. J. Lambert and D. L. Middleton, *Anal. Chem.* **62**, 1585 (1990).
31. J. C. Reijenga, G. V. A. Aben, T. P. E. M. Verheggen, and F. M. Everaerts, *J. Chromatogr.* **260**, 241 (1983).
32. T. Tsuda, K. Nomura, and G. Nakagawa, *J. Chromatogr.* **264**, 385 (1983).
33. X. Huang, J. A. Luckey, M. J. Gordon, and R. N. Zare, *Anal. Chem.* **61**, 766, (1989).
34. K. D. Altria and C. F. Simpson, *Chromatographia* **24**, 527 (1987).
35. P. Mukerjee and K. J. Mysels, Critical Micelle Concentrations of Aqueous Surfactant Systems, National Bureau of Standards (U.S.), Washington D.C., 1971.
36. M. Chen and R. M. Cassidy, *J. Chromatogr.* **602**, 227 (1992).
37. S. Hjerten, *J. Chromatogr.* **347**, 191 (1985).
38. C. S. Lee, W. C. Blanchard, and C. T. Wu, *Anal. Chem.* **62**, 1550 (1990).
39. C. S. Lee, D. McManigill, C. T. Wu, and B. Patel, *Anal. Chem.* **63**, 1519 (1991).
40. K. Ghowsi and R. J. Gale, *American Laboratory* **23**, 17 (1991).
41. H. F. Yin, M. H. Christmann, J. A. Lux, and G. Schomburg, *J. Microcolumn Sep.* (in press).
42. A. S. Cohen, A. Paulus, and B. L. Karger, *Chromatographia* **24**, 15 (1987).
43. A. Guttman, A. S. Cohen, D. N. Heiger, and B. L. Karger, *Anal. Chem.* **62**, 137 (1990).
44. J. A. Lux, H. F. Yin, and G. Schomburg, *HRC* **13**, 436 (1990).
45. J & W Scientific, High Resolution Chromatography Products, 1991 Catalogue, Folsom, CA.
46. H. F. Yin, J. A. Lux, and G. Schomburg, *HRC* **13**, 624 (1990).
47. M. Zhu, D. L. Hansen, S. Burd, and F. Gannon, *J. Chromatogr.* **480**, 311 (1989).
48. S. Nathakarnkitkool, P. J. Oefner, A. M. Chin, G. K. Bonn, and G. Bartsch, International Symposium on Capillary Electrophoresis, Poster No. 8, August 22–24, York, UK, 1990.

49. S. Nathakarnkitkool, P. J. Oefner, A. M. Chin, G. K. Bonn, and G. Bartsch, *Electrophoresis* **13**, 18 (1992).
50. S. Hjerten, *Chromatogr. Rev.* **9**, 122 (1967).
51. J. W. Jorgenson and K. D. Lukacs, *Anal. Chem.* **53**, 1298 (1981).
52. F. E. P. Mikkers, F. M. Everaerts, and T. P. M. Verheggen, *J. Chromatogr.* **169**, 11 (1979).
53. P. Jandik, W. R. Jones, A. Weston, and P. R. Brown, *LC GC* **9**, 634 (1991).
54. A. Weston, P. R. Brown, P. Jandik, W. R. Jones, and A. L. Heckenberg, *J. Chromatogr.* **593**, 289 (1992).
55. S. Terabe and T. Isemura, *Anal Chem.* **63**, 650 (1990).
56. P. Gebauer, M. Deml, P. Bocek, and J. Janak, *J. Chromatogr.* **267**, 455 (1983).
57. F. Foret, S. Fanali, A. Nardi, and P. Bocek, *Electrophoresis* **11**, 780 (1990).
58. P. Gozel, E. Gassmann, H. Michelsen, and R. N. Zare, *Anal. Chem.* **59**, 44 (1987).
59. S. H. Kuhn, A. Paulus, E. Gassmann, and H. M. Widmer, *Anal. Chem.* **63**, 1541 (1991).
60. T. Kaneta, S. Tanaka, and H. Yoshida, *J. Chromatogr.* **538**, 385 (1991).
61. Y. Walbroehl and J. W. Jorgenson, *Anal. Chem.* **58**, 479 (1986).
62. S. Terabe, K. Otsuka, K. Ichikawa, A. Tsuchia, and T. Ando, *Anal. Chem.* **56**, 111 (1984).
62a. J. Vindevogel and P. Sandra, Introduction to Micellar Electrokinetic Chromatography, Hüthig Verlag, Heidelberg (1992).
63. H. Nishi, N. Tsumagari, and S. Terabe, *Anal. Chem.* **61**, 2434 (1989).
64. E. Gassmann, J. Kuo, and R. Zare, *Science* **230**, 813 (1985).
65. R. Weinberger and I. S. Lurie, *Anal. Chem.* **63**, 823 (1991).
66. A. Weston, P. Jandik, and P. R. Brown, unpublished work.
67. S. Terabe, H. Ozaki, K. Otsuka, and T. Ando, *J. Chromatogr.* **332**, 211 (1985).
68. S. Fanali, *J. Chromatogr.* **474**, 441 (1989).
69. A. Nardi, S. Fanali, and F. Foret, *Electrophoresis* **11**, 774 (1990).
70. B. B. VanOrman and G. L. McIntire, *J. Microcol. Sep.* **1**, 289 (1989).
71. A. Zhu and Y. Chen, *J. Chromatogr.* **470**, 251 (1989).
72. H. K. Jones and N. E. Ballou, *Anal. Chem.* **62**, 2484 (1990).
73. D. A. Skoog and D. M. West, *Principles of Instrumental Analysis*, Holt, Rinehart and Winston, New York, 1971.
74. J. C. Giddings, *Separation Science* **4**, 181 (1969).
75. H. M. Wadsworth, *Statistical Methods for Engineers and Scientists*, McGraw–Hill Publishing Company, New York, 1990.
76. H. Engelhardt, *High Performance Liquid Chromatography*, English Edition, Springer Verlag, New York, 1977.
77. J. W. Jorgenson and K. D. Lukacs, *Anal. Chem.* **53**, 1298 (1981).
78. S. Hjerten, *Chromatogr. Rev.* **9**, 122 (1967).

79. J. H. Knox, *Chromatographia* **26**, 329 (1988).
80. A. E. Jones and E. Grushka, *J. Chromatogr.* **466**, 219 (1989).
81. M. Jansson, A. Emmer, and J. Roeraade, *HRC* **12**, 797 (1989).
82. E. Grushka and R. M. McCormick, *J. Chromatogr.* **471**, 421 (1989).
83. S. Hjerten, *Electrophoresis* **11**, 665 (1990).
84. F. E. P. Mikkers, F. M. Everaerts, and T. P. E. M. Verheggen, *J. Chromatogr.* **169**, 1 (1979).
85. W. R. Jones and P. Jandik, *Amer. Lab.* **22**, 51 (1990).
86. W. R. Jones and P. Jandik, *J. Chromatogr.* **546**, 445 (1991).

CHAPTER

3

Instrumentation for Capillary Electrophoresis

The brief discussion of elements of CE systems given in Section 2.1 provides an introductory description of modules for readers without previous experience in the method. In the following sections of this chapter, the operation of CE systems is discussed in more detail. The prerequisites for optimal operation have only recently become sufficiently clear to allow a comprehensive discussion. Selected questions on instrument design are also presented. Stress is placed on the more common types of instruments. The experimental approaches to various CE modules are reviewed briefly, and literature references are given for those design solutions deemed most promising by the authors.

3.1 Sample Introduction

The choice of operating parameters for sample introduction in CE influences the final results to a much greater degree than for sample injection in HPLC. Sample size in CE is determined by the seconds of duration of the sample introduction procedure. Consequently, most CE operators do not know the volume of sample introduced for a given separation. Yet, as we shall see, too large a sample volume can affect the efficiency of a CE separation. On the positive side, CE sampling, if performed properly, can increase the detectability by one or two orders of magnitude. The effect responsible for sample preconcentration under optimized conditions is known as *electrostacking*. The large difference between HPLC and CE is properly indicated by the use of word *introduction* instead of *injection* for the latter of the two separation methodologies. Actual injection of samples with syringes is much less frequently utilized in CE than with the other separation techniques.

The following is a list of sample introduction techniques in CE:

1. Hydrodynamic sample introduction;
2. Electromigrative sample introduction;
3. Alternative approaches to sampling (split flow injection, sample loop injectors, sample gating, electric sample splitting, etc.).

Before discussing the design features and proper functioning of the sample introduction modes listed, we should acquire a degree of understanding of electrostacking as a unique feature of capillary electrophoresis, shared only with some other electrokinetic methods such as isotachophoresis, gel or paper electrophoresis. In one form or another, electrostacking can be observed with all sample introduction methods in capillary electrophoresis.

3.1.1 Electrostacking

Let us consider the phenomenon of electrostacking as reported by Mikkers, Everaerts, and Verheggen in 1979.[1] The authors noticed a striking difference between two electropherograms of identical levels of analytes from two different sample matrices.

The separation on the left side of Figure 3.1, obtained from a dilute aqueous solution, has much higher and sharper peaks than the other electropherogram of the sample dissolved in the carrier electrolyte. Similar effects have been reported for peptide samples[2] and for amino acid derivatives.[3]

These observations are somewhat unexpected for anyone familiar only with standard procedures common in HPLC. In liquid chromatographic techniques, it is a good practice to dissolve sample constituents in the mobile phase (eluent, buffer, etc.) whenever possible. The procedure minimizes matrix peaks and uncovers possible solubility problems before the actual sample injections into the system. Experienced HPLC operators expect improved "appearance" of chromatographic recordings after injecting samples prepared in eluents. Such expectations are often extended to sample preparation for CE. Without an understanding of basic principles of electrostacking, an operator with LC experience would automatically dissolve samples in carrier electrolyte. The results would then in many cases (i.e., not for small sample volumes) be similar to those in the electropherogram on the right-hand side of Figure 3.1.

Improved peak shapes and sensitivity in capillary electrophoresis from more dilute sample solutions can be explained with the help of the theoretical framework presented in Chapter 2. According to Equation (2.2), the electrophoretic velocity, and thus also the rate of migration under conditions of capillary electrophoresis, increases in direct proportion to the applied electric field.

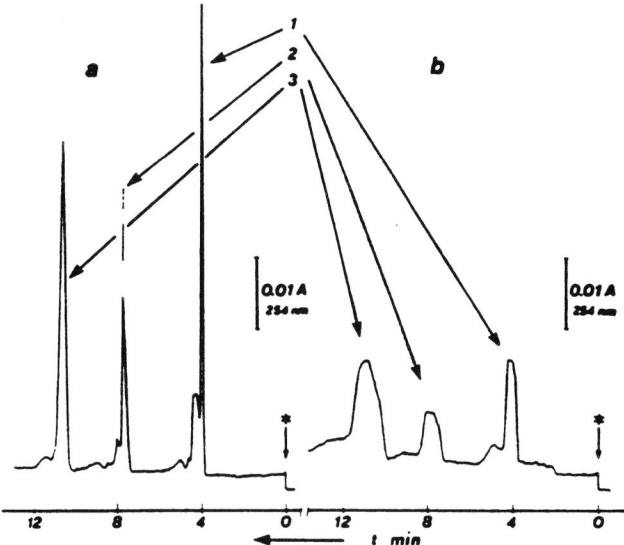

Figure 3.1 Two separations of identical concentrations. (a) Constituents dissolved in water. (b) Constituents dissolved in carrier electrolyte. Carrier electrolyte: 0.1 M γ-aminobutyric acid/acetate, 0.1% hydroxyethylcellulose, pH 4.00. Capillary: PTFE, 200 μm. Sample introduction by a sampling valve with a sampling volume of 0.7 μl. Driving current: 100 μA. UV detection. Peak identities: (1) sulfanilic acid, 70 picomoles, (2) 5-bromo-2,4-dihydroxybenzoic acid, 140 picomoles, (3) adenosine-5'-monophosphoric acid, 35 picomoles. Reproduced with permission.[1]

Prior to the application of the separation voltage, the sample components are distributed in a relatively broad segment filled predominantly by sample solvent or by other components from the sample matrix (Figure 3.2A). The electrical resistance of a dilute sample is higher than that of the carrier electrolyte inside the capillary. After application of the separation voltage, the electric field across the sample segment is much higher than across the carrier electrolyte ($U = RI$), resulting in a considerably higher velocity for sample ions in the sample segment. Rapidly moving sample ions inside the sample zone reach the electrolyte boundary as the sample ions that are already inside the electrolyte begin their migration toward the detector (Figure 3.2B). Once inside the electrolyte, the sample ions are slowed down to the much slower velocity determined by the lower electric field and shared by all other ions of the same chemical identity (Figure 3.2C). In most CE separations, the width of the solvent peak is indicative of the original length of the sample plug. In Figure 3.2C, the sample zone is shown moving behind the more rapidly migrating zone of ionic analyte. The mobility of the solvent zone is usually determined by the electroosmotic flow.

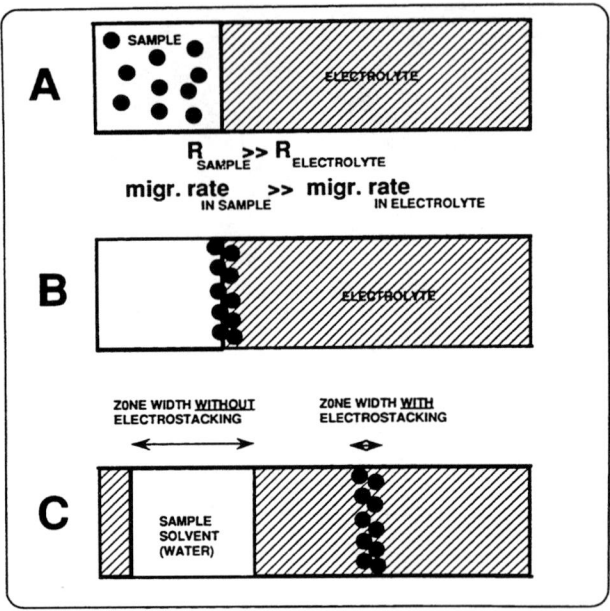

Figure 3.2 The role of electrostacking in capillary electrophoresis.

Under ideal conditions, the zone of ionic sample components retains its narrow character, yielding high separation efficiencies (see Section 2.5), during its migration through the capillary. However, it is also possible for analyte zones to be deformed and/or broadened during their migration, as discussed in Section 2.7.

Quantitative treatment of the sensitivity gain by electrostacking is more involved than the simplified picture presented in Figure 3.2. The field enhancement factor,[4] the ratio between the respective field strength across the sample zone (E_1) and across the electrolyte (E_0), is given by:

$$E_1/E_0 = \gamma/[\gamma x + (1 - x)] \quad (3.1)$$

where L[cm] is the total length of the capillary, x is the ratio of sample plug to L, and γ the ratio of conductivities of the electrolyte versus the sample zone. The complexity of the electrostacking mechanism arises from the fact that the enhancement ratio does not remain constant. It changes rapidly due to changes of ionic concentrations caused by electromigration and electroosmosis. The most complete mathematical treatment of electrostacking to date can be found in the report by Chien.[4]

3.1.2 Sample Introduction by a Pressure Differential

The introduction of samples into fused silica capillaries by means of differential pressure has become the most popular method in capillary electrophoresis. There are three main techniques of creating a pressure differential, and all of them have been incorporated in commercial CE systems.

The hydrodynamic, siphoning, or hydrostatic method is depicted in Figure 3.3A. For sample introduction, a sample vial is raised to a defined level (Δh) above the electrolyte vessel at the other end of the capillary. Provided that the capillary contains an uninterrupted column of electrolyte fluid, the hydrostatic pressure derived from the weight of liquid in a length Δh of the raised capillary leads to a sampling rate $S_{(\Delta h)}$[cm/sec]:

$$S_{(\Delta h)} = \rho g r^2 \Delta h / 8 \eta L_t \tag{3.2}$$

where ρ is the electrolyte density [g/ml], r the capillary internal radius [cm], $g = 9.81$ cm/sec^2 the standard acceleration of gravity, $\eta = 0.01$ P the viscosity of water at 20°C, and L_t the total length of capillary in centimeters. The volume sampled is then determined by the time interval for which the sample container is held at the elevated level Δh. Equation (3.2) is a classical relationship describing the flow of fluid in a tube and is reproduced here in

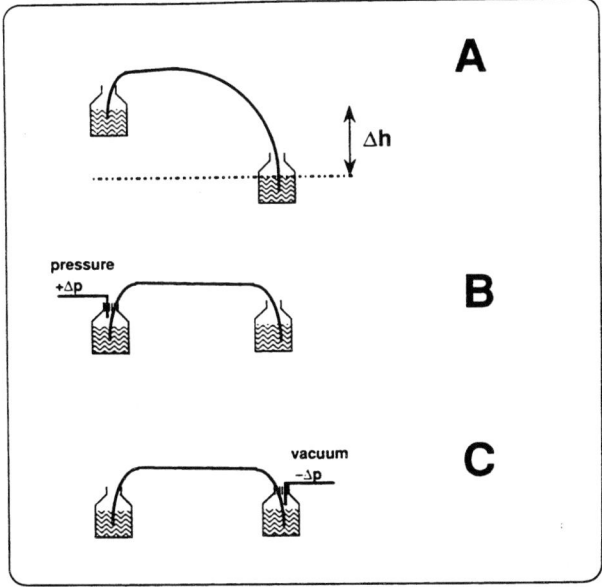

Figure 3.3 The three most common ways of creating a pressure difference for sample introduction in CE. The line connecting two vials signifies a capillary.

the same form as chosen by Rose and Jorgenson[5] in their 1988 article. The same authors also discussed another standard relationship allowing the prediction of the amount (Q in moles) introduced by a known hydrostatic pressure into a capillary with a defined geometry:

$$Q = [\rho g \pi r^4 \Delta h \, t_i / (8\eta L_t)] \, C \tag{3.3}$$

where t_i signifies the sample introduction period in seconds and C is the molar concentration.

Before discussing other differential pressure methods, we should make a note of another important use of Equation (3.2). During any CE separation, the liquid levels inside the electrolyte vessels should be at an identical level to that indicated by the dashed horizontal line in Figure 3.3A. If the two levels are not adjusted properly, the result may be a hydrodynamic flow affecting the separation efficiency and also the reproducibility of the migration times. As discussed in Section 2.5, the high efficiency of CE separations is based on the flat profile of the electroosmotic flow.

If the contribution of hydrodynamic flow is allowed to become significant, the flow profile becomes increasingly parabolic, contributing to the spreading of separated analyte zones. Combined use of Equations (2.2) and (3.2) makes it possible to compare relative rates of electroosmosis and hydrodynamic flow in same units [cm/sec]. Another easy calculation yields the value of maximum height difference that can be tolerated between two electrolyte levels.

Two other ways of generating controlled hydrodynamic flow of sample into the capillary are either by pressurizing the sample vial, Figure 3.3B, or by an application of vacuum to the electrolyte vessel at the opposite end of capillary, Figure 3.3C. The sample vial to which the Δp is directly applied is hermetically sealed (sample vial in Figure 3.3B, electrolyte vessel in Figure 3.3C). The sample rate $S_{(\Delta p)}$ [cm/sec] is then expressed as:

$$S_{(\Delta p)} = (\pi \Delta p \, r^2) / (8\eta L_t) \tag{3.4}$$

where Δp is the pressure differential [$g/(cm \, sec^2)$] : 1 atm = 14.696 psi = 1.013×10^6 $g/(cm \, sec^2)$ = 1.01×10^5 Pa.

In many CE instruments the pressure differential is applied not only for sampling, but also to purge capillaries and fill them with a new portion of carrier electrolyte. Unlike in HPLC, where the operator almost always has to wait for all sample components to elute from the column, in CE it is possible to interrupt the separation after detecting the peaks of interest. Since purging can be done at flow rates far exceeding the electrophoretic migration and electroosmosis, considerable savings in the time required for analysis can be accomplished. Equation (3.4) can be used to calculate purging flow rates and rinsing volumes in CE instruments equipped with the vacuum purging mechanism.

Without the benefit of electrostacking, as for example when the sample is first dissolved in the carrier electrolyte, the length of the introduced sample zones is severely restricted. Otsuka and Terabe[6] calculated the maximum lengths of sample zones under nonstacked conditions. For a 50 cm/50 μm capillary, they obtained 0.56 mm of injection segment length assuming that a separation efficiency of 500,000 plates not be decreased by more than 10%. A large number of CE reports utilize injection times of 30 sec with $\Delta h = 10$ cm. In a 50 μm/50 cm capillary this generates sample plugs ~ 5 mm long, Equation (3.1). Yet, even with the plug lengths far exceeding the maximum admissible values derived for nonstacking conditions, the separation efficiencies are within the maximum theoretical range of 100,000 to 500,000 plates. Electrostacking is thus widely relied upon to make possible sampling volumes that would otherwise not be possible due to restrictions imposed by the expected separation efficiency.

The only peak not narrowed by electrostacking is that belonging to the sample solvent (i.e., water). The solvent peaks are usually not recorded under the conditions of coelectroosmotic capillary electrophoresis (see Section 2.4). Such peaks, marking the electroosmotic mobility, are found migrating far behind the peaks belonging to the analytes of interest. In coelectroosmotic capillary electrophoresis, the separation is usually interrupted before the solvent peaks reach the detector. The zones containing the sample solvent are subsequently removed by vacuum purging of capillaries in the process of replacing the spent carrier electrolyte by a portion of new one.

The effective length of sample zone is determined not only by sampling time, but also by diffusion and inadvertent hydrodynamic flow from the misadjustment of electrolyte levels on the two sides of the capillary.[7] Immediately after the establishment of the concentration boundary between the sample zone and the electrolyte column inside the capillary, the sample begins to diffuse into the capillary. Sample introduction by diffusion can assume significant proportions even in the absence of inadvertent hydrodynamic flow. Dose and Guichon[7] calculate that in 9 cm/50 μm I.D. capillaries the amount of sample entered by diffusion can be as much as 10% of that contained in a 1 mm long sample zone; fortunately, most current CE operators use capillaries longer than 30 to 40 cm. Longer capillaries also reduce inadvertent hydrodynamic flow, which can have a positive or negative effect on sample volume; hydrodynamic flow decreases linearly with increasing capillary length. Uncontrolled sample introduction by diffusion may cause problems in the future with expected miniaturization of CE systems, but does not represent a serious handicap under current conditions. Another factor making diffusional problems less important is the typical length of sample zones (between 5 and 10 mm), made possible by analyte zone narrowing due to electrostacking.

3.1.3 Electromigrative Sample Introduction

Virtually all CE systems are equipped with the electromigrative sample introduction mode. From the point of view of instrument makers, that sampling technique is more easily implemented than the pressure differential methods discussed in Section 3.1.2. To initiate electromigration, the sampling end of the capillary is placed in the sample vial, and from that point on the procedure does not require any moving parts such as pumps or stepper motors for lifting the sample tray to a prescribed level. Transport of sample ions is accomplished by applying an electric field. The same source of high voltage can be used for sampling as for actual separations. Electromigration as a sample introduction represents one of the major differences between HPLC and CE. If performed properly, it can improve separation efficiency and detectability significantly. On the other hand, if electromigration is applied without proper understanding of underlying principles, quantitative results may be in error by a large margin, or in other cases, the separation efficiency may deteriorate strongly.

Let us consider what happens after the sampling end of a capillary has been taken from the electrolyte vessel and immersed into the sample solution containing two different ions (ii and i) and a nonionic compound (N) as shown in Figure 3.4A.

Immediately after the application of the sampling voltage, the ionic species begin to migrate across the concentration boundary between the sample and the electrolyte (Figure 3.4B). Although there are no systematic studies supporting the preferential choice of lower voltages for sampling, the sampling voltage is usually chosen at levels below 5 kV. Typical voltages for separation lie in the range between 15 and 30 kV for capillary lengths (L_t) between 30 and 70 cm. Electromigration is normally performed with a polarity that allows ions of interest to migrate from the sample toward and across the sample/electrolyte boundary.

With this choice of polarity and if the conductivity of sample segment is lower than that of the carrier electrolyte, electrostacking and trace enrichment of ionic components are achieved—see Section 3.1.2. Three other things should also be noticed at this stage. First, the ions of opposite charge to the analytes are migrating away from the capillary and do not even enter the capillary unless brought there by electroosmosis. Second, the nonionic components do not cross the sample/electrolyte boundary, and third, analyte ions migrate across the sample/electrolyte boundary at different rates. The difference in migration rates is tied to the so-called transference numbers t_i. If i^+ and i^- are the respective cationic and anionic currents, it is possible to calculate the exact portion of the total current i contributed by anions and cations:

$$t^+ = i^+/(i^+ + |i^-|) \text{ and } t^- = i^-/(i^- + |i^+|) \tag{3.5}$$

where t^+ is the total cationic and t^- the total anionic transference number. From this it follows that $t^+ + t^- = 1$. The total transference numbers for a given ionic charge are given by a simple summation of transference numbers for individual ions (t_i^+ or t_i^-). An individual transference number is deter-

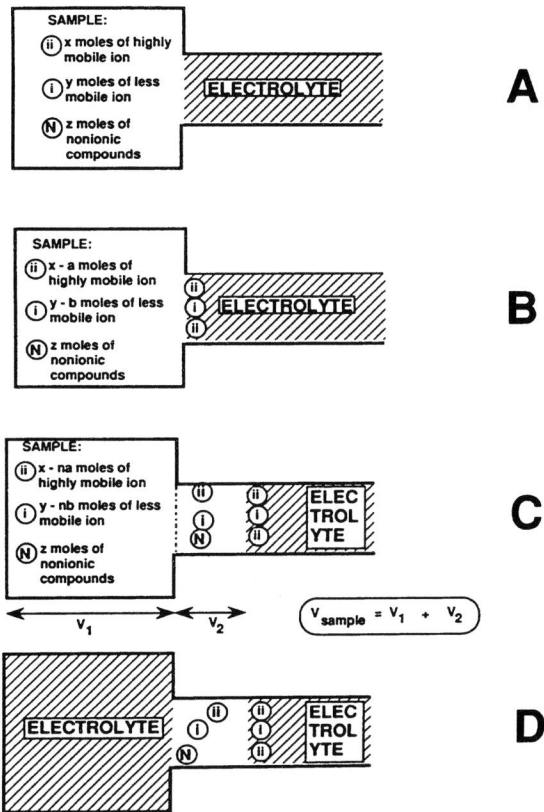

Figure 3.4 Four stages of sample introduction by electromigration. (A) Sampling end of capillary and the corresponding electrode are removed from the electrolyte vessel and placed in the sample. (B) First time segment after application of sampling potential, electrophoretic sampling prevails. (C) Electrophoretic sampling continues, electroosmotic contribution becomes apparent. (D) The sampling potential is turned off. The sampling end of capillary, along with the electrode, is removed from the sample and placed in the carrier electrolyte. See discussion in the text.

mined by a ratio of individual ionic equivalent conductance and *equivalent conductance of an ionic solution* (Λ_{eq}):

$$t_i^+ = \lambda^+/\Lambda_{eq} \text{ and } t_i^- = \lambda^-/\Lambda_{eq} \tag{3.6}$$

In Equation (2.4) we have seen that electrophoretic mobilities are directly proportional to equivalent conductances. Therefore we can write:

$$\mu_{sampling} = \lambda_i/(\Lambda_{eq}F) \text{ and } v_{sampling} = \lambda_i E/(\Lambda_{eq}F) \tag{3.7}$$

This equation indicates that the sample components are sampled into the capillary at different rates, depending not only on the individual equivalent conductances but also on the total conductance of the sample solution. A problem arises from the fact that a sample composition is an unknown by definition and therefore it is not possible to obtain an actual value of Λ_{eq} in many cases. As we shall see, however, practical solutions to this question can be found. The difference in sampling rates is indicated in Figure 3.4B for the two analyte ions of different mobilities.

Under most sampling conditions, another contribution to sampling needs to be considered. It is the contribution due to the electroosmotic flow. It is clear that for sampling purposes the electroosmotic flow should be directed away from the sample into the capillary, even though it may be of advantage for some matrix elimination procedures to proceed against that rule (i.e., elimination of excessive levels of low-mobility matrix components). Electroosmosis begins to be effective at the same time as the electrophoretic migration of analyte ions.

Stages B and C in Figure 3.4 will be observed only if the rates of electrophoresis exceed the rate of electroosmosis, as is certainly the case for most low-molecular-weight ions. Larger molecules (proteins, peptides, and DNA fragments) with rates of migration comparable to or lower than electrophoresis would than be drawn into the capillary mostly by electroosmosis, and consequently stage B could not be observed. As soon as the electroosmotic sampling becomes effective, the sample introduction becomes more complex. The different ionic components (ii and i) are now also introduced at a rate that is proportional only to their sample concentration, counteracting the purely electromigrative mechanism. The originally "preferential"[5] rates become similar once electrostacking occurs for the second time in stage D. Before shifting our attention to that step, we should notice that electroosmosis has also introduced the nonionic component into the capillary. An additional possibility, not illustrated in Figure 3.4, is that some of the slower, oppositely charged ions with electrophoretic mobilities lower than the rate of electroosmosis would move into the capillary at a sampling rate that is directly proportional to the net differential of electrophoretic and electroosmotic mobilities.

Disregarding that, the concentration of ions inside the electrolyte is now na and nb for the fast and slow ion, respectively. The ratio a/b is directly proportional to the ratio of corresponding transference numbers. The factor n stands for the number of time intervals between steps B and C. The concentration of both ions and the neutral component inside the volume V_2 is probably approximately equal to that in volume V_1. Ions transferred by electrophoresis from volume V_2 into the electrolyte are replaced by the ions migrating at a comparable rate from volume V_1 to V_2.

After interrupting the sampling potential, the capillary is placed back into the carrier electrolyte, and the separation potential can be applied. At the beginning of this last stage, we can thus observe two distinctly different sample segments inside the capillary. Provided that the sample conductivity is lower than that of the carrier electrolyte, there will be a narrow layer of preconcentrated (electrostacking) ionic sample components at the sample/electrolyte boundary. The ionic components will also be present in the V_2 segment inserted by electroosmosis. Their levels there will be close to concentrations in the original sample. Once the separation potential is applied, the electrostacking occurs for the last time. The ionic components inside the sample plug experience a higher field strength, and their higher rate of migration allows them to catch up with the sample ions already in the electrolyte. This last occurrence of stacking may under certain circumstances (i.e., electroosmotic flow rate comparable to electrophoretic migration rates or a relatively long sampling time) compensate partly or completely for the preferential rates of sample introduction by electromigration.

The four steps illustrated in Figure 3.4 and discussed in the text can thus be regarded as a combination of two limiting situations in electromigrative sample introduction. The two limits are defined by the relative magnitude of electrophoretic and electroosmotic mobilities. Step C will be eliminated under conditions where the rate of electrophoresis far exceeds the rate of electroosmosis. Analogously, the total contribution by step B will be negligible, if electroosmosis exceeds the electrophoretic migration rates. The dual contributions to electromigrative sampling can also be expressed in the following relationship for the total amount Q [moles] introduced by electromigration[5]:

$$Q = (\mu_{sampling} + \mu_{EOF})\pi r^2 V_i t_i \, C/L_t \qquad (3.8)$$

All symbols are explained in connection with Equation (3.2) and also in Section 2.3. Clearly, Equation (3.8) can be written in two limiting forms, omitting either the sampling or electroosmotic mobility in accordance with the discussion in the previous paragraph.

As indicated by Equation (3.7), the quantitative interpretation of analytical results in CE requires special consideration, if electromigration is used for sample introduction. Huang, Gordon, and Zare[8] have developed a two-step approach to the problem. The "bias" caused by differences in ionic equivalent conductances (λ_i) is treated separately from the influence of equivalent conductance in an unknown sample (Λ_{eq}). Typical results from a comparison of hydrodynamic and electromigrative introduction from an identical sample are listed in Table 3.1.

Note that the migration times and areas in Table 3.1 are listed as ratios to one of the components in sample mixture (Rb) and not as absolute values. As we shall see in Section 3.4.5, such referencing of peak areas and migration times is the simplest method of improving reproducibility in CE. Zare and co-workers noticed the difference between the reference ratios by electromigration versus hydrodynamic sample introduction, which they choose to call a "bias of electrokinetic injection." However, they also highlight the similarity of the column 1 to column 2 ratios with the corresponding refer-

Table 3.1. COMPARISON OF RESULTS BY ELECTROMIGRATION AND HYDRODYNAMIC SAMPLE INTRODUCTION. REPRODUCED WITH PERMISSION FROM REFERENCE 8

Peak Pair	Ratio of Peak Areas, Electromigration (column 1)	Ratio of Peak Areas, Hydrodynamic Meth. (column 2)	Column 1 to Column 2	Inverse Migr. Time Ratio
Rb^+/K^+	1.0	0.94	1.06	1.04
Rb^+/TMA	2.08	1.33	1.57	1.57
Rb^+/Li^+	1.17	0.69	1.70	1.73
Rb^+/DEA	6.91	3.93	1.76	1.81
Rb^+/arg	4.34	1.92	2.26	2.31

Conditions: 1 sec and 10 kV electromigration, 10 sec and 10 cm hydrodynamic sampling. All analytes at 5×10^{-5} M. Complete separation conditions are given in Reference 9.

ence ratios of migration times. From this observation, the authors develop the recommendation to use the ratios of migration times for correction of the observed bias. Such a procedure would certainly bring the ratios of peak areas by electromigration closer to those obtained by hydrodynamic sampling. A more practical solution can, however, be found in separate calibration for each of the sample introduction modes. The equality of peak area ratios between the two sampling methodologies appears to be more of an esthetic concern, having only minimal consequences for the practical utility of the technique.

A more practically relevant quantitation problem is that termed "bias between different samples caused by electrokinetic injection" (i.e., sampling utilizing electromigration). Huang et al. designed an experiment comprising the CE analysis of four different samples having identical concentration of two cations (Li, K) in solutions of differing conductivity. The differing conductivity was adjusted by means of different levels of morpholine ethanesulfonate/histidine buffer[9] concentrations. The peak areas by electromigrative and hydrodynamic sampling procedures are shown in Figure 3.5.

While the hydrodynamic sampling results in Figure 3.5 are constant and independent of sample conductivity, the results by electromigration vary in a linear fashion. The linear dependency of sampling rates on sample conductivity is given by Equation (3.7). From Figure 3.5 and in agreement with Equation (3.7), it is possible to design an appropriate calibration procedure for the majority of practical situations. Such a procedure consists in preparing two calibration plots. The first of the two is the standard calibration curve showing peak area versus concentration from standard solutions having identical resistance. The second calibration will be a plot of two or three standard solutions with constant levels of analytes, but with different levels of total resistance. The resistance/conductivity is adjusted by appropriate

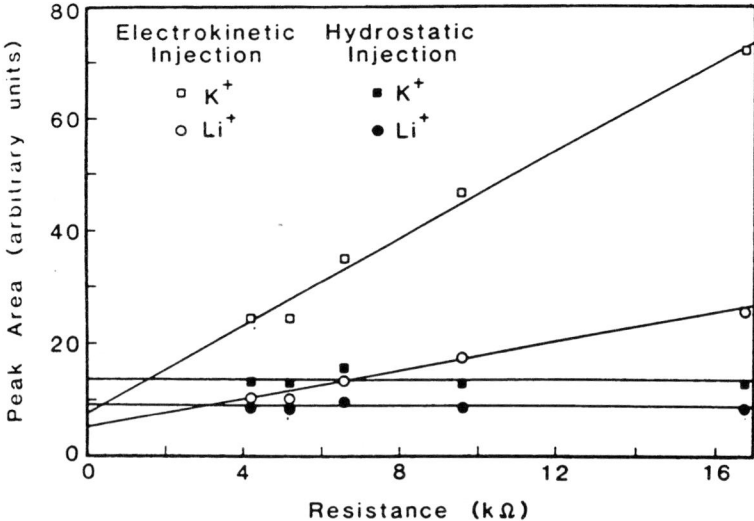

Figure 3.5 Plot of peak areas versus resistance of sample solution. The resistance was adjusted by varying levels of a buffer solution. The concentrations of cations were held constant in all experiments. Hollow squares and circles indicate the results by electromigration. Full black circles and squares indicate the results by means of hydrodynamic sample introduction. Reproduced with permission.[8]

concentrations of a suitable salt. Consideration should be given to the choice of chemistry of conductivity-adjusting compound. It should be free of ions to be analyzed, and its ionic components should have equivalent conductances sufficiently different from those of analytes to avoid close comigrations with peaks of interest. The plots of area change with known resistance/conductivity of standards yield the correction factor for unknown samples. A reliable quantitation is then possible for electromigration from samples of known resistance/conductivity. Before calculating quantitative results, the analyst has to obtain the value of conductivity for the measured sample and calculate a correction of peak area for a given conductivity level. Only the corrected peak area can then be used in calculations of unknown sample concentrations.

As a reminder: Some of the possible complications discussed here and in Section 2.7 (diffusion, inadvertent hydrodynamic flow, conductivity mismatch, etc.), leading to peak spreading and other effects, can also be observed in conjunction with electromigration under nonoptimum conditions.

3.1.4 Alternative Approaches to Sampling

Hydrodynamic and electromigration procedures are clearly the most frequently used sampling methods in CE at the present time. That notwithstanding, several alternatives have been proposed in the recent literature.

This chapter offers a brief overview of these techniques promising to supplement the currently prevalent sampling routines.

First we shall look at two methods of injection by syringes emulating the most frequently used procedure in gas and liquid chromatography. The extremely small sample volume (10–100 nl) dictated by the required dimensions of capillaries (25 to 100 μm I.D. and 10 to 100 cm total length) precludes the utilization of GC and LC sampling valves. The small sample volume requirement has been solved in capillary GC by split-flow injection.

Split-flow sample introduction for CE is described in a recent report.[10] The design of a split-flow injector shown in Figure 3.6 has become commercially available from one of the manufacturers (Isco, Inc., Lincoln, NE).

The volume (V_c) introduced into the capillary by split-flow injection is determined by the following equation.

$$V_c = (d_c/d_s)^4 (L_s/L_t) V_{inj} \tag{3.9}$$

where d_c and d_s are the capillary and venting tube diameters, respectively, L_s is the length of venting tubing, and L_t is the total capillary length. V_{inj} is the volume injected from the HPLC syringe connected to the split-flow injector.

An obvious advantage of the split-flow design is the ease of use and its similarity to devices common in more conventional separation methods. Tsuda and Zare[11] evaluated several designs for split-flow injectors for use with rectangular capillaries. Such capillaries are fragile and cannot be used in conjunction with conventional hydrostatic sampling in which the sampling end of a capillary has to be moved from one level to another. Split-flow design makes it possible to introduce samples without any movement of the fragile capillaries. A disadvantage of split-flow sampling is in the possibility that any obstruction of the flow either in the venting tubing or in the capil-

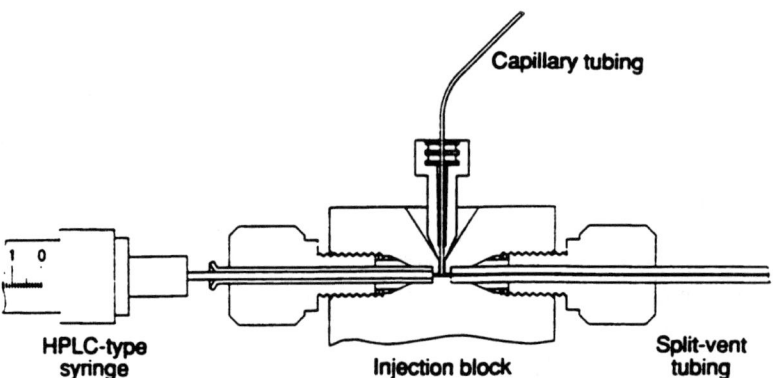

Figure 3.6 Split-flow injector for capillary electrophoresis. Reproduced with permission.[10]

lary can completely distort the ratio of hydrodynamic resistances and lead to questionable reproducibility of sample volumes, but this can probably be solved by the use of internal standards.

Tsuda and co-workers[12] proposed a modified design of a rotary injector for capillary electrophoresis. Rotary injectors are widely used in HPLC. The possibility to adapt that popular design to CE is thus of great interest. The first prototype of an injection valve design illustrated in Figure 3.7 included metallic parts in contact with the electrolyte. The authors reported problems with that type of injectors consisting in bubbles being generated at the metal/liquid interface. Portions of the valve between tubes 5 and 7 represent a segment of L_t and are exposed to electric fields during separations. That made it necessary to make all injector parts that are in contact with the liquid from nonreactive polymers. In its final version, the valve resembled the rotating sample loop injectors common in HPLC. In the initial position, Figure 3.7A, the sampling tubing 6 is connected to a cylindrical cavity inside the injector defining the size of the sample (part of the black rectangle reaching inside the rotor). To inject the sample introduced into the valve in the initial position, the actuator 8 is moved 90 degrees by hand. The sample volume inside the rotor is now placed in line with the separation capillary, Figure 3.7B. After voltage application the sample components are removed from their initial position by a combination of electrophoresis and electroosmosis.

As a main disadvantage, the authors mention the safety concerns stemming from the necessity to actuate a part of a high-voltage circuit manually.

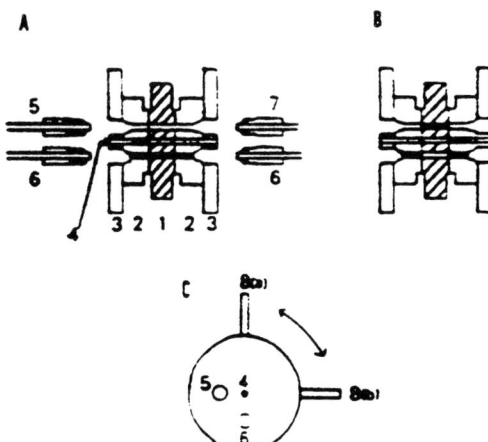

Figure 3.7 Rotary injector for capillary electrophoresis. (1) rotor, (2) stator, (3) plate aligning stators and rotor, (4) central pin holding rotor and stator in correct position, (5) fused silica capillary connecting injector to one of the two electrolyte containers, (6) tubing for sample introduction, (7) fused silica capillary, (8) lever actuator for switching between load and inject positions. Reproduced with permission.[12]

Increasing automation of all existing commercial CE instruments has made this concern rather less important. Manual actuation of parts of CE apparatus is undesirable for other reasons, such as reproducibility, contamination, etc. Rotary valves have also been used in the coupling of HPLC and CE.[12] The coupling of the two separation techniques with very different run times (slow and fast) enables a realization of two-dimensional separations. Two-dimensional separations were shown to yield great gains in separation power.[13,14] Rotary valves are very likely to become integrated in commercial instruments at some point in the future. Before this happens, however, suitable construction materials have to be found exhibiting a minimum of corrosion after a prolonged exposure to high temperatures and strong electric fields.

Another proposed approach to sample introduction, the electric sample splitter shown in Figure 3.8, could also be potentially utilized for HPLC CE coupling. A sample amount n_1 in Figure 3.8 is introduced through a conventional HPLC valve into a carrier electrolyte at a suitable distance from the point where a vertical CE capillary joins the broader sampling capillary. The system is equipped with one high-voltage electrode positioned at the left side of the sampling tubing and grounding electrodes at the respective ends of the CE and sampling capillaries. The sampling voltage is applied after the sample injection by means of a sampling syringe through a sampling valve into the sampling tubing. The total sample volume in the actual experiment was 1 µl. After the application of voltage, the injected sample plug moves through the sampling capillary toward the separation capillary. At the branching point, the original sample amount splits into two segments n_2 and n_3, which then continue to move toward the two grounding electrodes. The amount "trapped" in the separation capillary is directly proportional to n_1 and to the ratio of total currents ($n_2 = n_1 I_2 / I_3$) in the two types of capillaries. On the other hand, the sampled amount is independent of the concentrations and mobilities of sample components, and therefore the sample in the capillary is a true aliquot of the original sample.

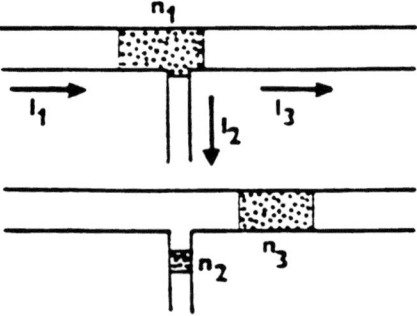

Figure 3.8 Principle of electric splitter. Reproduced with permission.[15]

Capillary electrophoresis is ideally suited for the analysis of microscopic samples. Volumes of 10 to 100 μl are sampled and analyzed easily with the help of commercially available hardware. Two reports present evaluations of sampling devices for samples smaller than 10 μl.

Schomburg and co-workers[16] experimented with highly polished titanium electrodes and conically shaped sampling vials. The evaluated microinjector enabled easy handling of samples in the range of 1–2 μl. Wallingford and Ewing[17] developed a microinjector designed to sample volumes as small as 430 picoliters. Miniaturization of CE will require procedures for the introduction of even smaller sample volumes (fractions of picoliters) from larger volumes. Monnig and Jorgenson[18] describe a "gating" procedure utilizing lasers to modulate the size of an introduced sample. The sample components are first tagged with a fluorescent molecule and then pumped continuously into the separation capillary. Tagged sample components are continuously "detagged" by a laser beam upon entering the separation capillary. The fluorescent portion of the analyte is photodegraded, becoming undetectable by a fluorescence detector positioned at the other end of the separation capillary. An analyzable sample amount is created by a controlled, very brief interruption of the laser beam. During that time interval a small amount of sample passes the gating zone with its tag intact. Its components are easily detected after a separation at the other end of the CE capillary. According to the authors, such optical introduction exceeds in flexibility any mechanical sampling procedure.

3.1.5 Sample Preconcentration in Capillary Electrophoresis

Sample preconcentration in capillary electrophoresis has developed into an exciting field of research. Recent developments have rendered previous concepts regarding the sensitivity of CE obsolete. Until about 1990, CE was considered inherently less sensitive than HPLC. While it was recognized that absolute sensitivity in CE (μg or μmoles injected) might be relatively high, the concentration sensitivity (moles per liter, ppm, ppb) was deemed to be poor. Some hope was seen in the capabilities of laser-induced fluorescence detection or in amperometric detection. Only for fluorescent and electroactive compounds was it believed that a significant sensitivity could be achieved.

It is actually quite surprising that the belief in the insensitivity of CE could last so long. Improvements of sensitivity under the conditions of electrostacking were recognized already in one of the early reports on CE.[1] Burghi and Chien[3] quote altogether seven articles on electrostacking published between 1964 and 1990. However, electrostacking is only one of several preconcentrating methods available to CE users which are:

1. Electrostacking;
2. Isotachophoresis (off line or on line);

3. Preconcentration capillaries (capillaries incorporating a segment of retentive packing).

3.1.5.1 Preconcentration by Electrostacking

The introduction to electrostacking presented in Section 3.1.1 focuses mostly on the effects of electrostacking on separation efficiency. Sample components from a relatively broad sample segment can be "compressed" into very narrow zones before the actual separation begins (Figure 3.2). CE separations performed under optimum conditions (see Chapter 2.5) generate much less zone spreading than conventional HPLC. It is interesting to note that, in many cases, improved separation efficiency improves detectability. Under optimal conditions, the separation efficiency in CE approaches 10^6 plates even in routine applications. Assuming that peaks without electrostacking exhibit separation efficiencies only in 10^3 plates and that these peaks are Gaussian in shape, it can be shown[19] that the improved efficiency in electrostacking leads to approximately thirty times higher peaks:

(peak height with stacking/peak height without stacking) = $(10^6/10^3)^{1/2} = 31.6$

The peak height enhancement thus calculated applies directly to hydrodynamic sampling. For electromigrative sample introductions the sensitivity improvement is accomplished by maintaining a high field ratio between sample and carrier electrolyte zones in accordance with Equation (3.1). Chien and Burghi[20] describe possible sensitivity enhancements by as much as several hundred times for low-concentration samples dissolved in pure water and introduced by electromigration into a capillary filled with 100 mM morpholine ethanesulfonate/histidine carrier electrolyte.

Improvements of that order of magnitude cannot be explained only by a stacking effect inside a narrow sample plug drawn into the capillary. The improved sensitivity is probably based to a certain degree also on the ability to draw a significant amount of analyte ions from the bulk of the sample solution. The authors also report that the enhancement is much smaller than predicted by Equation (3.1), if one moves the capillary directly from the electrolyte vial into the sample. They speculate that the carrier electrolyte boundary at the end of the capillary may be disturbed during the transfer and the electric field in the initial stages of sample introduction may not be applied properly. To avoid such a problem, they developed a modified procedure relying on the introduction of a short water plug between the electrolyte and the sample. The capillary is moved first into a vial containing pure water. A hydrostatic injection is performed and only then is the capillary dipped into the sample. Using that approach assures that the field enhancement ratio approximates that calculated by Equation (3.1). Figure 3.9 shows a comparison of electropherograms for identical analyte levels obtained with electromigration under three different conditions. The peak heights from

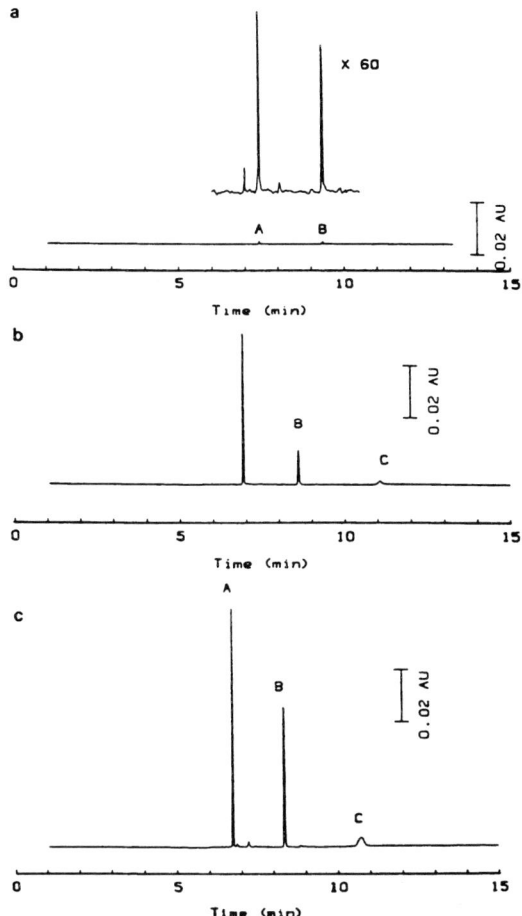

Figure 3.9 All three separations were carried out from the samples containing identical levels of PTH-arginine (Peak A), PTH-histidine (Peak B), and neutral marker (Peak C). The sampling conditions were 5 kV and 10 sec. Separation was performed at 30 kV. The separation buffer is 100 mM MES-HIS. (a) Electropherogram after electromigrative sampling from a sample dissolved in carrier electrolyte. (b) Electropherogram obtained with electromigrative sample introduction from the sample dissolved in pure water. (c) Electropherogram obtained with electromigration carried out with a water plug between the carrier electrolyte and the sample. Reproduced with permission.[20]

these separations are shown in Table 3.2, referenced to a 10 sec hydrodynamic injection of the same sample.

As already noted, sampling from solutions having equal conductivity as the carrier electrolyte is quite counterproductive. Figure 3.9a is another example of unsatisfactory results obtained by that approach. The sensitivity im-

Table 3.2. COMPARISON OF PEAK HEIGHTS FOR THE TWO IONIC ANALYTES FROM FIGURE 3.9 USING FOUR DIFFERENT SAMPLE INTRODUCTION METHODS. ALL PEAK HEIGHTS ARE NORMALIZED IN RESPECT TO 10 SEC HYDROSTATIC INJECTION. REPRODUCED WITH PERMISSION[20]

Method	PTH-arginine	PTH-histidine
Hydrostatic injection 10 sec at 7.6 cm	1	1
Electromigration (a) sample dissolved in 100 mM MES-HIS	0.311	0.225
Electromigration (b) sample in water	16.96	3.38
Electromigration (c) as (b) but with a water plug between electrolyte and sample	28.84	13.44

proves visibly by dissolving the same sample in pure water. Better detectability is due to field enhancement stemming from the increased resistance of the sample segment, Figure 3.9b. However, the greatest increase in peak height is accomplished only after the introduction of a water plug into the sampling end of capillary prior to the transfer into the sample vial, Figure 3.9c.

The hydrostatic injection to which the values of peak heights in Table 3.2 are referenced was carried out from a sample dissolved in the carrier electrolyte. The separation efficiencies achieved in all electropherograms were comparable. The changes of peak heights are thus due only to changes of sample to electrolyte field ratios. We can see an enhancement factor of ~ 100 between the two extreme forms of electromigrative sample introduction. The optimized electromigration, on the other hand, is approximately thirty times more sensitive than hydrostatic sampling, requiring about the same duration of time (10 sec).

In counterelectroosmotic CE (see Section 2.4) there is a frequent need to introduce both cations and anions into the capillary by electromigration. Under conditions where the polarity is unchanged during electromigration, only cations or anions are sampled. The oppositely charged analytes are brought into the capillary only when the ionic mobility in the opposite direction is overcome by the electroosmotic flow. If such sampling of oppositely charged ions occurs, it is at a much lower rate leading to poor detectability of those analytes. A modified electromigrative method has been developed,[21] making it possible to sample ionic species of opposite charges by switching the polarity of the voltage applied during the sample introduction step.

3.1.5.2 Preconcentration by Isotachophoresis

Isotachophoresis is a technique closely related to capillary electrophoresis. The similarities and differences between the two are illustrated in Figure 3.10.

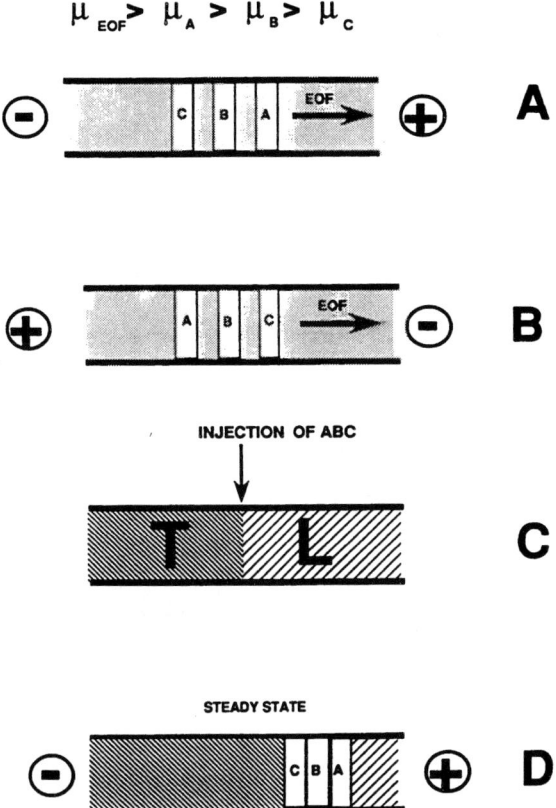

Figure 3.10 Comparison of anion separations by capillary electrophoresis and isotachophoresis. A, coelectroosmotic CE; B, counterelectroosmotic CE; C, injection in isotachophoresis; L and T stand for leading and terminating electrolyte, respectively; D, isotachophoretic steady state.

The main similarity lies in the fact that both techniques separate ionic analytes according to their electrophoretic mobilities. On the other hand, unlike isotachophoresis, CE is able to separate ions in at least two different ways, depending on whether the coelectroosmotic (A) or counterelectroosmotic (B) approach is utilized. Isotachophoresis uses two electrolytes, compared to only one in CE, and the sample has to be injected at the concentration boundary dividing the two electrolyte zones, Figure 3.10C. Once a potential of correct polarity is applied, the sample components are separated and migrate toward the detector. As shown in Figure 3.10D, isotachophoretic zones migrate in contact with each other, inserted between the leading and terminating electrolytes. They are not separated by regions of electrolyte as in CE. The important prerequisite for successful isotachophoretic separations is to find suitable ions for the leading and terminating electrolytes. The main component of the leading electrolyte has to have the highest electrophoretic mobility of all ions of like charge in the system. The terminating electrolyte has to contain ions of lowest mobility in the same system. The analyte ions are thus bracketed by the respective mobilities of ions of same charge in the leading and terminating electrolytes. A common counterion is then chosen for the two isotachophoretic electrolytes. The term *isotachophoresis* is based on the observation that all zones in the system migrate at the same speed ("iso tacho" in greek), which is dictated by the lowest mobility of the main ionic component of the terminating electrolyte. For further information on isotachophoresis, the reader is referred to a comprehensive text[22] and a review.[23]

An important difference between isotachophoresis and CE and other separation techniques (HPLC, GC) is the way of determining sample concentration from the recorded separation. In chromatographic techniques and in CE, the concentrations inside the separated zones and thus the corresponding signal from a detector are directly proportional to the original sample concentration. In contrast, in isotachophoresis it is the length and not the concentration of a separated zone that is proportional to the original sample concentration.

The concentration in isotachophoretic zones is adjusted by the concentration of the leading electrolyte [see the discussion of Eq. (3.10) in the following paragraph]. Since the zone concentration is always constant, different amounts of analyte ions express themselves in different lengths of separated zones. Zone length is thus the actual analytical signal related to the sample concentration in isotachophoresis. In the adjustment of sample zones to a concentration approximating that of the leading electrolyte lies the powerful preconcentrating effect of isotachophoresis. Three stages of the adjustment are illustrated in Figure 3.11.

The ratio of sample ion concentration c_A to leading electrolyte ion concentration c_L can be calculated from the Kohlrausch Regulating Function.[24]

$$c_A/c_L = [\mu_A(\mu_L + \mu_R)/\mu_L(\mu_A + \mu_R)] \qquad (3.10)$$

where μ denotes the ionic electrophoretic mobilities and R denotes the counterion in the leading and terminating electrolytes. In order to calculate possible preconcentration gains, for example, for low-molecular-weight anions (ionic equivalent conductivity $\sim$ 40 Siemens cm^2 equiv^{-1}), we can generate a plot of c_A/c_L versus leading electrolyte and counterelectrolyte mobilities. Because of the direct proportionality given by Equation (2.4), it also possible to use limiting equivalent conductivities instead of electrophoretic mobilities. The plot in Figure 3.12 is calculated for leading electrolyte anion conductivities between 41 (carboxylates) and 198 (hydroxide) Siemens cm^2

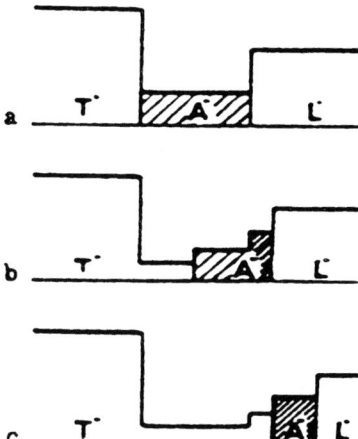

Figure 3.11 Preconcentration effect in electrophoresis. The ordinate indicates concentration, and the abscissa shows time. T, A, and L denote terminating, analyte, and leading electrolyte anions, respectively. (a) Situation after the sample introduction, prior to the application of electric field. (b) Situation shortly after the application of electric field. The adjustment of zone concentration begins. The terminating electrolyte anion replaces the portion of analyte ions that has moved forward during the adjustment. (c) The adjusted zone of analyte is formed. The terminating electrolyte replaces the analyte concentration during all steps (only two of many possible steps are indicated in the figure). Reproduced with permission.[23]

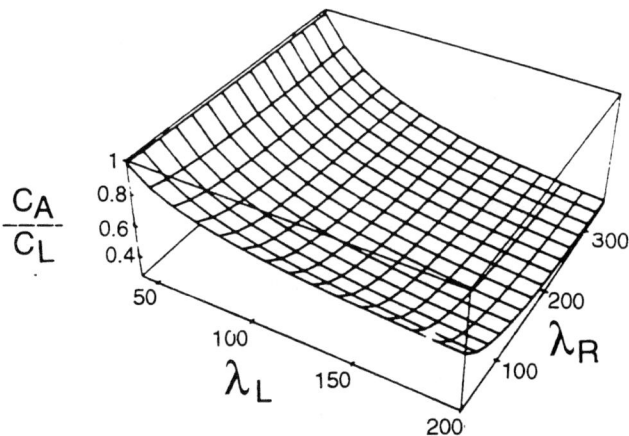

Figure 3.12 Plot of Kohlrausch Regulating Function. Limiting equivalent conductivities (λ) are used instead of electrophoretic mobilities. See further discussion in the text. Reproduced with permission.[25]

equiv^{-1}. The countercation limiting equivalent conductivities are varied between 35 and 350 Siemens cm^2 equiv^{-1}, covering the entire range between alkylammonium ($\sim$ 30) and hydronium ions (394.7).

For the ionic range illustrated in Figure 3.12, the sample zone during the isotachophoretic steady state adjusts to a level between 25 and 100% of leading electrolyte ion concentration. For some very dilute samples (i.e., 10^{-6} to 10^{-9} M), this represents a very useful trace enrichment factor, with leading electrolytes being mostly in the 10^{-3} M range. Under optimal conditions the analyte zone can thus be preconcentrated up to 10^6 times.

The great preconcentrating power of isotachophoresis was recognized earlier,[26] and a corresponding module was incorporated in a commercial instrument.[27] On-line coupling of isotachophoresis and CE is of a more recent date. Kaniansky and Marak[28] investigated nitrophenols and 2,4-dinitrophenyl-labeled amino acids coupling a commercial isotachophoretic instrument[27] to a home-made CE system using tubing made of fluorinated ethylene–propylene copolymer. In this arrangement they were able to achieve the inherent preconcentration effect of isotachophoresis for their model compounds. They also report another advantage of the coupling of the two different electrokinetic separation methods—the elimination of excessive matrix concentrations. Sample constituents that would interfere with any of the two separation modes, if used separately, could be removed during the isotachophoretic preconcentration, matrix elimination step. Similar observations were also made by two other groups of authors.[29,30] Several steps during a typical isotachophoresis—CE coupling procedure are illustrated in Figure 3.13.

Figure 3.13a–e shows the procedure that would have to be employed for very difficult analyses with two zones of interfering matrix components removed. In many cases, for example, for preconcentration only, one or both matrix eliminating steps may be avoided.

In some instances, it is possible to carry out the isotachophoretic trace enrichment even without any specially designed isotachophoretic modules, directly in a CE instrument.[25] A good example of where this simplification becomes possible is the analysis of trace levels of anions in pure water. A recently introduced method for low-molecular-weight anions[31] utilizes chromate as the carrier electrolyte. The chromate anion exhibits a relatively high electrophoretic mobility. For a majority of analyte anions, chromate thus can fulfill the function of an isotachophoretic leading electrolyte. A terminating electrolyte is made by adding a suitable anion of relatively low electrophoretic mobility to the sample. Sodium octane sulfonate[25] was used as the additive. A side benefit in using that anion is that in coelectroosmotic CE, it comigrates closely with the water peak and does not appear in the separation recording since the broad water peak along with other sample components of low electrophoretic mobility is usually removed by vacuum purge after the detection of the last peak of interest in coelectroosmotic CE.

INSTRUMENTATION FOR CAPILLARY ELECTROPHORESIS 103

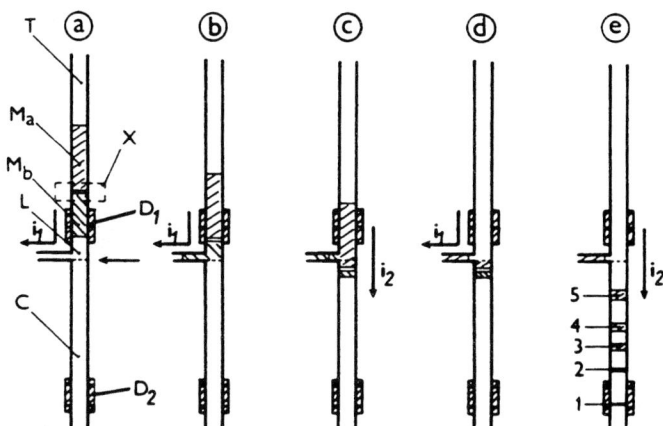

Figure 3.13 Schematic illustration of the separation steps in combined isotachophoresis (ITP) and CE modules. (a) ITP separation in the isotachophoretic module, ITP Step 1, (b) Removal of fast-migrating matrix constituents M_b from the separation compartment, ITP Step 2, (c) transfer of the sample fraction containing the analytes X into the CE module, CE Step 1, (d) removal of slow-migrating matrix constituents M_a from the separation compartment, ITP Step 3, (e) CE separation in the capillary electrophoresis module, CE Step 2. L, T, and C stand for leading, terminating, and carrier electrolytes, respectively. D1 and D2 are the detectors in the ITP and CE modules. Arrows for i_1 and i_2 indicate the direction of the driving currents and also the direction of migration in the respective stages of the coupling procedure. Numbers 1 through 5 denote the separated analytes in CE Step 2. Arrow on the right of (a) indicates the bifurcation point. Reproduced with permission.[28]

It is possible to assume that in the presence of suitable leading and terminating electrolytes an electromigrative sample introduction may include an isotachophoretic step, such as shown in Figure 3.14.

Stages A and B in Figure 3.14 correspond to the conventional electromigrative sampling discussed in Section 3.1.4. Again, the analyte anions in the sample experience a higher level of electric field than in the carrier electrolyte, the consequence of which is a narrow layer of ionic sample constituents at the sample electrolyte concentration boundary. If the terminating electrolyte is added in an excess relative to the analyte ions, it is possible to accomplish the isotachophoretic steady state depicted in Figure 3.14C. There, the analyte ions become sufficiently depleted at the entrance into the capillary, so that a situation analogous to that observed after an injection in isotachophoresis (Figure 3.10C,D) may develop. Even if this occurs for only a very brief interval of time, high preconcentration ratios can be accomplished. Nanomolar concentrations of ions can be detected after the capillary is put back into the carrier electrolyte (Figure 3.14D) and after a CE separation. (See Figure 3.15.)

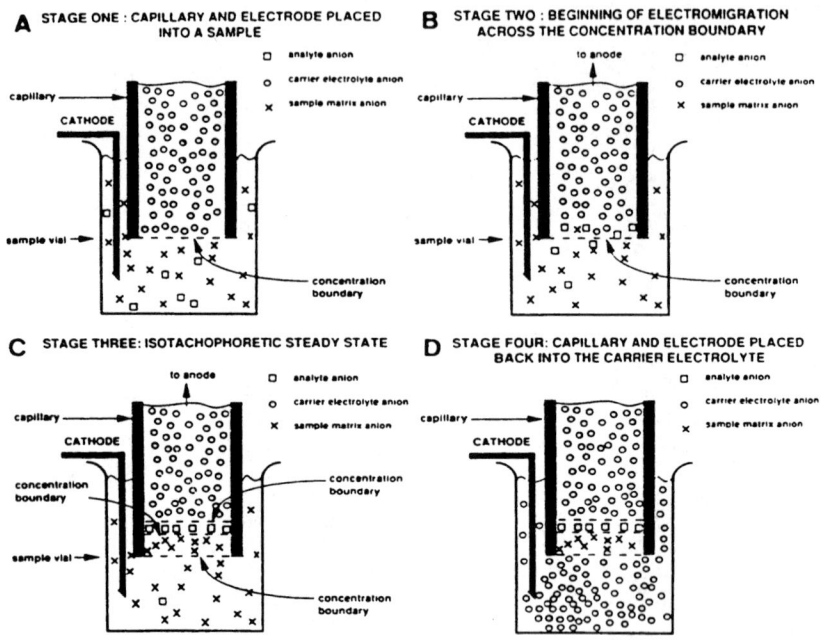

Figure 3.14 Four stages of simplified isotachophoretic sample preconcentration. Reproduced with permission.[25]

A more detailed discussion of the applications of simplified isotachophoretic trace enrichment is offered in Chapter 5.3.

3.1.5.3 Preconcentration in Capillaries Incorporating a Segment of LC Packing

The electrokinetic trace enrichment methods (electromigration and isotachophoresis) discussed in the previous two sections are eminently suitable for highly mobile ions. Their efficiency tends to decrease, however, with the decreasing mobility of analyte ions. A preconcentration technique analogous to trace enrichment in HPLC would thus be a method of choice for such slowly migrating sample components. Special capillaries containing a segment of LC packing (e.g., C18 silica or polymeric reverse phase, ion exchange resin, etc.) have been shown to be useful in these cases. (See Figure 3.16.)

Prior to use, the preconcentrator capillary is equilibrated with the carrier electrolyte. Following this, a known volume of a sample is pumped through the segment filled with chromatographic packing and into the rest of the capillary. Partially or fully ionized analytes can also be introduced into the

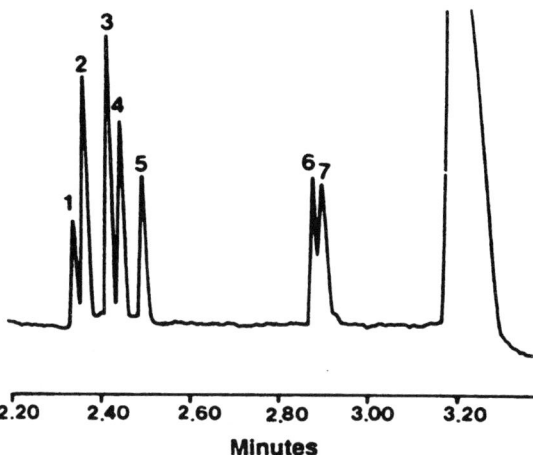

Figure 3.15 CE separation of inorganic anions after trace enrichment by isotachophoresis performed directly inside a CE capillary (60 × 52 cm × 75 μm). Chromate electrolyte was used at 5 mM concentration and pH 8.0. The sampling and separation voltage were −5.0 and −20 kV, respectively. The sampling was carried out for 45 sec. The added concentration of sodium octane sulfonate in the sample was 18 μM. Indirect UV detection was performed at 254 nm. The peak identities, concentrations (ppb, i.e., ng/ml) and detection limits (three times the noise in 10^{-9} M) were as follows: (1) bromide, 4 ppb, 13.6 nM, (2) chloride, 4 ppb, 13 nM, (3) sulfate, 4 ppb, 25.4 nM, (4) nitrite, 4 ppb, 25.4 nM, (5) nitrate, 4 ppb, 24 nM, (6) fluoride, 2 ppb, 19.8 nM, (7) phosphate, 8 ppb, 17.8 nM. The large peak at ~3.2 min belongs to carbonate. The levels of carbonate were not controlled under the conditions of the experiment. Reproduced with permission.[25]

capillary by electromigration. The objective of the first step is to strip the sample volume that is being pumped through the preconcentrator capillary of all analytes of interest. After the preconcentration period, the sample solution is replaced by the carrier electrolyte in the whole capillary. If the chemistry of the preconcentrator segment was chosen correctly, the analytes

Figure 3.16 Preconcentration capillary filled with C18 silica. Reproduced with permission.[32]

of interest remain on the preconcentrator and are not removed by the stream of the electrolyte. The actual transfer of the analytes from the preconcentrator segment into the following portion of the capillary for electrophoretic separation is carried out by introducing a short plug of a strong chromatographic eluent mixed with electrolyte for conductivity into the capillary and by forcing that segment by pressure or by electroosmotic flow through the preconcentrator packing. Once the preconcentrated sample constituents leave the chromatographic packing, they can be easily separated from the segment of the strong eluent used for their transfer. When the separation voltage is applied, the segment of the strong eluent, usually an organic solvent (acetonitrile, THF, etc.), migrates with the velocity of the electroosmotic flow, whereas the analytes exhibit their own electrophoretic mobility. Since the eluent zone is of much lower conductivity than the carrier electrolyte, it is safe to assume that an electrostacking occurs at the eluent/electro-

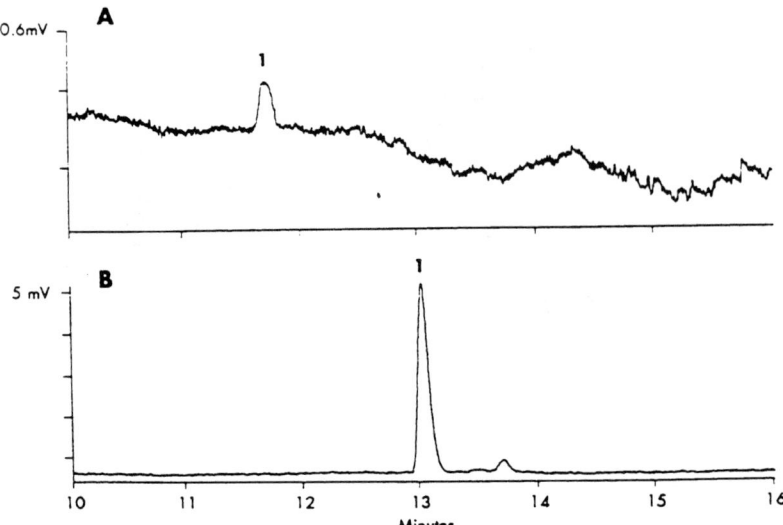

Figure 3.17 (A) Electropherogram of 0.5 ppm Doxepin (an antidepressant drug) after electromigrative sample introduction at +5 kV for 20 sec. (B) The same concentration of Doxepin as in A, preconcentrated for 15 min by electroosmotic pumping at +5 kV on a segment of polymeric reverse phase packing (PRP) in the preconcentrator capillary (Waters, AccuSepC/PRP). The elution from the PRP packing was carried out with an acetonitrile segment (95/5 acetonitrile/carrier electrolyte) introduced electroosmotically (18 sec and 5 kV). The acetonitrile segment was subsequently transported by the electroosmosis trough the preconcentrator bed after the application of the separation voltage. Common conditions for both electropherograms: Carrier electrolyte, 25 mM sodium citrate, pH 4.0. Separation voltage, +15 kV. Direct detection at 214 nm. Capillary dimensions: 60 × 52 cm × 75 μm. The 1.0 mm long segment of PRP material is located approximately 1.0 mm from the sampling end of the capillary. Reproduced with permission.[33]

lyte boundary, contributing additionally to the improvement in detectability. Analogously to HLPC, the preconcentrator capillaries permit the use of a much larger sample volume than would otherwise be possible by a direct introduction with the help of a pressure differential. Typical sample volumes with the introduction by pressure differential methods discussed in Section 3.1.6.1 are between 10 and 100 nanoliters. The preconcentrator capillary makes possible the use of sample volumes in the microliter range. As already mentioned, the preconcentrator capillary gives improved results even over electromigration for most compounds of very low electrophoretic mobility. This advantage is illustrated in the example of Doxepin in Figure 3.17.

Swartz and Merion[33] report an approximately thirtyfold improvement in detectability with the preconcentrator capillaries in comparison with the standard electromigrative sampling. They also discuss a possible use of their capillaries for determination of organic molecules from urine and serum. Undesirable sample components are either not retained on the preconcentrator segment or can be removed prior to the analysis by suitable rinsing steps.

3.2 Selection and Handling of Capillaries

In Section 2.3 we discussed chemical properties of capillaries and the origins of the electroosmotic flow. The conclusion reached was that there are developments in the properties of capillaries that are analogous to developments in the early stages of HPLC columns. One of the essential functions of capillaries, dissipation of Joule heat, was discussed in Section 2.6. Various reasons for narrow internal capillary diameters were explained in Sections 2.5 and 2.6.

The following text contains additional information regarding capillary geometries. Also discussed are some routine and as well as some specialized steps to prepare capillaries for sensitive UV detection.

Cylindrical capillaries were already used in the first reported application of CE,[34] and cylindrical, polyimide-coated, fused silica capillaries (Figure 2.15) are used by the majority of CE users today. Although cylindrical geometry is common, there are some interesting points about capillaries with square or rectangular cross sections. Some of the early work in CE was done in rectangular capillaries designed originally for isotachophoresis. One such capillary consisted of a groove in the organic glass block, covered by a PTFE foil.[35] The capillary dimensions in this 1983 report were: 0.2×1.0 mm, $L_t = 200$ mm.

The advantages of rectangular profiles of capillaries were first pointed out in a theoretical study.[36]

Figure 3.18 compares the combined influence of geometry and capillary materials. The calculated temperature gradients between the center of the

108 CAPILLARY ELECTROPHORESIS OF SMALL MOLECULES AND IONS

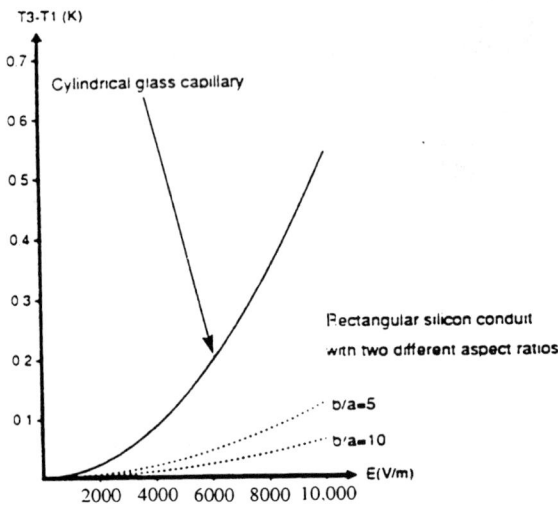

Figure 3.18 Temperature differential in a rectangular (dotted lines) and cylindrical conduit as function of applied electrical field strength. T_3 and T_1 are the respective temperatures at the center and at the outside wall of a capillary. An identical electrolyte with thermal conductivity of 0.4 W m^{-1} K^{-1} was used for all calculations. The respective thermal conductivities for silicon and borosilicate glas were taken as 149 and 1.08 W m^{-1} K^{-1}. Reproduced with permission.[36]

capillary and the outside capillary wall are five to ten times lower for the rectangular silicon conduits than for cylindrical glass capillaries. Table 2.6 shows that the thermal conductivities for fused silica (quartz 1.40 W m^{-1} K^{-1}) and borosilicate glass (1.0 W m^{-1} K^{-1}) are similar, and thus the conclusions from Figure 3.18 can be extended to include not only the borosilicate glass but also the fused silica capillaries. Improved heat dissipation in rectangular silicon capillaries thus makes possible the use of higher electrical field strengths. The ability to use higher field strengths leads to higher separation efficiencies [Eq. (2.24)]. Higher separation efficiency means in turn better resolution and reduced analytical run times. The authors[36] predict as much as ten times shorter analytical run times in rectangular capillaries made from silicon than in cylindrical borosilicate capillaries. This finding indicates a great potential for miniaturized capillaries produced by photolithographic etching techniques commonly utilized in semiconductor manufacturing.

Tsuda, Sweedler, and Zare[37] investigated the use of rectangular (20 × 200, 30 × 300, 50 × 500 and 50 × 1000 μm) and square (50 × 50, 100 × 100 μm) borosilicate glass capillaries. The capillaries are marketed (Wilmad Glass Co., Buena, NJ) as microvials and are not protected against breaking by a protective polymeric layer. Thus hydrostatic injection (Section 3.1.3) is impossible since the capillaries cannot be moved. The researchers have utilized a split-flow injector (Section 3.1.5) to overcome that problem. The results in Figure 3.19 compare signal intensity in cylindrical, square, and rectangular cross sections, and an improved sensitivity in noncylindrical capillaries is evident.

The separation efficiency remained virtually unchanged for all three geometries: $N = 210,000$ cylindrical, $N = 280,000$ square, and $N = 210,000$ rectangular. The sensitivity enhancement can be explained by the gradual increase in the length of the optical path in going from experiment (c) to experiments (b) and (a) in Figure 3.19. The enhanced peak height with the rectangular geometry is obtained by irradiating the sample across the longer of the two dimensions, that is, 340 μm.

There are at least two other ways of extending the optical path lengths in CE capillaries. The first approach[38] makes a certain portion of the length of a cylindrical capillary a part of the optical path.

The rectangle in Figure 3.20 represents a capillary holder. Prior to placing the capillary, a short straight piece of tubing made of steel or a suitable polymeric material is inserted into the capillary holder. The capillary will be placed inside this tubing, which provides a snug fit for the middle part (11) of the capillary (9) inside the holder and helps to center it properly. Before the installation of a conventional fused silica capillary, a length of the protective polyimide tubing has to be removed between points (12) and (13) by one of the methods described in the final portion of this chapter. A capillary with a length of removed polyimide tubing is then inserted into the tubing

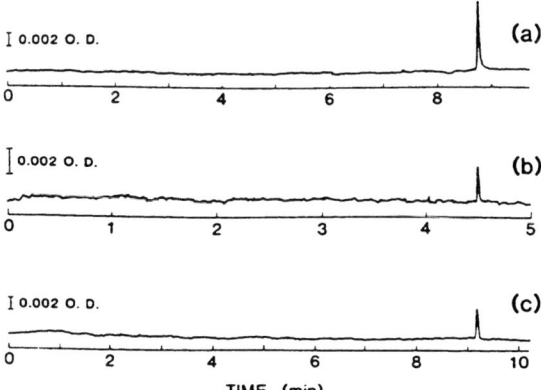

Figure 3.19 Electropherograms of a constant amount of pyridoxine in three different types of capillaries. (a) 27 × 340 μm rectangular, $L_t = 83$ cm, $L_d = 63$ cm. (b) 52 μm square, $L_t = 60$ cm, $L_d = 43$ cm. (c) 51 μm I.D. cylindrical, $L_t = 53$ cm, $L_d = 34$ cm. The carrier electrolyte contained 5 mM phosphate pH 6.8 and 5% ethylene glycol. The separation voltage was +15 kV for all three experiments. Reproduced with permission.[37]

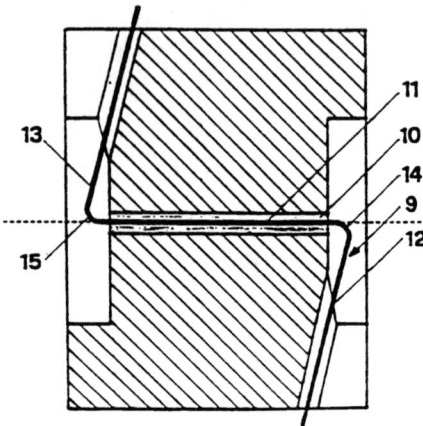

Figure 3.20 Drawing of a Z-shaped capillary cell for improved sensitivity of optical detection. See text for explanation of numerical descriptors. The numbering system is identical with that used in the original reference. Reproduced with permission.[38]

(10) positioned inside the capillary holder. Following this step, the capillary is still essentially straight and protrudes from both sides of the capillary holder in the direction of dotted line in Figure 3.20. The portion of the capillary from which the opaque polyimide coating had been removed to enable optical detection is relatively brittle and would easily break if an attempt were made to bend it at this point. The two bends (14) and (15) constituting the entrance and exit windows for the light beam can be made only after the capillary was heated to a high temperature. The two grooves, upward from point (13) and downward from point (12), provide support for the capillary in its bent shape after cooling down and during the subsequent handling. In accordance with Beer's Law, the sensitivity gains in a 50 μm I.D. Z-shaped capillary can exceed a factor of 100 for an optical path of 5 mm. An obvious limitation of Z-shaped cells and of other approaches that increase the length of the capillaries used for detection is the deteriorating resolution of closely migrating zones. In the worst case, two or more separated peak zones can be inside the optical path at the same time. In such a case, the separate peaks would be detected and recorded as one continuous peak zone. However, using the theory from Section 2.3, a minimum difference in migration time resolvable by a given optical path can be easily calculated. In a slightly more time-consuming approach, a separation can be evaluated first with a conventional capillary to make the quantitative evaluation of closely comigrating peaks with the Z-shaped cells more reliable. Separations can also be modified with the help of principles outlined in Section 2.4 to maximize resolution.

The second approach[39] to optical path length extension by a capillary modification employs an egg-shaped or bubble-shaped cell fabricated directly into a fused silica capillary. This approach extends not only the path length for irradiation, but it also increases the optical flux. While the sensitivity increases with a multiple of the path length according to Beer's Law, it is also improved by the square root of the path's cross-sectional area due to the improved photon flux.[39] The total improvement is actually greater than indicated, due to improvements in the light scattering characteristics of the egg-shaped cell. A smaller portion of the light beam is scattered, and a larger portion of the monochromatic light is thus available for measurement. Thus bubble cells offer a combined contribution with three factors: light path, area, and reduced light scattering.

Consider the following example. A simultaneous increase of optical path and cross-sectional area in a bubble cell with 450 μm I.D. made from a fused silica capillary having 50 μm I.D. equals $9^{0.5} \times 9$ or 27 times. To achieve a similar sensitivity improvement by a Z-shaped cell, the optical path length would have to be ~ 1.3 mm long. In summary: The bubble cell can enhance sensitivity without endangering the resolution of closely comigrating peaks.

In Figure 3.21 a beam of monochromatic light (64) is directed toward the capillary (10) and the center of the egg-shaped cell (32) so that the direction of the radiation is perpendicular to the longitudinal axis of the capillary (10). The total length of the optical cell is indicated by the dimensional bracket

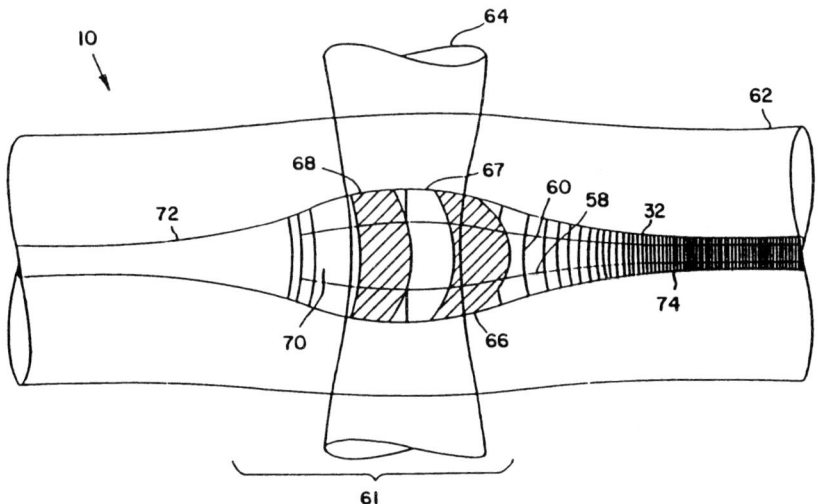

Figure 3.21 Cross-sectional diagram of a bubble cell. The numbering system is identical with that used in the original reference. See text for explanation of numbers in this figure. Reproduced with permission.[39]

(61). The outside circumference of the original capillary is indicated by the numeral 62. The lines 58, which are linear and parallel outside and curved inside the bubble cell, illustrate the electric field lines. Vertical lines 60 indicate equal potential. Also depicted are the shapes of two separated analyte zones. Zone 68 has just reached the detection zone, while zone 66 is leaving it. Figure 3.21 also indicates a certain amount of longitudinal peak spreading evident from the comparison of the middle portions of the two analyte zones 66 and 68. While this may well be the case, the author[39] reports that the total decrease of separation efficiency due to the egg-shaped cell is negligible for practical purposes. Also evident in Figure 3.21 is the possibility of lacking resolution for two closely migrating peak zones. The problem of resolution deterioration is considerably less important in bubble cells than with the Z-shaped cells, due to a much narrower length of the former type compared to latter design.

An interesting example of the great flexibility of CE is another manipulation of capillary geometry. In this case, capillaries are primarily manipulated not for a gain in detection sensitivity, but for a possible improvement in separation efficiency. Note: Detection sensitivity usually improves with improved separation efficiency as well; the peak heights increase by the square root of the ratio of two plate counts.

As indicated by Equation (2.24), the separation efficiency increases proportionally with the strength of electrical field. Given that for reasons of safety the maximum applicable separation voltage is limited to ~ 20 kV and that most commercial instruments would not handle capillaries shorter than ~ 20 cm, one could incorrectly assume that the field strengths could be optimized within a relatively narrow range only. This is certainly not the case, as demonstrated recently by Monnig and Jorgenson.[18]

By inserting a short segment of an extremely narrow capillary (e.g., 10 or 20 μm I.D.) between two longer segments of 50 or 75 μm I.D. capillary, it is possible to combine the easy handling of long capillaries and improved sensitivity of detection due to broader capillary diameters with the enhancement of separation efficiency provided by capillaries of narrow diameters (see Figure 3.22). A suitable instrument for making connections between two pieces of fused silica capillaries is commercially available (e.g., Fused Silica Manipulator from Power Technology, Inc., Mabelvale, AK).

As mentioned in Section 2.3, the polyimide-coated fused silica capillaries are currently utilized by the majority of users of CE instrumentation. The polyimide coating renders the fused silica capillaries very flexible, and it is possible to wind them in narrow coils, for example, around a finger. While the polyimide coating makes the capillary virtually unbreakable, its opaqueness prevents the use of optical "on-capillary" detection methods. So far, all instrument designs rely on "on-capillary" detection in order to avoid the challenging requirements of a dead-volume free microscopic "off-capillary" detector cell. Another factor contributing to the popularity of the use of portions of fused silica capillaries for optical detection is the ideal optical

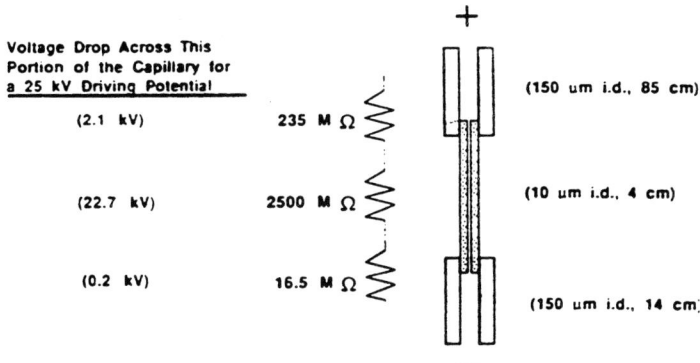

Figure 3.22 Diagram illustrating the coupling of capillaries of different diameters for enhancing separation efficiency. The figures in kV and resistor equivalent circuit to the left of the capillary list the voltage drops and resistances calculated for each section of the capillary. Reproduced with permission.[18]

properties of fused silica, which approach those of pure quartz. In order to make the optical "on-capillary" detection possible, it is necessary to remove a few millimeters of polyimide coating from the portion of the capillary to be used as a detector cell. This can be accomplished by either of the following ways:

(a) Thermal Method: The simplest way to remove a portion of the polyimide protective layer is to use an open flame, for example, from a cigarette lighter. After holding the capillary for a few seconds above the flame, it is possible to observe the decomposition of the polyimide coating. Closer inspection under a magnifying glass usually reveals some small black fragments of the original brown coating still clinging to the colorless fused silica surface. To assure a proper degree of translucence, the newly prepared fused silica window should be gently treated with a soft cloth dipped in methanol, followed by a repeated visual inspection under a magnifying glass. The portion of the capillary without its protective coating immediately makes the handling of the originally unbreakable capillary more demanding. Even though the fused silica material obtained by a carefully controlled heating and cooling of the original raw quartz is not inordinately brittle, local overheating from the flame can easily result in a decreased homogeneity due to crystallization, which makes the material increasingly brittle. This increasing brittleness, along with the fact that a relatively broad section of the capillary (5 to 15 mm) is affected by the treatment, makes the removal of coating by a flame considerably less attractive than other thermal techniques.

A simple device described by Lux, Haeussig, and Schomburg[40] (Figure 3.23) allows a more controlled exposure of the polyimide coating to heat. The detection window can be made as narrow as 200 μm, and overheating of fused silica material is prevented. It is also possible to adjust the temper-

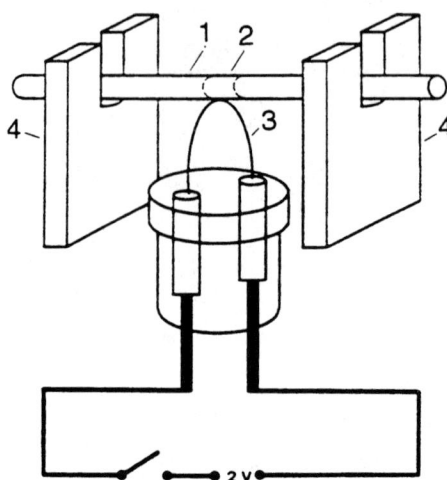

Figure 3.23 Electrically heated device for production of detection windows. 1, Polyimide-coated fused silica capillary; 2, detection window; 3, hot filament; 4, metal blocks. Reproduced with permission.[40]

ature of the filament (3 in Figure 3.23) below the softening point of fused silica to avoid the distortion that can occur with the open flame method.

(b) Chemical Method: With chemically modified inside capillary walls, thermal treatment may lead to undesirable reactions or even destruction of the capillary chemistry. Changes in electroosmotic flow and irreversible adsorption of analytes on the capillary walls may ensue. Chemical decomposition of polyimide polymer can be considered as a useful alternative to the thermal method for capillaries with chemically derivatized inside walls. A manufacturer of capillaries[41] recommends the application of concentrated sulfuric acid at 100°C for polyimide removal. Such treatment is claimed not to cause any brittleness of fused silica. The same reference advises against the use of strong bases for the removal of protective coating, since it can produce brittle sections in the capillaries. A successful application of etching of polyimide coating on capillaries by KOH solutions was reported in a recent article.[42]

(c) Mechanical Method: Even in their most refined forms, thermal and chemical methods are always likely to affect the capillary wall to a certain degree. The refinement of the techniques can only succeed in keeping the effect of heat and chemicals below an acceptable threshold. The only techniques potentially capable of removing the protective coating without any effect whatsoever on the capillary walls are mechanical methods (Figure 3.24). McCormick and Zagursky[43] described a lathelike instrument that makes it possible to scrape away the polyimide coating with a chisel-pointed knife.

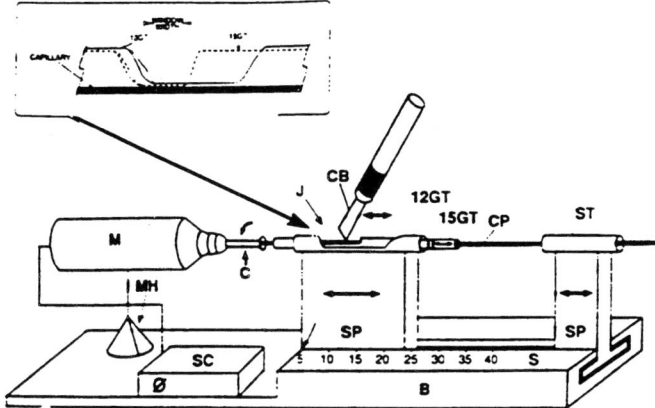

Figure 3.24 Capillary stripper apparatus for polyimide removal from fused silica tubing. The capillary is held in a jig (J) formed from two concentric sections of hypodermic tubing mounted on a sliding support standard (SP). One end of the capillary (CP) is attached to a motor (M) powered by a speed controller (SC) using a force-fit chuck (C) attached to the motor. Milled-out regions in the two concentric sections of the jig assembly define the region where the window is to be made. The cutout on the 23 cm long section of the 12-gauge tubing (12GT) is 20 mm long and 1.5 mm deep. The milled-out region on the 30 cm long section of the 15-gauge tubing (15GT) is 10 mm long and 1.2 m deep. The length of the support tubing (ST) is 4 cm. The motor is attached to a Lucite base (B) using a Moto-Tool Holder (MH), which is positioned so that the axes of the motor and sections of hypodermic tubing are aligned. The location of the window on the capillary is determined by positioning the jig an appropriate distance from the motor using the scale (S) on the base. An X-Acto chisel blade (CB) removes the polyimide from the capillary in a back-and-forth motion with the point of blade resting on the bottom of the cutout in the 15-gauge tubing. (A second support tube may be positioned between the motor and the jig to provide additional stability when windows are being formed near the center of long (75–125 cm) capillaries. Reproduced with permission.[43]

All their advantages notwithstanding, the mechanical and chemical methods are usually more laborious than the thermal procedures. The latter techniques are thus very likely to remain the predominant approach by the users of CE instrumentation. The chemical and mechanical approaches will probably remain in use solely by the equipment manufacturers, who will utilize them in handling gel-filled and chemically derivatized capillaries.

Fused silica capillaries are also easily modified to any length by simple cutting devices. As shown in the three steps in Figure 3.25, the only necessary precaution is to hold the capillary downward to minimize a possibility that microscopic fragments of the coating material could obstruct or even plug the opening of the capillary. A brief inspection of freshly cut ends of capillaries under a magnifying glass may also be very useful.

Commercial suppliers of fused silica capillaries are listed in Table 3.3.

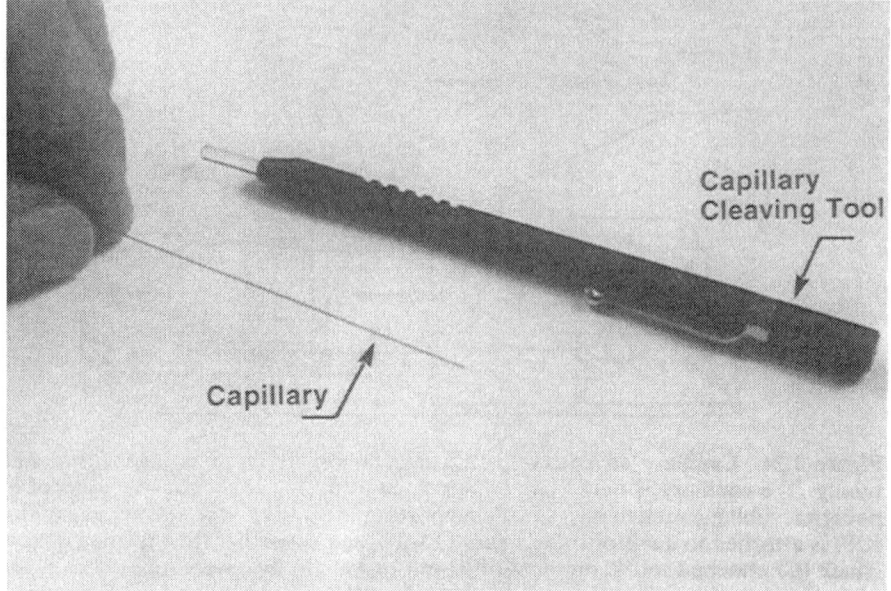

A

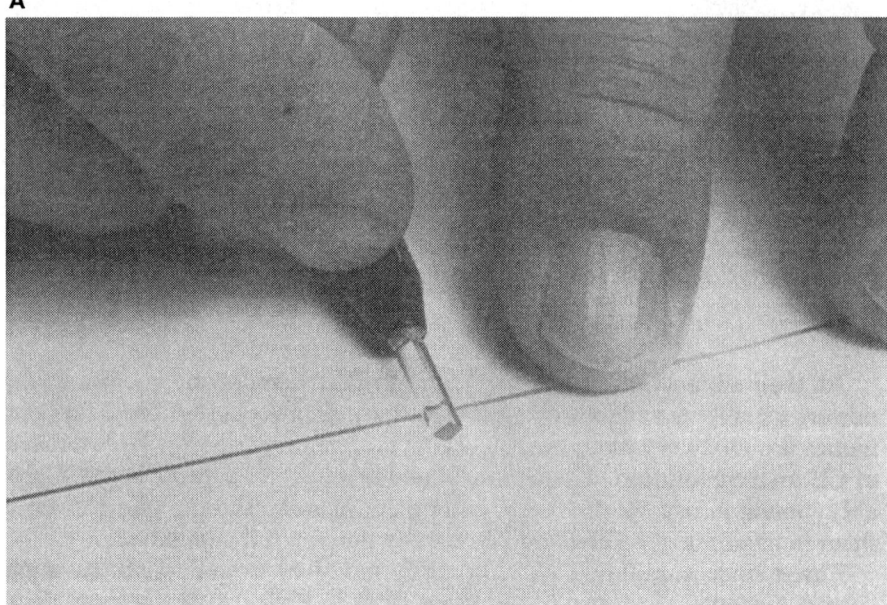

B

Figure 3.25 Cutting of fused silica capillaries.

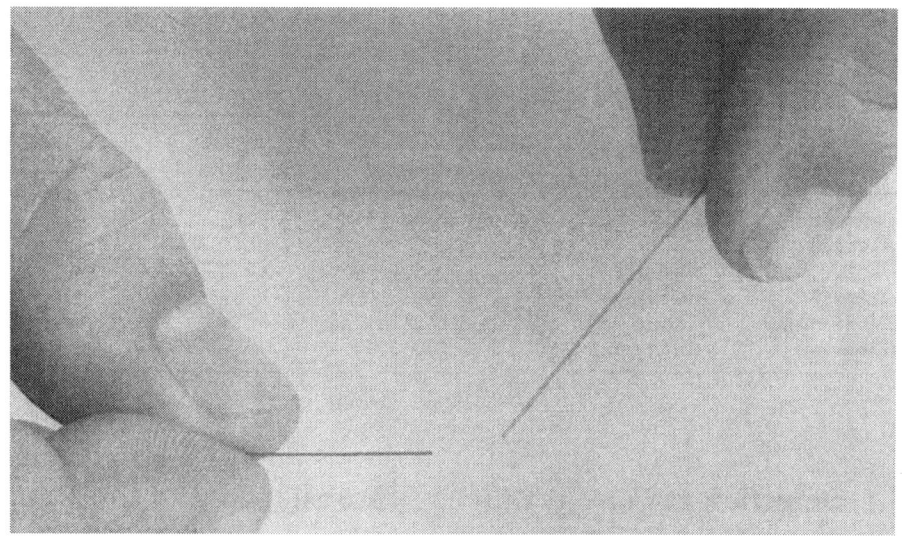

C

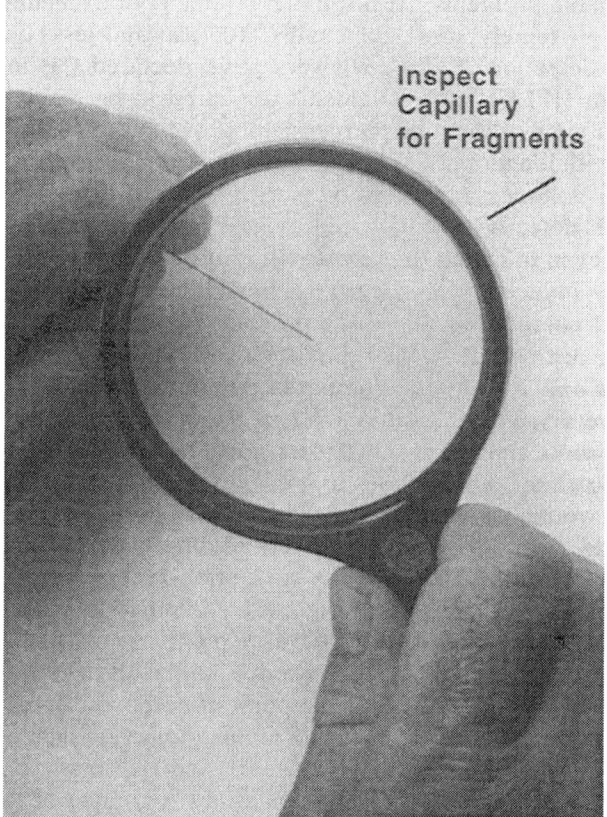

D

Figure 3.25 (*continued*)

Table 3.3. COMMERCIAL SUPPLIERS OF FUSED SILICA CAPILLARIES

Company	Location	Cylindrical Capillaries	Rectangular Capillaries	Chemically Modified
Polymicro Technologies	Phoenix, AZ	yes	yes	no
SGE	Austin, Tx	yes	no	no
Supelco	Bellefonte, PA	yes	no	yes
J&W Scientific	Folsom, CA	yes	no	yes
Sarasep	Santa Clara, CA	yes	no	yes

3.3 Detection in Capillary Electrophoresis

At first glimpse, detection in capillary electrophoresis seems to impose many formidable problems, with sample volumes not exceeding 100 nanoliters and the extremely short light paths (100 μm and less) encountered in on-capillary detection. Early reviewers have declared CE inherently less sensitive than HPLC. Such judgments appeared to be eminently justified. Especially in view of the much larger sample volumes in HPLC—10 μl and more—and with longer optical path lengths in HPLC detectors—5 to 10 mm. With sample volumes and detector path lengths at least a hundred times smaller, a CE detector, it was stated, would have to be about ten thousand times better even to match the sensitivity routinely accomplished in modern HPLC. Laser-induced fluorescence has been cited frequently as a technique with the best potential to overcome the lack of sensitivity. Less expensive lasers will have to be developed, but even so, the improved sensitivity will be applicable only to a limited number of compounds with excitable fluorescence. As we discuss in Section 3.3.3, indirect laser-induced fluorescence detection is more universal, but lasers must become more stable for this technique to achieve greater sensitivity.

The odds would have been weighted heavily against the emerging analytical technique, were it not for one essential difference between HPLC and CE: Unlike CE, HPLC is basically a dilution technique. Many times in HPLC, the sample components, originally contained in about 10 μl of volume, give rise to peak zones of one or more milliliters by the time they reach the HPLC detector cell. As a consequence, the analyte concentration is as much as a hundred times lower at the point of detection than in the original sample. On the other hand, due to the almost complete lack of longitudinal diffusion in optimized CE (Section 2.5), the analyte zones arrive at the detector in essentially the same volume as at the beginning of the separation. Moreover, due to electrostacking (Section 3.1.1) that initial volume can be a factor of ten to one hundred smaller than the original sample volume. Even if larger volume injections (more than 200 μl) may be possible in HPLC in conjunction with peak volumes under 1 ml in certain cases, we see the originally assumed ten thousand–fold disadvantage of CE (100 × smaller sample, 100 × shorter optical path) disappear, for the majority of applications, in view of the possibilities of electrostacking (analyte amount contained in a 100× smaller volume at the point of detection) and due to the effect of longitudinal diffusion in HPLC (analyte contained in 100× larger volume during the detection). Given that the introduction of extended path length detector cells (Section 3.2) may frequently put CE at an advantage by about a factor of ten (50 to 500 μm in a bubble or rectangular cell).

In some cases, such as, for example, with isotachophoretic preconcentration preceding a CE separation (Section 3.1.5.2), it is possible to increase the analyte concentration between the sample and the point of detection as much as a hundred thousand to a million times. Similar enrichment is of course possible with preconcentrating columns even in HPLC. Only, the isotachophoretic trace enrichment requires ~ 30 to 45 sec; the typical trace enrichment times in HPLC are ~ 10 min.

Based on these facts, the discussion in this chapter focuses mainly on direct and indirect UV detection for capillary electrophoresis. Laser-induced fluorescence detection will be merely reviewed along with several other interesting experimental techniques.

3.3.1 Optical Detection at UV and Visible Wavelengths in Capillary Electrophoresis

Measurements of absorbance [Eq. (3.11)] of UV or visible light are usually classified as a "solute property" detection. The designation comes from the assumed degree of selectivity obtained by the evaluation of a property of one of many components of a liquid sample. Another approach is then termed "bulk property" detection, evaluating a fundamental property of a liquid, such as, for example, the refractive index or conductivity. It is interesting to note that the application of this classification to instruments rather than to methods can be misleading at times. It is difficult to classify indirect UV detection as a solute property detection method because it has more of the characteristics of bulk property detection. On the other hand, conductivity, if used with chemical suppression,[44] can hardly be described as a bulk property measurement as the analytes of interest are detected against a background that is close to zero for all practical purposes. Other examples can be found where an identical detector can be used in both modes of measurement. Uninformed readers can be misled by some authors' classifications of "bulk property detectors" as relatively insensitive or by statements of high sensitivity of "solute property detectors" due to minimal levels of background signal. In summary: While the attributes "bulk" and "solute" are very useful in method description, they should not be applied to instruments utilized for the execution of the methods.

120 CAPILLARY ELECTROPHORESIS OF SMALL MOLECULES AND IONS

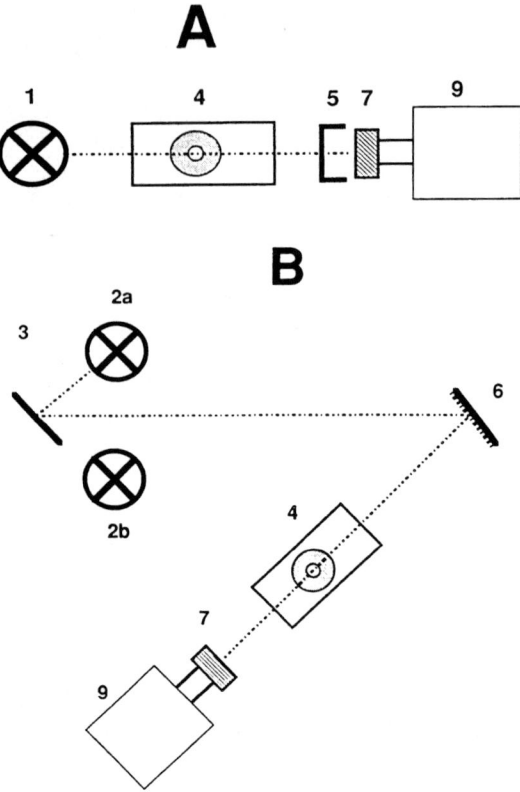

Figure 3.26 Four most common UV/visible detectors in capillary electrophoresis. (A) Fixed-wavelength detectors, (B) variable-wavelength detectors, (C) scanning monochromator detector, (D) photodiode array detector. See the text for explanation of numerals.

All commercially available CE systems are equipped with one of the four basic types of UV/visible detectors. The functional principles of the most common UV/visible CE detectors are shown in Figure 3.26.

The fixed-wavelength detectors are noted for their simplicity and ruggedness. They consist of only a minimal number of components and do not utilize any moving parts. In addition to ruggedness and simplicity, they offer to the user considerable increases in sensitivity, especially at detection wavelengths below ~ 230 nm. The source of light (1) is either a deuterium or more often an atomic vapor lamp. Typical elements utilized in the latter type of lamps are zinc, cadmium, and mercury. The rectangular component (4) in Figure 3.26 represents the detection cell assembly consisting usually of a section of fused silica capillary made translucent by the removal of the protective layer of opaque coating (Section 3.2) and of other components

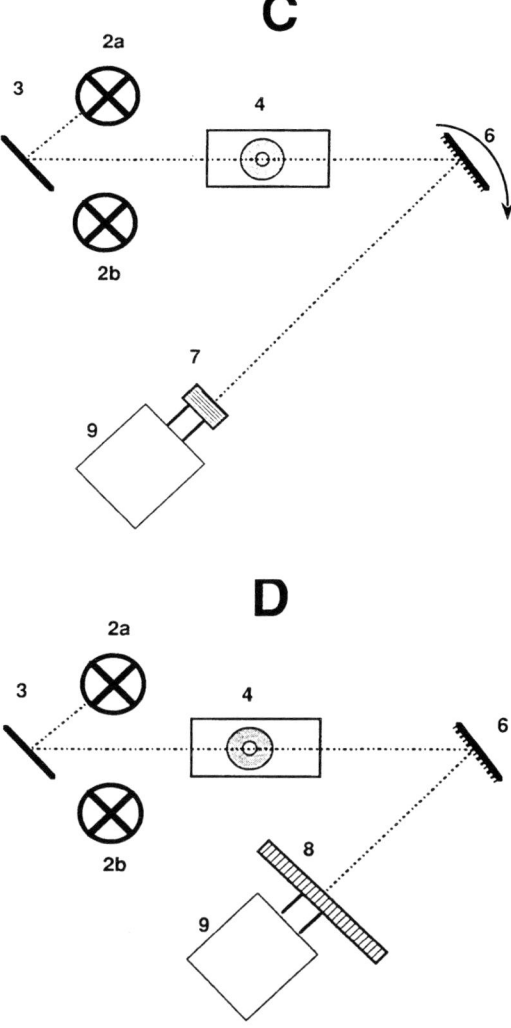

Figure 3.26 (*continued*)

that will be discussed in the subsequent paragraphs. Wavelength selection is carried out by an optical filter (5), and a decrease in light intensity due to light absorption in the sample is then sensed by a photodiode or a photomultiplier[45] (7) connected to the electronics (9) of the instrument. The instrument electronics continuously compares the light intensity with zero sample concentration I_o to the intensity after passage of the light through a cell filled with a sample I. The term I_o is obtained in most HPLC and CE instruments by measuring the light intensity of a reference beam in the air, without going through the detection cell. The ratio I/I_o is called the *transmittance* and is not used for output since it does not yield a linear relationship to sample concentration. The instrument electronics are designed to evaluate the logarithm of the reciprocal value of the transmittance, called the *absorbance A*. According to Beer' Law, also called the Lambert–Beer Law in some countries, there is a linear relationship between c, the concentration of a dissolved compound, and absorbance:

$$A = \log_{10} (I_o/I) = kLc \tag{3.11}$$

where k is the molar absorptivity coefficient (the value of absorbance for 1 M concentration of a solute) and L is the optical path length. As already discussed in Section 3.2, Beer's Law is a one-dimensional approximation of a somewhat more complex relationship. It does not account for the cross-sectional area of a beam transmitting through a sample and reaching the photosensitive element. In a strict sense, Beer's Law is thus valid only for light beams having identical geometries. Only for identical beams will the absorbance increase linearly with multiples of the path length. According to reference 39, if the cross-sectional area of the light beam and the light path increase simultaneously, a useful approximation is to calculate the change of absorbance as: $(d_2/d_1)^{1/2}(L_2/L_1)$, where d is the diameter of the cross-sectional area of the cell.

Some variable-wavelength detectors are equipped with two different lamps (see Figure 3.26B). The usual combination consists of a deuterium (2a) and a tungsten (2b) lamp. The two lamps are necessary because of the low energy output of the deuterium light source in the visible range of the electromagnetic spectrum. The choice between the two lamps is made by a proper positioning of a mirror (3). After passing through the optical cell assembly (4), the polychromatic light reaches the monochromator (6) responsible for the selection of a detection wavelength reaching the photodiode or photomultiplier (7). In most modern optical instruments the function of a monochromator is fulfilled by a diffraction grating.[45] The inclusion of a monochromator in the detector greatly increases the convenience of switching between various wavelengths. Unlike in fixed-wavelength instruments, where the operator replaces one wavelength kit (usually filter plus lamp) for another, the change of wavelengths in variable instruments is as simple as turning a dial knob, connected to the diffraction grating, from one position to another. On the other hand, even with variable detectors the wavelength is seldom changed during an analytical run, where speed of change is of importance. The changes of detection wavelengths are usually implemented between analytical runs. Furthermore, because UV spectra are typically very broad, stretching over hundreds of nanometers with relatively flat maxima, the convenience of wavelength change becomes less important in some cases.

For many applications the advantage in sensitivity held by the fixed-wavelength instruments may be of greater importance. Even though there have been great improvements in the ability of variable instruments to carry out detection at wavelengths below 230 nm (approximate energy maximum of a deuterium light source), the sensitivity of fixed-wavelength instruments in the range between 185 and 254 nm still remains unsurpassed. This is especially important since absorption coefficients increase in the direction toward lower wavelengths for the vast majority of compounds.

A number of factors are responsible for degrading the performance in the low UV range with variable wavelength instruments. The output of deuterium lamps typically used in variable detectors falls off sharply below the energy maximum at 230 nm (see Figure 3.27).

In contrast to deuterium lamps, the pen-ray lamps first applied to CE by Wahlbroehl and Jorgensen[47] offer a sufficient optical throughput for a sensitive detection at several different wavelengths between 254 and 185 nm. Note that at 254 nm the energy of the mercury lamp actually exceeds the plotted range. Even at the range where the energy output of the deuterium lamps is close to maximum, the greater energy of the mercury vapor lamp yields a more sensitive detection. Another factor in favor of fixed-wavelength instruments is that the efficiencies of the monochromator gratings are usually not optimized for wavelengths below 220 nm.[48]

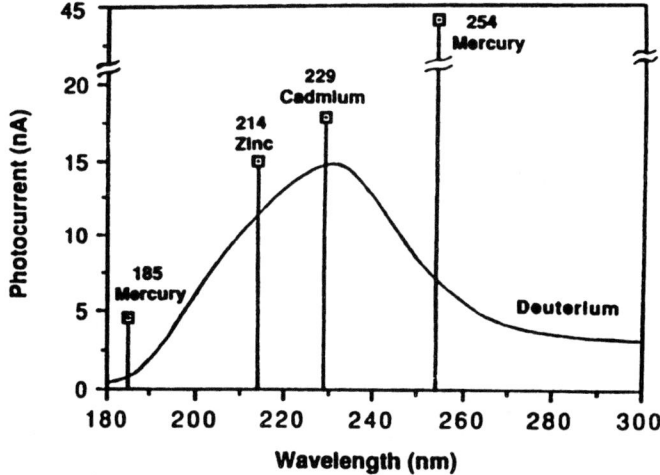

Figure 3.27 Comparison of optical throughput (light intensity sensed by a photodiode) for fixed and variable detectors. The continuous curve indicates the energy distribution stemming from a deuterium lamp. The perpendicular lines give the energy value sensed by the photodiode in combination with wavelength filters and atomic vapor lamps. Reproduced with permission.[46]

Unlike fixed-wavelength instruments, the monochromator detectors have substantial path lengths through the air (see Figure 3.28). The absorbance of oxygen is low in the near UV, but it increases sharply toward shorter wavelengths.[49] Further compounding the energy loss is the generation of ozone by the UV light in the air. Ozone has a broad absorption maximum at 250 nm, with a much higher molar absorptivity coefficient than oxygen.

It is interesting to comment on the relative importance of detection at low UV wavelengths in HPLC and in CE. Solvent transparency is a major issue in HPLC, making sensitive detection increasingly difficult with lower wavelengths. Most common solvents and buffers cannot be used at 200 nm. Water, acetonitrile, and hydrocarbons such as hexane are among the few remaining useful choices. In contrast, in CE the separation medium is mostly aqueous, much lower solvent and salt concentrations are used, and the on-column detection generates path lengths shorter by a factor of one hundred or more than in HPLC. Solvent transparency is thus much less an issue in CE than in HPLC.

The greater feasibility of 185 nm UV detection in CE makes it possible to improve detectability for a large number of analytes. An example of such an improvement is illustrated by the separation of antibiotics shown in Figure 3.29.

Unlike in HPLC, where the fixed-wavelength detectors have been gradually replaced by variable and scanning instruments, in CE the former type of detector will remain highly useful for some time into the future. The latter two types of detectors will become comparable in sensitivity to the fixed-wavelength type only after additional improvements in the efficiency of optical elements and with shorter optical path lengths through the air.

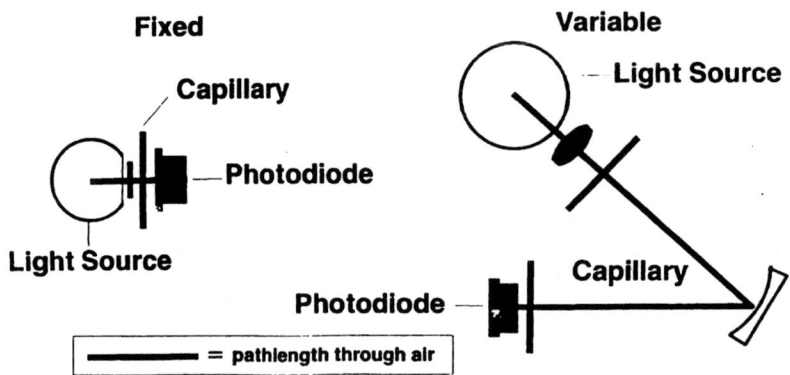

Figure 3.28 Comparison of optical layouts of fixed and variable detectors on the same scale. The path lengths through the air differ by a factor of 10. Reproduced with permission.[46]

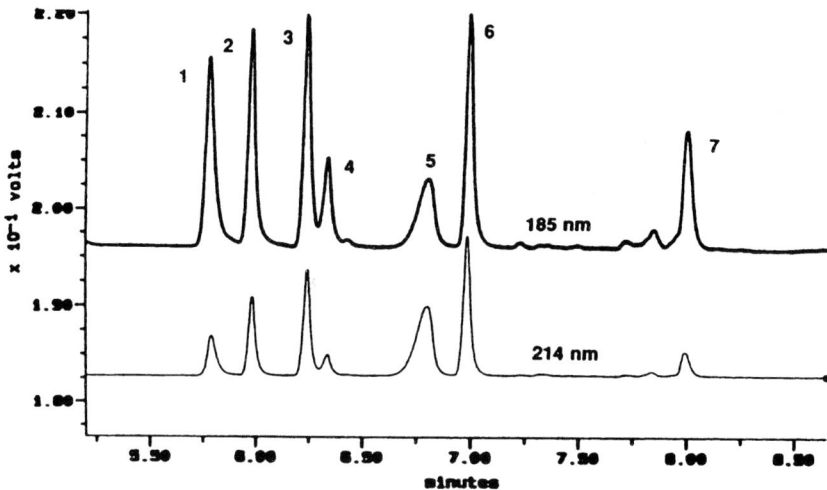

Figure 3.29 Increased sensitivity of detection at 185 nm in comparison with 214 nm. Peak identities: (1) amoxicillin, (2) oxacillin, (3) cloxacillin, (4) 6-amino penicillanic acid, (5) nafcillin, (6) dicloxacillin, (7) ticarcillin. A 10 sec hydrostatic sample introduction from a standard solution containing 10 μg/ml of each analyte. Capillary dimensions: 75 μm and 60 × 52 cm. Carrier electrolyte: 20 mM sodium phosphate/sodium borate, 50 mM SDS, pH 9.0. Separation voltage: +18 kV. Reproduced with permission.[46]

Currently the term *scanning* is used rather narrowly in optical instrumentation to signify the evaluation of an optical signal over a certain range of wavelengths. However, the word *scanning* may be also utilized to describe signal evaluation in three-dimensional space, across a certain segment of an area, or over a certain length in only one dimension. In the early days of capillary electrophoresis, Hjerten[34] had experimented with a version of scanning detection consisting of UV absorbance measurements over the entire length of a separation capillary.

Polychromatic scanning or detection within a range of wavelengths can be carried out with two different techniques. First, it is possible to design instruments capable of carrying out well-defined, rapid, and precise movements of the diffraction grating (6) in a similar configuration as in Figure 3.26C. Unlike most variable detectors, however, the polychromatic light is first directed by the mirror 3 through the optical cell assembly 4, and the change of detection wavelengths is carried out only after the passage through the sample. Such an arrangement is usually referred to as *reversed optics* to stress the principal difference between the variable (tunable) detectors on one hand and the scanning detectors on the other.

For separations having peaks broader than, for example, 10 sec, the movement of a monochromator in 0.5 sec from one end of the wavelength range to another makes it possible to collect twenty or more measurements for a single peak. As discussed in Section 3.4.5, twenty data points describe a peak sufficiently. However, with analysis times dropping to a few minutes for separations containing tens of peaks, total peak widths less than 0.5 to 1.0 sec are becoming increasingly common. Prior to collecting the data with instruments utilizing scanned gratings, a user should always verify that the scanning rate is sufficient to provide enough data points for a given peak width.

Scanning detection of very narrow peaks can be carried out with the help of photodiode array instruments (Figure 3.26D). At least theoretically, it is possible to collect twenty and more data points for the narrowest of CE peaks. In practice, many commercial instruments have maximum scanning rates of 10–20 scans per second. Realistically, the practical limit to maximum scan rates resides with the data storage devices used for the acquired data. A storage capacity on the order of magnitude of several gigabytes has to be made available for a routine data processing of three-dimensional electropherograms lasting more than 1 min. Similarly to the variable instruments, a photodiode array detector may also employ more than one lamp (2a and 2b in Figure 3.26D) for optimal energy output over the entire span from low UV across the visible range. As in the monochromator-equipped, rapidly scanning detectors, the polychromatic light is first directed by mirror 3 through the optical cell assembly 4, and the change of detection wavelengths is carried out after the passage through the sample. The optical element 6 can, for example, be a holographic grating providing a dispersion of the monochromatic light across the array of photodiodes 8. As can be expected, the electronics part 9 of a photodiode array detector is considerably more complex than the electronics module in a variable- or fixed-wavelength detector. In the majority of cases, it has to be supplemented by a personal computer with an adequate data processing and storage capacity. All these factors in combination make photodiode array detectors considerably more expensive than other UV detectors.

The three-dimensional electropherograms obtained with the help of scanning UV detectors are potentially of high practical value, since the third dimension of wavelength improves the degree of certainty by which the identity of an unknown compound can be determined. By itself, a two-dimensional recording of signal intensity versus migration time does not suffice to provide positive proof of identity. The situation in this respect is quite similar to HPLC, but different from gas chromatography, where the retention time is frequently utilized as a proof of identity. In the same way as two-dimensional HPLC and CE, a UV spectrum alone cannot be used for identification. However, if the migration times are available in combination with the corresponding UV spectra, a three-dimensional electropherogram, such as the one in Figure 3.30A, becomes a useful identification tool.

The three-dimensional chromatograms in Figure 3.30 were obtained with the help of a fast scanning monochromator detector. The schematic diagram of such a detector is presented in Figure 3.26C. The spectra were collected at the rate of 3.69 per sec and in 5 nm intervals, corresponding to 25 wavelengths between 195 and 320 nm. The single wavelength electropherograms can be distinguished in Figure 3.30 as straight lines or lines with one or several maxima. In contrast, a typical photodiode array detector can collect spectra at ~ 1–2 nm intervals and across a much wider range of wavelengths. A scan between 190 and ~ 600 nm is possible with most commercial instruments.[51]

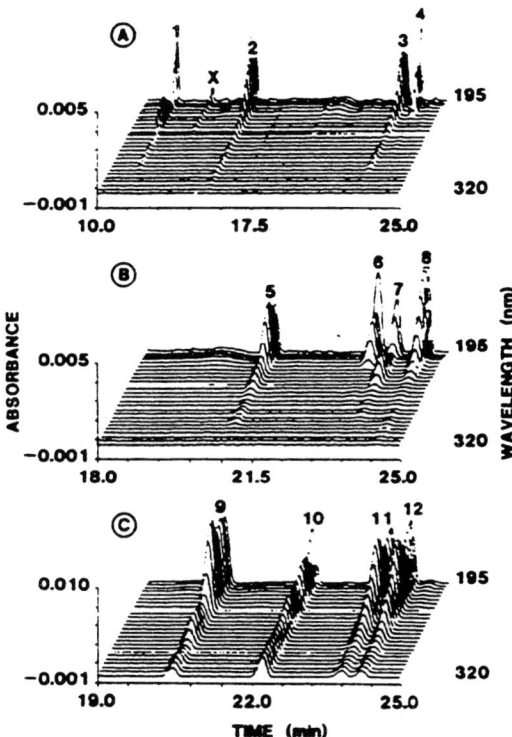

Figure 3.30 Three-dimensional electropherograms of three model mixtures of different drugs. The separation voltage was 20 kV in all three cases across a 75 μm × 70 × 90 cm fused silica capillary. Hydrostatic sample introduction was done at the height of 34 cm for 5 sec. All sample components were at 2 mg/ml. The carrier electrolyte contained 75 mM sodium dodecyl sulfate, 6 mM sodium tetraborate, and 10 mM disodium phosphate, pH 9.1. The peak identities were: (1) benzoylecgonine, (2) morphine, (3) heroin, (4) methamphetamine, (5) codeine, (6) amphetamine, (7) cocaine, (8) methadone, (9) methaqualone, (10) flunitrazepam, (11) oxazepam, (12) diazepam, (X) benzoic acid impurity from benzoylecgonine. See text for discussion of detection method. Reproduced with permission.[50]

128   CAPILLARY ELECTROPHORESIS OF SMALL MOLECULES AND IONS

For identification purposes, the spectra of the compounds in each of the peaks from an unknown sample can be compared with the reference spectra stored by the instrument from the previous injections of single compounds or standard mixtures of known composition. The comparison is facilitated by a software algorithm that can adjust (normalize) the intensity of the sample spectrum to the same level as that of the reference spectrum. Comparison of peak spectra obtained from a urine specimen with those generated in Figure 3.30 is shown in Figure 3.31. The separation conditions for the analysis of the urine sample were the same as those in Figure 3.30.

From the four compounds examined in Figure 3.31, morphine (Figure 3.31B) is a metabolite of heroine; another illicit drug, codeine (Figure 3.31C) does not metabolize readily and is excreted in the urine unchanged. The three-dimensional electropherograms represent a powerful instrumental ap-

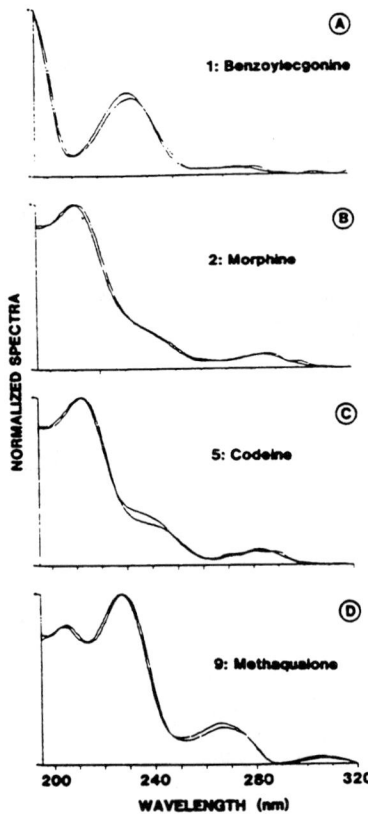

Figure 3.31 Comparison of UV spectra from a standard of known composition with UV spectra obtained from a urine sample. Reproduced with permission.[50]

proach to the analysis of illicit drugs and their metabolites. The authors[50] recommend their techniques for confirmation testing following a positive response of a toxicological screening procedure.

Up to this point we have considered only the general optical schemata of the instruments utilized in UV/visible detection. The detailed arrangement of optical elements in the immediate vicinity of the capillary segment used for detection (module 4 in Figure 3.26) will be discussed next. Some of the possible geometries of CE detection cells are discussed in Section 3.2. The simple calculation of sensitivity increase in connection with bubble capillaries is based on a spherical approximation of the detector cell cavity.

The theory of the transmission of light through a cylindrical cell was developed by Hjerten[34] and by Poppe et al.[52] Both treatments assume, somewhat unrealistically, collimated incident light (i.e., parallel rays). Their derivation of Beer's Law for cylindrical cells formed by segments of fused silica capillaries from which the opaque protective coating has been stripped (Section 3.2) is based on the representation of the optical cell assembly given in Figure 3.32.

In Figure 3.32, I_0 indicates the intensity and direction of the incident light, a is the length of the translucent capillary segment illuminated by the incident light beam, l is the path length of an infinitely small element of the total cell volume. Analogously, as in the derivation of Beer's Law in one dimension,[45] the three-dimensional derivation starts with the equation defining the transmission through an infinitely small segment having the length l and the width dx.

$$dT = (1/2s) \exp(-2\varepsilon' c\{r^2 - x^2\}^{1/2}) \, dx \qquad (3.12)$$

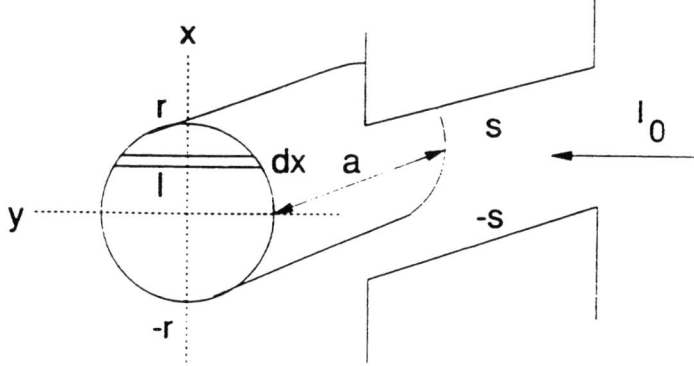

Figure 3.32 Schematic representation of an optical cell assembly consisting of the inner cyllindrical cavity (radius r) of a fused silica capillary and an aperture having a total width $s + (-s)$. Additional explanation of the symbols is provided in the text. Reproduced with permission.[52]

where dT is the contribution to the total transmission T by the segment I times dx, ε' is ln 10 times the molar absorptivity of the medium filling the inner cavity of the capillary. By integrating Equation (3.12) over the entire aperture width between $+s$ and $-s$ in Figure 3.32, and for effective absorbance A less than 0.1, we can obtain:

$$A = \varepsilon c[\{r^2 - s^2\}^{1/2} + (r^2/s) \arcsin\{s/r\}] \tag{3.13}$$

For all cases where the aperture width is adjusted to match exactly the inside diameter of the capillary, the expression inside the brackets is simplified to $0.5\pi r$, and Beer's Law for cylindrical cells becomes:

$$A = 0.5\pi r \varepsilon c \tag{3.14}$$

Matching of the capillary I.D. by the aperture not only yields the simplest form of Beer's Law, it also helps to reduce the noise levels of the detection signal. The noise is increasing if only a portion of the light beam energy is utilized for the absorbance measurement and if too much of the light energy is not directed through the cell by too large an aperture width or is lost to diffraction and reflection. Additional increases of noise are due to stray light. The stray light is the small portion of the light outside the narrow wavelength range defined by the monochromator.[53]

Poppe and co-workers[52] carried out modeling calculations of light intensity losses by diffraction and reflection, the results of which are shown in Figure 3.33.

Figure 3.33A shows an experiment where the aperture width matches the inner diameter of the capillary. As can be seen, only a relatively insignificant dispersion, due to different refractive indices of the capillary material and sample, is observed. By increasing the aperture width, while keeping the capillary inner diameter constant, as in Figures 3.33B and C, the refractive dispersion soon exceeds acceptable levels. Inclusion of focusing lenses may lead to an energy-efficient utilization of relatively broad light beams (Figure 3.33D), if the focusing lens can be placed at a proper distance from the inside cavity of the capillary. If, however, for some reason, it is not possible to position the focusing lens properly, reflective and refractive losses can again increase over an acceptable level (Figure 3.33E). A misalignment of optical elements of a CE cell thus frequently leads to a decreased magnitude of the absorbance signal.

On the right-hand side of Figure 3.34 are depicted three configurations of a 50 μm I.D. and 375 μm O.D. capillary with the different optical elements. The configuration containing the 0.5 mm aperture is comparable with the modeling experiment in Figure 3.33C and represents a gross mismatch of the aperture relative to the inner diameter of the capillary. The corresponding calibration curve exhibits the lowest value of slope of the three calibration curves displayed. Inclusion of a properly designed and positioned lens can correct for a wide aperture, as shown by the uppermost drawing, and the

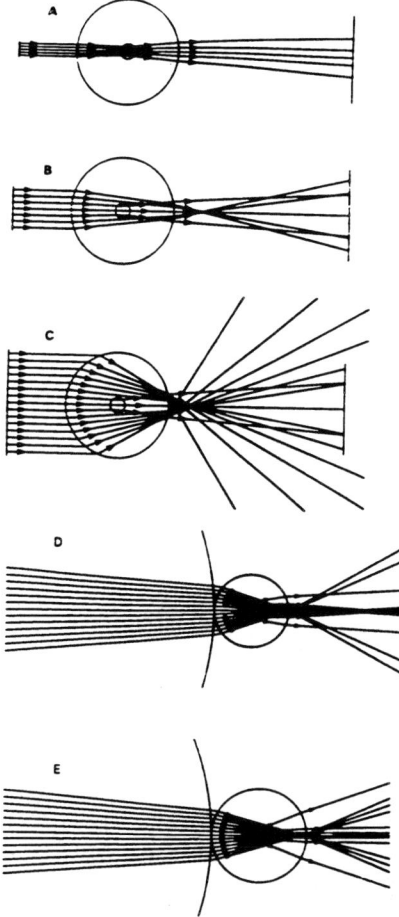

Figure 3.33 Results of modeling experiments with various configurations of optical elements in relation to a cylindrical capillary. Configurations A, B, and C are combinations of different aperture widths (A: 50 μm, B: 145 μm, C: 350 μm) for three capillaries having an identical I.D. of 50 μm. The two last experiments utilize identical focusing lenses with capillaries differing in I.D. as well as O.D. D: 75 μm I.D. and 275 μm O.D., E: 50 μm I.D. and 350 μm O.D. Reproduced with permission.[52]

corresponding calibration curve with the largest slope in Figure 3.34. In those cases, where the aperture width on one hand and lens design and positioning on the other hand are not combined properly, the inclusion of a lens does not lead to an optimal strength of the absorbance signal for a given capillary, for example, the calibration curve with an intermediate value of

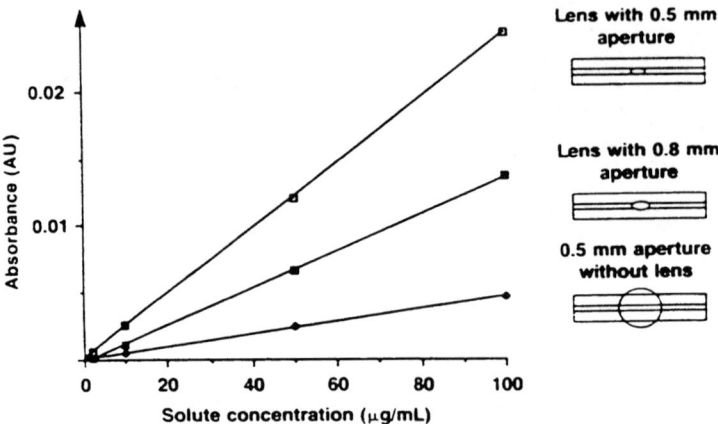

Figure 3.34 Changes in detector sensitivity at 200 nm for different lens and aperture combinations. Several different concentrations of peptide dynorphin in 20 mM sodium citrate at pH 2.5 were used in static measurements. Reproduced with permission.[54]

the slope in Figure 3.34 obtained with an 800 μm aperture in combination with the same lens as used for the 500 μm wide aperture.

The dependency of signal-to-noise ratios on the size and geometry of apertures defining the shape of the light beam, before its passage through a fused silica capillary, was also evaluated by Wang, Hartwick, and Champlin.[55] The authors observed up to sixfold increases in signal-to-noise ratios between 1 mm and 50 μm aperture widths used in conjunction with a 50 μm I.D. capillary. They also report an expansion of the linear range of detection with a properly matched capillary–aperture combination.

The primary approach to improving detection sensitivity is to expand the path lengths of the optical cell. As discussed in Section 3.2, in capillary electrophoresis, optical paths may be expanded by different cell geometries (e.g., bubble and Z cells). Another interesting approach consists of forcing a light beam to travel several times through a relatively narrow segment of modified fused silica capillary.

The multireflection cell shown in Figure 3.35 was prepared from a 75 μm I.D. and 364 μm O.D. capillary by burning off about 1 cm of the protective polyimide coating. Subsequently, a silver layer was deposited on that opening by redox reaction of $Ag(NH_3)_2^+$ and glucose. The silver coating was then covered by a layer of black paint to protect it from physical damage. The two windows in the silver layer and black protective coating were 0.8 mm (D1) apart and about 0.35 mm wide. The critical parameter was the incident angle Θ, by which the laser beam (this type of cell only works with a laser) was directed to enter the capillary. The output light intensity was at a maximum with an incident angle of about seven degrees. The authors calculated

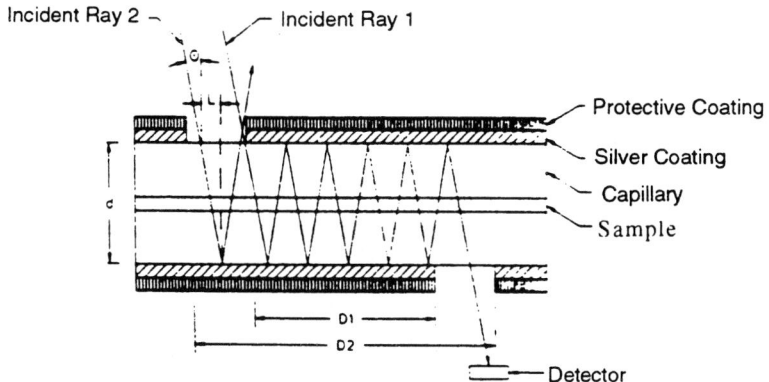

Figure 3.35 Multireflection cell. Reproduced with permission.[56]

about 30–40 reflections of the light inside a capillary at the optimally adjusted angle Θ. This was found to be in good agreement with the observed ~ 40-fold sensitivity improvement in going from a single-pass to multiple-reflection cell.

The design and manufacturing of optical cell assemblies for capillary electrophoresis can be greatly simplified (no lenses and apertures required) and improved by an incorporation of optical fibers as waveguides. Bruno et al.[42] evaluated a number of possible configurations for including optical fibers in the CE cells and formulated guidelines for achieving an optimum performance. The first of these guidelines prescribes the maximum possible diameter of the optical fiber providing the incident beam for the capillary cell, the source fiber in Figure 3.36.

The second guideline for the optimum design of cells incorporating optical fibers specifies the minimal diameter of the collecting fiber relative to the internal diameter of the fused silica capillary. Guidelines one and two help to minimize those portions of total light energy lost to various reflections and to refractive effects. Different light paths of portions of the original incident beam are illustrated in Figure 3.36. The rays numbered 1–8 represent those portions of incoming light lost to reflection at the outside surface of the capillary. Rays 9–12 are also lost and useless for the detection, since they run only through the fused silica capillary material without entering the inside cylindrical cavity containing the electrolyte or sample zones. Ray number 13 is deflected at the capillary's innermost surface and represents another portion of light that is lost. In order to maximize the sensitivity of detection, the cell design has to provide a maximum yield of light passing through the full length of the sample and accepted by the collecting fiber—rays numbered 14–30 in Figure 3.36. The various fates for the four groups of rays just described were calculated with the help of Snell's Laws[45] and from the resulting angles of incidence at the interfaces represented in Figure

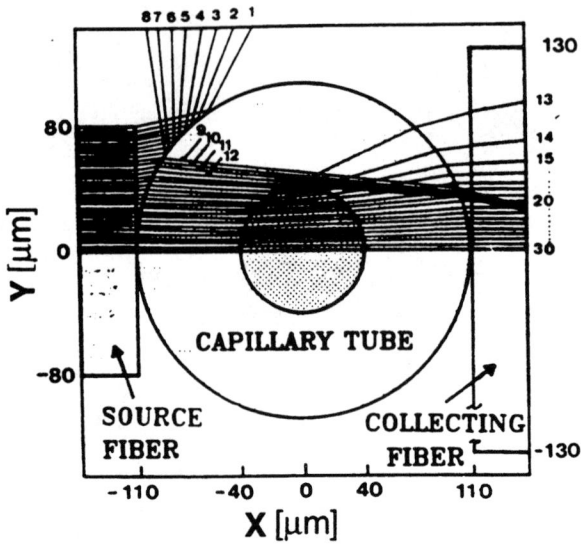

Figure 3.36 Optical tracing diagram illustrating several possible paths for fractions of the incident light beam. Reproduced with permission.[42]

3.36. The utility of a properly designed optical cell assembly incorporating optical fibers is not limited just to CE. They have already been found useful in HPLC detectors and hold an equally great promise for UV/visible and fluorescence detectors.

3.3.2 Indirect UV Detection in Capillary Electrophoresis

Indirect detection measures the decrease in a relatively high level of background signal in a separation medium caused by zones of analyte having only a negligible or no response. Indirect UV detection makes use of compounds exhibiting a relatively high level of absorption of UV light for the separation medium. Under the most suitable conditions, the concentration of the UV absorbing substance in the separation medium (e.g., eluent in HPLC or carrier electrolyte in CE) and the detection wavelength are chosen to maximize the UV absorbance and to minimize the noise. The absorbance can of course be increased by simply increasing the concentration of the UV background-providing compound. However, at too high concentrations of such a compound, the signal noise usually increases beyond an acceptable level. In extreme cases the change of detector signal is no longer a linear function of the concentration of background-providing substance or may even become independent of that concentration. Providing that suitable conditions could be found, indirect UV detection can be carried out according to the schema illustrated in Figure 3.37.

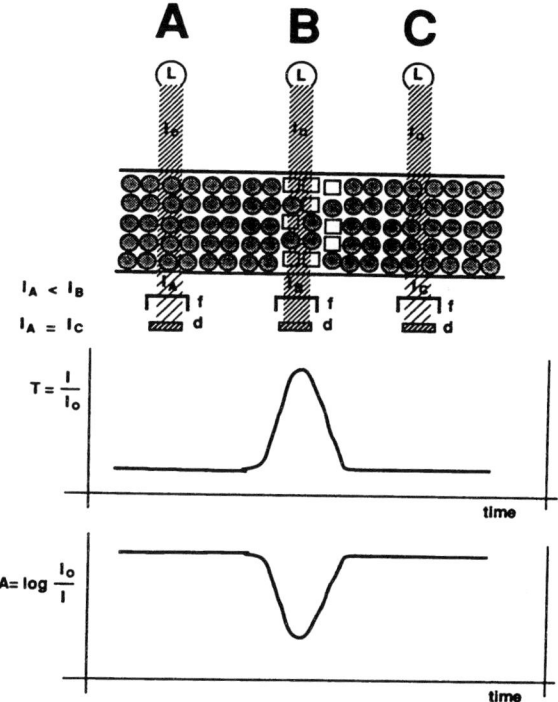

Figure 3.37 Indirect UV detection in a separation medium. L, light source; dark circles, UV background-providing compound; squares, analyte; f, optical filter; d, photodiode; I_o, intensity of the incident light; I, light intensity after the passage through the streaming medium; T, transmission; A, absorbance. See text for additional explanations.

In the absence of an analyte, as illustrated in Figure 3.37A, the absorbance of the separation medium is large, as only a small portion of the light from the light source penetrates through the fluid and reaches the photodiode detection element. If a sample zone passes through the detector cell, the originally high level of the absorbance signal is decreased due to the dilution of the high UV absorbing compound by the more translucent analyte molecules (Figure 3.37B). A larger portion of the light than in the previous case passes through the flowing medium, now containing the sample molecules, and reaches the sensor. The transmission of light through the fluid is at a high level. After the sample zone has left the detector cell, the light path becomes less translucent again, and the absorbance background has returned to its original high level from the time interval before the passage of the sample zone through the detector cell (Figure 3.37C). As before the passage of the sample zone, only a small portion of the incident light travels the entire path length between the light source and the sensing element.

Indirect UV detection was introduced for both HPLC[57] and ion chromatography[58] a relatively long time ago, and it was also utilized in CE[59,60] at an early stage. The theory of indirect UV detection following liquid chromatographic separations of nonionic compounds is relatively complex and has been a subject of several studies, even relatively recently.[61] The mechanism of indirect UV detection after an ion exchange separation is relatively more straightforward. In the following paragraphs, we shall discuss indirect UV detection in both ion exchange chromatography and in capillary electrophoresis for a better understanding of the similarities and differences between the two methodologies.

In an ion exchange column at equilibrium with an ionic UV absorbing component in the mobile phase, the ionic analytes without a UV chromophore (A,B,C in Figure 3.38) displace the UV absorbing eluent component (X in Figure 3.38) at the first available segment of the ion exchange column. The sample components in the solvent zone are thus replaced by equimolar quantities of the UV absorbing eluent component (equivalent of analyte per equivalent of the ionic component carrying the same sign of charge from the

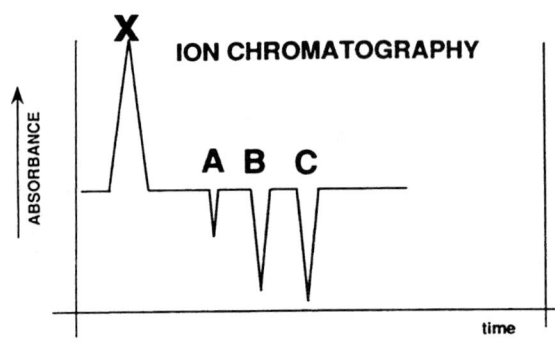

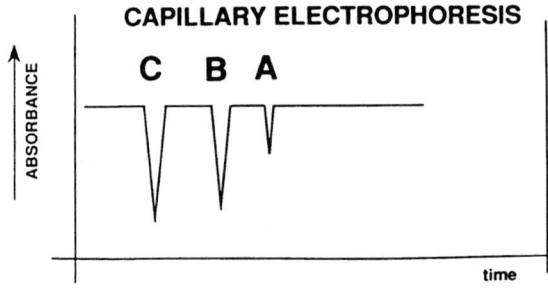

Figure 3.38 Indirect UV detection in ion chromatography and in capillary electrophoresis. The different sequence of peaks indicates the different selectivities of the two methods. See text for the explanation of symbols.

mobile phase). The sample solvent zone, with its now increased content of the UV absorbent, elutes unretained as a positive peak at the retention time corresponding to the dead volume of the ion exchange column (peak X in Figure 3.38A).

After their exchange onto the column, the sample components partition between the ion exchange groups and the surrounding eluent. The sample components are displaced from the ion exchange groups by the mass action of the eluent coions (ions having the same sign of charge). Whenever this happens, one equivalent of the UV absorbing eluent coion replaces one equivalent of one of the analyte in the stationary phase, and the separated analytes travel down the column, as a zone of a lower UV absorbance. On elution, the sample zone is detected as a negative peak. In the majority of cases, the area of the peak X is an exact sum of the peak areas of the sample components, A, B, and C, confirming the mechanism just described.

The main advantage of indirect detection is the ability to detect a large number of analytes that are undetectable by direct methods. Another positive aspect of the indirect UV detection in ion chromatography is the possibility of "universal calibration."[62] The concentration of an unknown analyte is indicated by the concentration decrease of always the same compound—the UV absorbing coion from the eluent. The concentration of all analytes can be calculated from a single calibration curve, since one ionic equivalent from a sample always displaces exactly one equivalent of eluent ion. The concentration can thus be determined even for the sample components of an unknown identity. Alternatively, it should also be possible to arrive at the analyte concentrations without any calibration at all.[62] In such cases, the analysis has to be performed sequentially with different UV absorbing coions in two different eluents.

In practical applications, however, indirect UV detection in ion chromatography is characterized by a frequent occurrence of system peaks. The system peaks may have several different causes (e.g., pH, ionic strength, solvent effects), always resulting in disturbances of the equilibrium between the stationary phase and the UV absorbing components from the eluent. Since the analyte signals may be frequently and unpredictably obscured by such system peaks, the practical utility of indirect UV detection in ion exchange chromatography is relatively limited.

A properly designed indirect UV detection in capillary electrophoresis, on the other hand, has all of the advantages of the technique's application in ion chromatography and none of its disadvantages. As in ion exchange chromatography, the nonabsorbing analyte replaces the UV background-providing coionic (same sign of charge) electrolyte component on an equivalent per equivalent basis. Unlike in ion chromatography, capillary electrophoresis does not employ any stationary phase. In consequence, there is no equilibrium between the electrolyte components and a stationary phase, a disturbance of which could lead to an occurrence of one or more system peaks. There is also no large peak X in the initial portion of the capillary electrophoretic separation (Figure 3.38B), as in the case of indirect detection following the ion exchange separation. If the question is asked: Where does the UV absorbing component being replaced by an analyte go? The answer is: As in ion chromatography, it goes into the sample solvent zone. Unlike chromatography, however, in capillary electrophoresis the operator has the option of choosing coelectroosmotic conditions to let the sample solvent zone migrate far away from the zones of analytes of interest. In coelectroosmotic capillary electrophoresis the sample solvent zone migrates at the rate of the electroosmotic flow, whereas the migration velocity of the analytes is given by the vector sum of the respective electrophoretic mobility and the electroosmotic mobility (Section 2.2).

The regular equivalent per equivalent exchange in capillary electrophoresis is enforced not by the ion exchange groups as in ion chromatography, but by a constant and oppositely directed stream of counterions and by the condition of electroneutrality. After an application of separation voltage across the capillary filled with carrier electrolyte and containing a volume of sample, the cations and anions are forced to migrate in the opposite directions. Even during the resulting migration, the counterion (ions of opposite charge relative to the analyte ions) concentration is constant and predetermined by the original electrolyte concentration. The sample counterions are removed from the capillary already during the electrostacking (see Section 3.1), and do not play a role during separation. Once the opposite migration begins after the application of the separation voltage, each electrolyte segment contains a constant number of counterions. Because of that, the sum of equivalents of coions and analyte ion also has to remain constant. The gap created by a forward-moving analyte ion has to be filled by the electrolyte coion from the segment into which the analyte ion is moving. The schematic representation of that process in Figure 3.39 does not begin with the sample still dissolved in the original sample solvent. The first step represents the situation after the last stage of electrostacking; all sample constituents had been removed from the zone of higher electric field into the zone of a lower electric field beyond the concentration boundary inside the carrier electrolyte. Only the movement of the highly mobile analyte (squares) is discernible at this stage. During the separation, the sample solvent zone (crosses in Figure 3.39) does not consist only of the nonionic solvent molecules; it also contains a small portion of the electrolyte coions moved there to compensate for analyte ions transferred into the electrolyte during the electrostacking. In summary, the constant flux of counterions in capillary electrophoresis fulfills a similar function as the constant ion exchange capacity in ion exchange chromatography; it maintains a constant number of opposite charges per volume segment of carrier electrolyte column inside the fused silica capillary. The distance between the analyte and solvent zones gradually increases with each of the steps in Figure 3.39. We should recall, however, that only with zero electroosmotic flow can the solvent zone be expected to remain in its original position. Under the coelec-

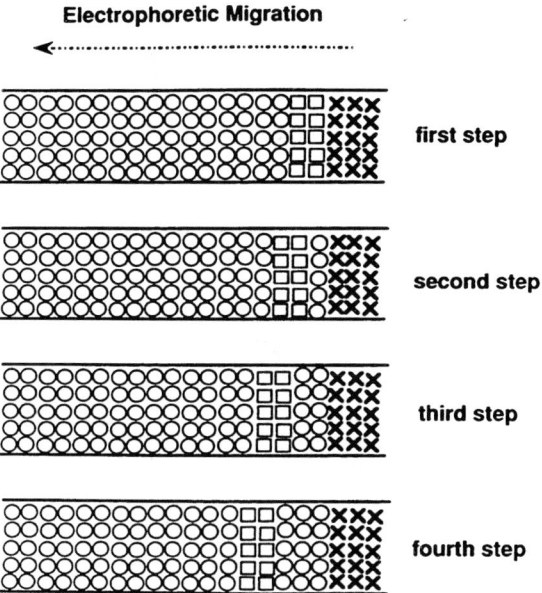

Figure 3.39 Movement of the zone of nonchromophoric analyte ions through the carrier electrolyte containing UV absorbing coions. The crosses represent the nonionic sample solvent. The squares and circles are the respective coionic analyte and carrier electrolyte components. See the text for additional explanations.

troosmotic conditions (Section 2.4), the inherent electrophoretic movement of the analyte zone is enhanced by the rate of electroosmosis. In such a case, the solvent zone will also move, but the relative distance between the sample and solvent zones will remain the same as in the absence of electroosmosis.

The basic principles used to predict and to discuss the performance of indirect detection techniques were introduced by Yeung.[62] The minimum detectable concentration c_{lim} can be expressed as a function of the concentration of the UV background-providing species c_m, transfer ratio TR, and dynamic reserve DR:

$$c_{lim} = c_m/(TR \times DR) \qquad (3.15)$$

The transfer ratio is the number of equivalents of the UV background-providing ion that is going to be displaced by each equivalent of analyte ions. For example, in a chromate carrier electrolyte, TR = 1 for sulfate and TR = 0.5 for chloride. The transfer ratio predicts that, with everything else being same, the molar sensitivity for divalent ions is going to be twice that for monovalent ions. Equation (3.15) also allows a conclusion that, with all other parameters being constant, the sensitivity of detection in monovalent coion electrolytes can be expected to be two times higher than in divalent coion electrolytes. A recent report[63] concludes from the Kohlrausch theory[24] that a regular displacement, indicated by the transfer ratio and discussed in connection with Figure 3.39, occurs only for "analytes having the same mobility as the background ion." The same report also states that ". . . a drawback for quantitative analysis using the (universal) indirect detection follows from the Kohlrausch theory. . . ." However, the conclusions by Nielen are valid only for "universal calibration"[62] (i.e., calibration using only one standard to quantitate several different compounds) performed in conjunction with hydrostatic sampling. Consider, for example, the calibration data for a variety of cations obtained with electromigrative sampling and shown in Figure 3.40.

In Figure 3.40, we observe slopes of calibration plots that are identical within experimental error for the monovalent cations on one hand and for

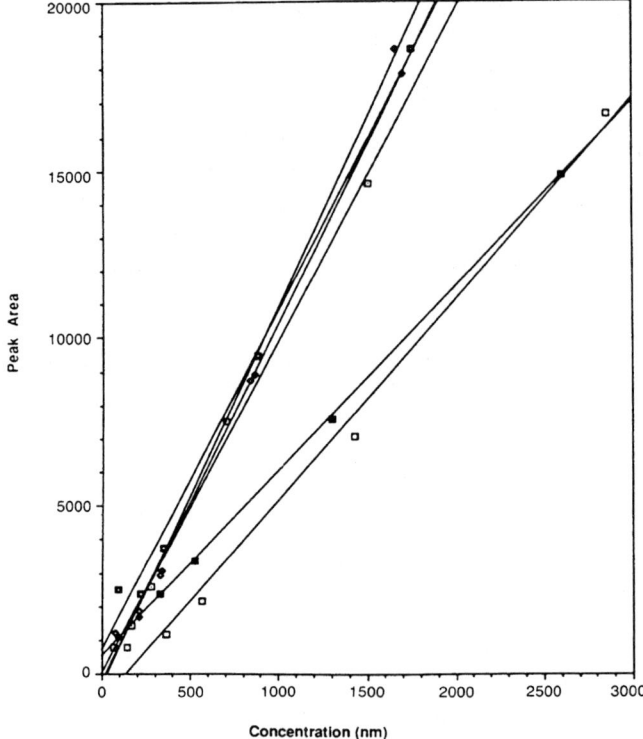

Figure 3.40 Calibration plots for alkali metal and alkaline earth cations. The complete experimental conditions used to generate the calibration data are described in the caption to Figure 2.37. The sample introduction for the experiments summarized in this figure was done by electromigration (10 sec, 5 kV) and not by the hydrostatic method described in Figure 2.37. Courtesy Andrea Weslon, University of Rhode Island. ⊡ = Barium, ♦ = Strontium, ■ = Calcium, ◇ = Magnesium, ■ = Sodium, □ = Lithium

the divalent cations on the other hand. Furthermore, the values of the slopes for the divalent analytes are different from those for the monovalent analytes by exactly a factor of two. Thus we see that a broad selection of divalent cations displaces exactly an identical amount of the "background ion." By no stretch of imagination can magnesium, calcium, strontium, and barium be described as having "the same mobility" as each other or as the background-providing cation. The different degrees and types of tailing for the cations in question, in the electropherogram shown in Figure 2.37, is another indication of varying differences of analyte mobilities relative to the mobility of the electrolyte cocation. Another strong indication of the regular one-to-one displacement in ionic equivalent terms is the fact that the monovalent cations are observed to displace only one half of the amount of the background ion in comparison to the divalent cations. According to the theory,[24,63] the ions of lower mobility should be replacing a larger number of carrier electrolyte coions. This should be discernible in the calibration plots as a larger value of slope for such slower ions. Nothing of that kind is seen in Figure 3.40. However, the regular appearance of calibration plots in Figure 3.40 is to a large extent due to the electromigrative method of sample introduction. As discussed in section 3.1.3, electromigrative sample introduction yields higher recoveries for high-mobility ions and vice versa. This "preferential" mode of sample introduction appears to have compensated for differences in transfer ratio between the higher- and lower-mobility ions calibrated in Figure 3.40. In simple terms: The term TR in Equation (3.15) is the same for all analyte ions, if and only if sample introduction is performed by electromigration. From a practical point of view, the term TR is nearly identical even in most electropherograms obtained after a hydrodynamic sample introduction. If the Kohlrausch ratio[24] is calculated for the anions in Figure 3.41, the change of that ratio and correspondingly the change of TR to be expected is smaller than 25%. The conclusions discussed by Nielen[63] are thus affecting, and only in a limited way, the possibility of "universal calibration" if performed after a hydrostatic injection. The validity of calibration in indirect photometric detection is not affected even if hydrostatic injection is used. The only restriction in such a case is the necessity to calibrate with one standard for each analyte to achieve better accuracy and to eliminate the trend to higher values of TR for low-mobility ions. As we shall see in Section 5.2, a universal calibration appears to be possible even for traces of anions preconcentrated by a newly developed electromigrative method.

The term DR in Equation (3.15) describes the so-called dynamic reserve, the ratio of the intensity of the background signal to the noise of that signal. For the indirect photometric detection, the dynamic reserve can be defined as follows[64]:

$$DR = (\varepsilon L c_m)/AN \tag{3.16}$$

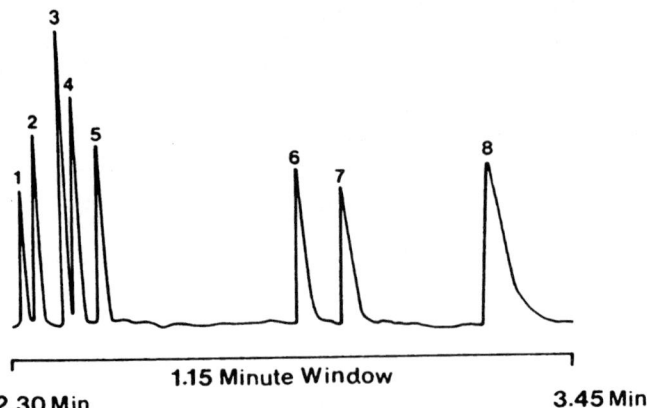

Figure 3.41 Separation of inorganic anions using indirect photometric detection at 254 nm. The carrier electrolyte contained 5 mM chromate and 0.5 mM electroosmotic flow modifier added to reverse the direction of the electroosmotic flow (OFM BT from Waters). The pH was adjusted to 8.1. The fused silica capillary dimensions were 75 μm I.D. and 52 × 60 cm. The separation voltage was −20 kV. The sample was introduced by hydrostatic injection lasting 30 sec at 10 cm height above the normal level. The peak identities, concentration injected (ppm), and detection limits (three times the noise in μM) were as follows: 1, bromide, 2 ppm, 4.8 μM; 2, chloride, 2 ppm, 4.2 μM; 3, sulfate, 2 ppm, 1.8 μM; 4, nitrite, 2 ppm, 7.2 μM; 5, nitrate, 2 ppm, 5.6 μM; 6, fluoride, 1 ppm, 5.2 μM; 7, phosphate, 4 ppm, 4 μM; 8, carbonate, 2 ppm, 2.2 μM. Reproduced with permission.[25]

where ε is the molar absorptivity [abs. units $\times$ M^{-1} $\times$ cm^{-1}], L is the light path [cm], c_m is the concentration of UV background-generating compound in the carrier electrolyte [M], and AN is the absorbance noise [abs. units]. Equation (3.16) describes the dynamic reserve as a combination of quality of design of the detector and a proper choice of the UV background-providing compound. The operator attempts to optimize the dynamic reserve by choosing a suitable compound (the electrophoretic mobility has to be matched with that of the analytes) with a maximum value of molar absorptivity and by preparing maximum possible concentration (the electrolyte has to be of low conductivity to minimize Joule heating) of that substance in the electrolyte. The operator's options will be limited by a ceiling imposed by the characteristics of the photometric detector. The background noise will only remain independent of the total value of the background absorbance within a certain range of c_m. Above some maximum level of the product $\varepsilon L c_m$, the noise will start to increase appreciably, counteracting in the denominator any increases of the value of the numerator. Clearly, the maximum value of the product $\varepsilon L c_m$ at which the noise is still constant and unchanged compared to low values of background absorbance can be used to evaluate the suitability of various detectors for indirect photometric detection.

By substituting the term for dynamic reserve from Equation (3.16) for DR in Equation (3.15) and assuming that TR = 1, we obtain:

$$c_{\lim} = AN/\varepsilon L \qquad (3.17)$$

In the relationship of Equation (3.17), the minimum detectable concentration can be calculated independently of the concentration of the background-providing electrolyte component. Another important point from Equation (3.17) is that in contrast to other indirect techniques, it is not necessary to use very low concentrations of background ions. This results in a wider dynamic range for indirect absorbance. Note that the discussed relationships also use only the standard linear version of Beer's Law. The results are thus not transferable between different geometries of optical cells. The best of the currently available UV detectors achieve noise levels of about 10^{-5} abs. units. With a background compound having $\varepsilon = 10^3$ (chromate anion according to Reference 53) and with the capillary I.D. = 75 µm, we can calculate a 1.33×10^{-6} M concentration limit of detection:

$$c_{\lim} = 10^{-5}(10^3 \times 75 \times 10^{-4}) = 1.33 \times 10^{-6} \qquad (3.18)$$

This approximate calculation has several underlying assumptions that cannot be expected to hold in all instances, but as we can see in Figure 3.41, it predicts rather correctly the order of magnitude of achievable sensitivity by indirect detection under optimized conditions. The agreement for sulfate is in fact very close. Recall that Equation (3.17) was derived using the transfer ratio TR = 1, as in the case of the divalent sulfate replacing the divalent chromate anion. Another close agreement between the theoretical and calculated values of $c_{\lim}$ is found for carbonate. The phosphate anion cannot be considered as a divalent species since the pH of the chromate carrier electrolyte is within one pH unit of the second dissociation constant for that anion. For the monovalent anions in Figure 3.41, the transfer ratio equals 0.5. According to Equations (3.15) and (3.17), the theoretical minimum detectable concentration for the monovalent anions in chromate is thus ~ 2.66 µM. The observed values are in the range of 4–7 µM, showing that the observed close agreement of the divalent ions should not be overrated. The values for noise and molar absorptivity in Equation (3.18) were used only as estimates giving merely the order of magnitude. No attempt was made to determine more accurate values within that estimated order of magnitude. There are also additional factors influencing the magnitudes of detection limits such as electrostacking, the increasing degree of peak asymmetry with longer migration times, and varying rates of migration of the analyte zones through the detector.

The term *dynamic reserve* or DR in Equations (3.15) and (3.16) should not be confused with another term that is frequently utilized to describe the range of concentrations for which calibration of the detection signal is feasible—the dynamic range. The dynamic reserve is useful for calculations of detectability and does not provide, by design, any information about the feasibility of calibration. DR is actually equal to the signal-to-noise ratio of the background UV signal.[62] The conceptual difference between the two definitions notwithstanding, there is also a degree of similarity that should not be overlooked. As with DR, the size of dynamic range depends on the concentration of the UV background-providing electrolyte coion as shown by the following evaluation.

The portion of the dynamic range within which the detector signals are a linear function of sample concentration and that is the actual portion of a calibration plot that can be easily used for a routine analytical work may be determined by plotting logarithms of detector signal versus logarithms of concentration. The calibration plots are found linear,[65] if the slope of a double logarithmic plot remains within the range of 0.98 to 1.02. (Note: Certain nonlinear, regular patterns of data points may yield the value of slope within the "linear range." Close visual inspection of calibration plots is an essential requirement.) An indirect UV detection method used in conjunction with a CE separation of inorganic anions in Figure 3.41 was evaluated with the help of the logarithmic linearity test.[25] The complete analytical conditions are given in the caption to Figure 3.41. The method is similar to that presented in Figure 2.10. It relies on coelectroosmotic conditions (Section 2.4) for an enhancement of separation speed. For the test of linearity, the standard mixtures of the seven anions from Figure 3.41 were prepared at eight different concentrations between 0.5 and 200 ppm and tested with three different concentrations of the chromate anion providing the UV background at 254 nm. Figure 3.42 shows a summary of the results of the linearity evaluation for the sulfate anion.

The experimental protocol in the discussed evaluation of linearity consisted of analyzing the standard mixtures of higher concentrations first, followed by the standard mixtures of the next lower concentration. For the two lower concentrations of the carrier electrolyte (5 and 7.5 mM chromate, full squares and circles in Figure 3.42, obscured by an almost complete overlap), the linearity of calibration was observed to deteriorate between 200 and 50 ppm. The slopes of the double logarithmic plots (log peak area vs. log ppm sulfate) were 0.90 and 0.81 for the 7.5 and 5 mM electrolytes, respectively. For that reason the evaluation was not extended to lower sulfate concentrations. The data points for 0.5 and 2 ppm sulfate are not included in Figure 3.42 for the two lower electrolyte strengths (5 and 7.5 mM). The slopes in the 10 mM chromate electrolyte, on the other hand, remained within the linear range of 0.98–1.02, not only between 5 and 200 ppm, but also in the entire concentration range between 0.5 and 200 ppm. The determined value of the slope of the double logarithmic plot for the 10 mM chromate and 0.5 to 200 ppm sulfate was 1.003. The data points for 10 mM chromate are recognizable in Figure 3.42 as empty squares positioned along the lowest of three calibration curves.

In another experiment of the same evaluation, the 5 mM chromate electrolyte "failed" the linearity test for five different, evenly spaced concentrations of sulfate between 2 and 30 ppm, giving a slope value of 0.97848. The

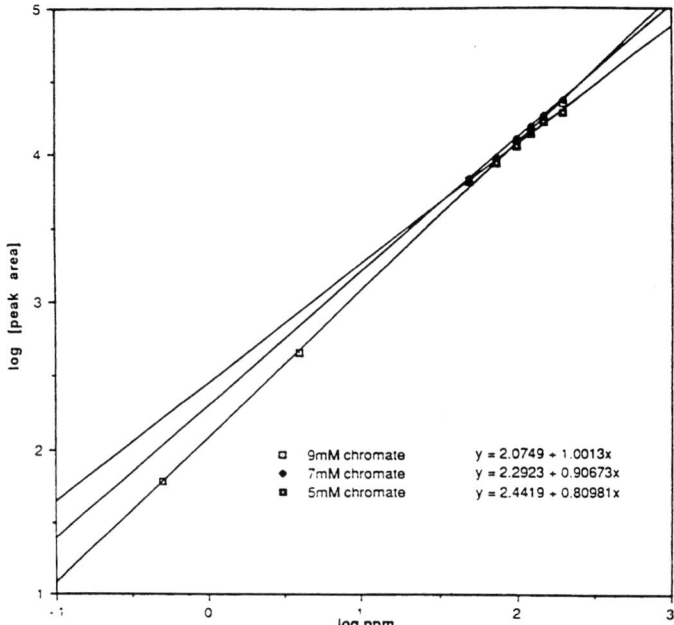

Figure 3.42 Linearity evaluation between 0.5 and 200 ppm sulfate for three different concentrations of the chromate electrolyte. The concentration of the electroosmotic flow modifier was held constant at 0.5 mM. See the text for further explanation. Reproduced with permission.[25]

7.5 mM electrolyte gave satisfactory results for the same set of standards, a slope of 0.99799. The best results were again obtained with the highest concentration of chromate, a slope of 0.99956 for six evenly spaced data points between 2 and 50 ppm.

Both parts of the discussed evaluation confirm the expected gradual improvement of linearity with an increasing concentration of the UV background-providing ion in the carrier electrolyte. It seems obvious and the experimental evidence discussed previously confirms that the detector signal in indirect photometric detection can be expected to change only as long as there are enough UV absorbing coions in the electrolyte that can be displaced by the analyte ions. As in many other detection techniques, the closeness of the limit of the dynamic range is observed not as a sudden failure to respond, but rather as a gradual deterioration of the linearity of calibration.

In yet another reference to Equation (3.17), we observe that the detectability is predicted to improve with increases of molar absorptivity. The higher the ε, the lower the value of c_{lim}. This is because the molar absorptivity improves sensitivity as defined, for example, by the slope of a calibration plot. Background ions of higher molar absorptivity generate more response for the same amount of analyte in indirect photometric detection. A displaced coion equivalent with a larger absorptivity coefficient will cause a larger signal deflection. This can be verified experimentally by determining the peak area for the same amount of an analyte in several different carrier electrolytes. For an easier correlation the peak areas and molar absorptivities can be normalized as in Figure 3.43. The relative molar absorptivity coefficients were obtained by dividing the absorbance of a given carboxylate by the absorbance of benzoate at the same wavelength and concentration (254 nm and 1 mM, respectively). The relative molar absorptivity of chromate on this scale equals 3.07. (For broader comparisons, recall that molar absorptivity of chromate is approximately 10^3 abs. units M^{-1} cm^{-1}, Reference 53.) In a similar way, all measured peak areas for 10 ppm propionate were divided by the peak area of propionate in the benzoate electrolyte. As shown, a very good correlation exists between the intensity of UV absorp-

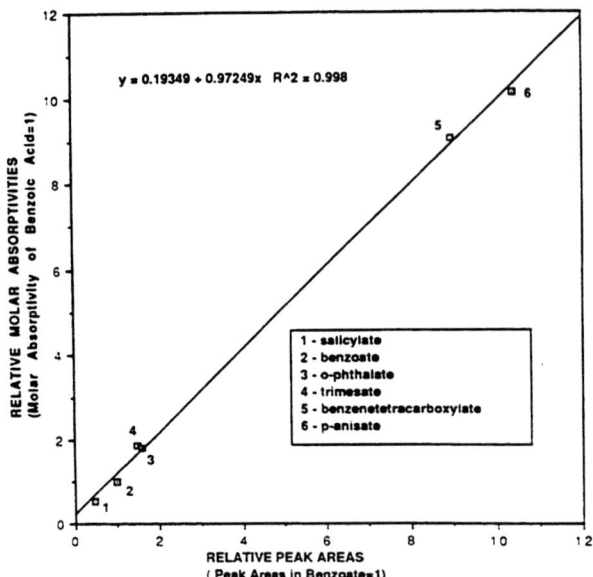

Figure 3.43 Correlation of relative molar absorptivities of six different background anions and the corresponding peak areas for 10 ppm propionic acid. The propionic acid standard was introduced by a 15 sec hydrodynamic sampling from a 10 cm height. Carrier electrolyte contained 5 mM of one of the six carboxylic acids in a mixture with 0.5 mM electroosmotic flow modifier. The resulting electrolyte solution was adjusted to pH 6.0. The separations were carried out at -20 kV in a 75 μm, 52 × 60 cm fused silica capillary. 1, Salicylate; 2, benzoate; 3, o-phthalate; 4, trimesate; 5, benzenetetracarboxylate; 6, p-anisate. Reproduced with permission.[25]

tion by the background ion in the carrier electrolyte and the sensitivity of indirect photometric detection evaluated as the peak area obtained for 10 ppm propionate standard.

The correlation in Figure 3.43 is of great practical use in the optimization of detection sensitivity. We can see, for example, that the sensitivity can be increased twentyfold on going from salicylate to *p*-anisate. Sensitivity improvements of an order of magnitude can be gained by changing from benzoate, a popular electrolyte among the early CE pioneers, to other, more suitable UV-absorbing components such as *p*-anisate.

In a final reference to Equations (3.15)–(3.17), we should note that those equations are valid only for cases where the analytes do not exhibit any measurable absorption of light at the wavelength selected for detection. Any UV absorption of analytes at the detection wavelength chosen for indirect UV detection decreases the dynamic reserve, yielding a worse than expected detectability. The linear range of calibration is also affected by a direct analyte signal during an indirect detection.

Many of the low-molecular-weight ions that are frequently considered non-UV absorbing actually exhibit a considerable absorbance at lower UV wavelengths. See, for example, tungstate, arsenate, molybdate, thiosulfate, iodide, bromide, and nitrate in Figure 3.44. Other ions that may be known to have considerable UV absorption may surprise one by the broad range of wavelengths at which the UV absorbance occurs—see vanadate, benzoate, phthalate, and trimesate in Figure 3.44.

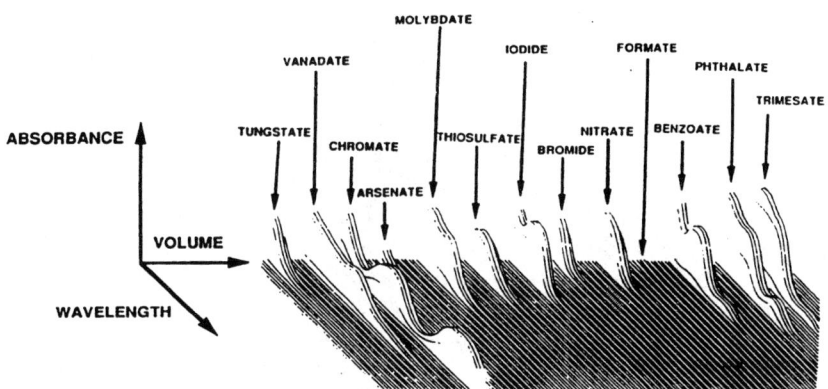

Figure 3.44 Typical UV spectra of some inorganic and organic anions. All anions were injected in a constant volume (100 μl) of 1 mM aqueous solutions. The spectra were recorded between 200 and 400 nm. The range of sensitivity was set at 0–1.1 absorbance units. The volume between any two sample zones was determined by selected time intervals between two injections into a flow-injection analysis apparatus and does not represent any separation effect. Reproduced with permission.[25]

The usefulness of chromate for indirect photometric detection in capillary electrophoresis is apparent from Figure 3.44. The anion has a broad absorption spectrum throughout the entire UV range. For any of the anions whose spectra are depicted in Figure 3.44, chromate can offer a suitable wavelength for indirect photometric detection, including the broadly absorbing anions such as vanadate and the aromatic carboxylates. The mobility of chromate anion also matches the mobility of a large number of low-molecular-weight anions. The possibility of a nearly universal detection is thus combined with low degree of asymmetry for a large number of low-molecular-weight anions (see also Figure 2.10).

In summary: An optimally designed indirect UV detection method makes possible a highly sensitive analysis of a large number of low-molecular-weight ions. The sensitivity of detection is optimized by choosing background ions with large values of molar absorptivity. The dynamic range and linearity of calibration can be expected to increase with increasing concentration of the UV-absorbing background ion. The ceiling for the optimization of linearity through increases of background ion concentrations is imposed by increasingly serious Joule heating problems at higher ionic strengths and by detector linearity. Even so, a linearity over more than three orders of magnitude is achievable in many cases. The underlying mechanism for indirect photometric detection is straightforward, offering a high degree of predictability. In contrast to ion chromatography, the practical value of indirect UV detection in capillary electrophoresis is not diminished by unpredictable system peaks.

3.3.3 Alternative Detection Methods in Capillary Electrophoresis

The predominance of UV detectors in the published reports and in the commercially available instruments has not deterred numerous researchers from developing a remarkable level of activity in the field of alternative detection methods for capillary electrophoresis. Interest in fluorescence detection alone generated a considerable number of publications. Among the alternative techniques, fluorescence is the only other technique besides photometric detection that has been included in several commercial capillary electrophoresis systems. A broader availability can probably also be expected soon for laser-induced fluorescence, mass spectroscopy, conductivity, and perhaps also amperometric detection. The following text offers a brief overview of all alternative detection techniques reported to date. The objective is to give only a brief introduction to the various detection modes, rather than an exhaustive treatment with a complete list of literature references. For a more in-depth study, the reader is referred to two of the most recent review articles published in the literature.[66,67]

3.3.3.1 Direct Fluorescence Detection

For the compounds containing fluorophores, fluorescence achieves a high level of sensitivity. For example, Wu and Dovichi[68] reported detection limits in the picomolar range (10^{-12}) for fluorescein isothiocyanate (FITC)-derivatized amino acids. In fact, most of the applications of fluorescence detection to date have been reported in connection with biochemically relevant separations of amino acids, peptides, proteins, and nucleic acids. The field of biochemical applications of capillary electrophoresis was reviewed authoritatively by Novotny and co-workers.[69]

As with photometric detection, fluorescence can be relatively easily adapted to the requirements of on-column detection in capillary electrophoresis. On-column detection avoids the design problems of an off-column CE detector cell fulfilling all requirements, that is, would not contribute to peak spreading and would thus not affect the efficiency of CE separations. The first successful realization of on-column fluorescence detection in capillary electrophoresis was reported by Jorgenson and co-workers.[70-72]

A schematic diagram of their detector design, utilizing a mercury–xenon lamp, a monochromator on the excitation side and optical filters on the emission side, is shown in Figure 3.45.

The 200 W, Hg-Xe arc lamp is configured with a collimating lens, L1, and a water-filled filter absorbing the IR radiation to prevent overheating of the monochromator. The light enters the monochromator slit after focusing in lens L2. On the other side of the monochromator, the light is collimated by lens L3 and then focused onto the capillary by lens L4. On the other side of the capillary is a neutral-density filter (ND) and a reference photodiode. The neutral-density filter prevents an oversaturation of the photodiode by the light transmitted through the capillary. The fluorescence emission coming from the capillary is collected by lens L5 at a right angle to the original direction of the light coming from the lamp. After passing through an optical filter, the light reaches the photomultiplier tube (PMT). The performance of

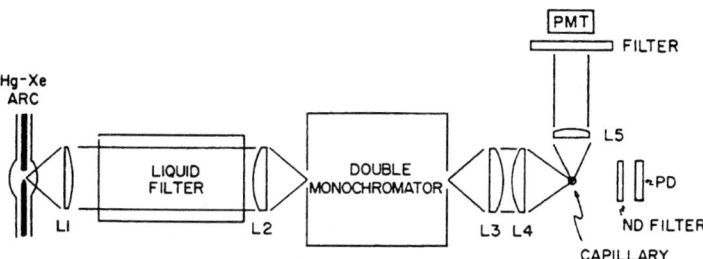

Figure 3.45 Optical layout of an on-column fluorescence detector. See text for a detailed explanation. Reproduced with permission.[72]

the detector is illustrated by the electropherogram of derivatized amino acids shown in Figure 3.46.

The detection limits of dansylated amino acids were found to be in the 10^{-7} M range. Linarity tests of the calibration plots according to Scott[65] confirmed a good linearity over three orders of magnitude. The high-frequency noise of the baseline in Figure 3.46 was attributed by the authors to the poor spatial stability of arc lamps. In conventional instruments, the constantly moving image of the arc can be easily kept on the relatively large areas of cuvettes exposed to the excitation light. However, the effect of the arc wander represents a serious problem with the small areas offered by the 50–75 μm I.D. fused silica capillaries. Saito and co-workers[73] identified another significant source of noise in on-column fluorescence detection, refraction and reflection of excitation light on cylindrical capillaries, as shown in Figure 3.47.

As illustrated in Figure 3.47, the proportion of the light reflected at the outer boundary can be minimized by focusing and positioning of the excitation beam. However, even with optimal focusing and positioning, a portion of the light hitting the internal surface of the fused silica capillary may undergo multiple reflection inside the capillary wall. Whenever the light strikes the outer boundary at an angle smaller than the critical angle, part of the light passes through the outer wall of the capillary. The excitation light is thus scattered in every direction, and a considerable portion of that scat-

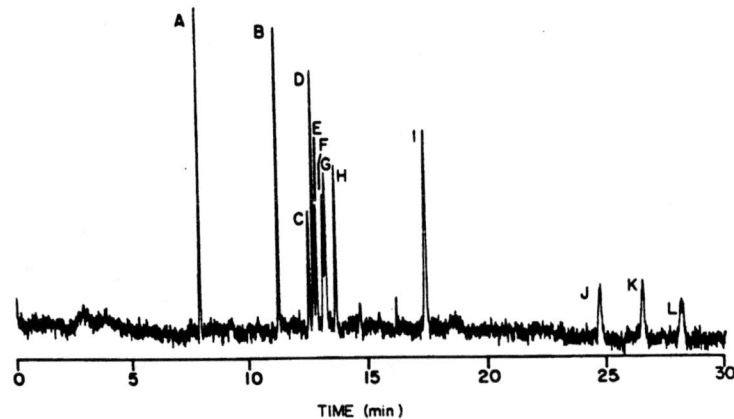

Figure 3.46 Counterelectroosmotic separation of twelve dansyl-amino acids in a 100 cm long, 75 μm I.D. fused silica capillary at +30 kV and with 0.0125 M phosphate, pH 6.86 as a carrier electrolyte. The sample introduction was by electromigration for 1 sec at +30 kV. Excitation wavelength 365 nm, emission wavelength 490 nm. Peak identities: A, ε-labeled lysine; B, doubly labeled lysine; C, isoleucine; D, methionine; E, asparagine; F, serine; G, alanine; H, glycine; I, doubly labeled cystine; J, glutamic acid; K, aspartic acid; L, cysteic acid. The concentration of each analyte was 5×10^{-6} M. Reproduced with permission.[72]

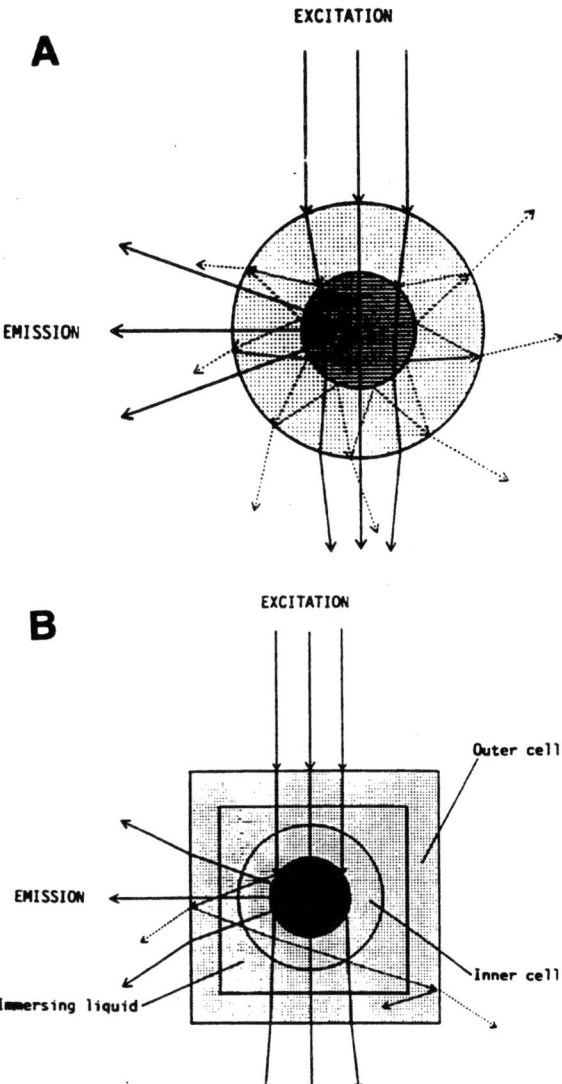

Figure 3.47 Reflection and refraction of light on two different on-column fluorescence detection cells. A, Internal reflection of a cylindrical cell. B, Immersed flow cell minimizing reflection and refraction of excitation light. Reproduced with permission.[73]

tered light may be reflected at a right angle and be added as additional background to the emission signal. To reduce scattering of light through refraction and reflection, the authors inserted a fused silica capillary, from which a section of polyimide coating had been removed, into a rectangular HPLC flow cell and filled the space between the capillary and internal wall of the flow cell with a liquid of suitable refractive index, Figure 3.47B. The intensity of the background signal diminishes, and the detectability improves the closer the immersing liquid matches the refractive index of fused silica. Using a xenon lamp as light source, the authors report detection limits in 10^{-9} M range. A more consequent realization of the principle of the immersed detection cell is the so-called sheath flow cuvette,[68,74,75] Figure 3.48.

The sheath stream, usually delivered by a conventional HPLC pump, surrounds the carrier electrolyte as it leaves the capillary forming a thin column in the center of the chamber. If the sheath stream's composition is similar to that of the carrier electrolyte flowing from the capillary, there is no refractive index boundary between the two streams and consequently also no light scattering at the interface. By restricting the illuminated field to the sample stream and by using a cuvette with flat windows, the background of the fluorescence measurement can be reduced to very low values. The CE electrical circuit is completed by a contact in the stainless steel upper body of the cuvette. In all reports so far the sheath cell has been used in conjunction with a laser. The lasers are frequently described as the most promising approach to increasing the sensitivity of detection in capillary electrophoresis. The two most widely used types of lasers are He–Cd, using the strong lines at 442 and 326 nm (Omnichrome, Chino, CA, or Liconix, Sunnyvale, CA), and Ar, operating at 488 nm (Control Laser Corp., Orlando, FL). There are, however, some drawbacks in using a laser as a source of excitation light in fluorescence detection. The most important drawback, at least for the time

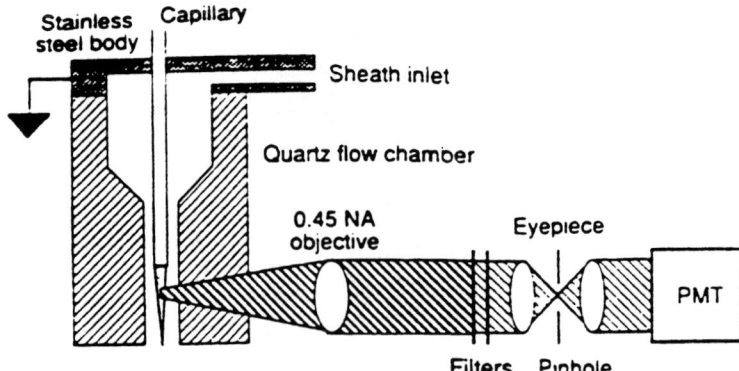

Figure 3.48 The sheath flow cuvette for capillary electrophoresis. Reproduced with permission.[74]

being, is the high cost of lasers. Another disadvantage is that a continuing variation of wavelength, such as, for example, with more conventional sources of light, is impossible with lasers.

On the other hand, for compounds with a fluorophore, laser-induced fluorescence is an "ultrasensitive detection method."[69] The reported picomolar, femtomolar, and attomolar limits of detection translate to amounts of samples of 10^{-18} moles (attomolar) or lower, 10^{-21} (zeptomolar) to 10^{-24} (yoctomolar). The detection of very low numbers of molecules, in the 10^{-24} mole range, is considered as a realistic possibility by the leading researchers in the field of capillary electrophoresis with laser-induced fluorescence detection.

Since the majority of the analytes do not contain any fluorophores, it is necessary to adapt some pre- and postcolumn derivatization techniques from HPLC for the purposes of capillary electrophoresis. It is also obvious that an adaptation of HPLC derivatization procedures will not lead in all cases to a sufficient sensitivity in capillary electrophoresis, and as a consequence new derivatization chemistries designed specifically for use with lasers have been reported. Still, the use of the terms post- or precolumn derivatization is more common among the workers in the field of capillary electrophoresis than the more accurate designation post- or precapillary derivatization. The main activity in developing derivatization procedures for fluorescence detection in CE has centered around amino acids, due to their biological importance, and also because any procedures developed for amino acids can be expected to work for related classes of compounds, for example, peptides and proteins. The need for chemistries optimized for laser detection was first pointed out by Nickerson and Jorgenson[76] in a study comparing detection limits with a He–Cd laser operated at 326 and 442 nm excitation wavelengths. Three different precolumn derivatization procedures were evaluated for phenylalanine. The authors obtained 20, 1.6, and 0.22 nM detection limits with ortho-phthaldialdehyde (OPA), naphthalene dicarboxaldehyde (NDA), and fluorescein isothiocyanate (FITC) reagents, respectively. They explain the improved detectability with NDA and FITC as compared with OPA by the greater energy output of the laser at 422 nm, the wavelength used for the excitation of fluorescence in NDA and FITC, but not in the detection of OPA-labeled amino acid. The excitation wavelength of 326 nm was used for the latter reagent. These results clearly show that the limited choice of laser excitation wavelengths brings an additional requirement to the development of laser-induced fluorescence detection—the necessity to find a reagent with a proper position of absorption maximum to match one of the wavelengths that can be delivered by a laser. Once that requirement is fulfilled, separation conditions have to be found for the new derivatives of amino acid to be separated by CE. FITC has the highest sensitivity of the three reagents, but Hsieh and Novotny[77] reported difficulties in their attempts to separate FITC derivatives of common amino acids. Separation problems with labeled compounds are quite common, and not just with fluorescence. Derivatization usually reduces the physical and chemical differences between analytes, and can often make separation difficult to achieve. For example, Wright and co-workers[78] reported virtually identical migration times for three dansylated (dansyl chloride: 1-dimethylaminonaphthalene-5-sulfonyl chloride) marine toxins. They could overcome the separation problem by choosing fluorescamine (4-phenylspiro[furan-2(3H)-1'phthalane]-3,3'-dione) as a labeling reagent. The fluorescamine-labeled toxins were easily separated but the observed sensitivity of detection was lower than in the case of dansyl derivatives. Another problem connected with the use of fluorescamine was the formation of multiply labeled derivatives, such as, for example, the dilabeled saxitoxin in Figure 3.49.

In Figure 3.49, the most interesting toxin is saxitoxin. It is a powerful neurotoxin that can render a variety of shellfish poisonous. The poisonous

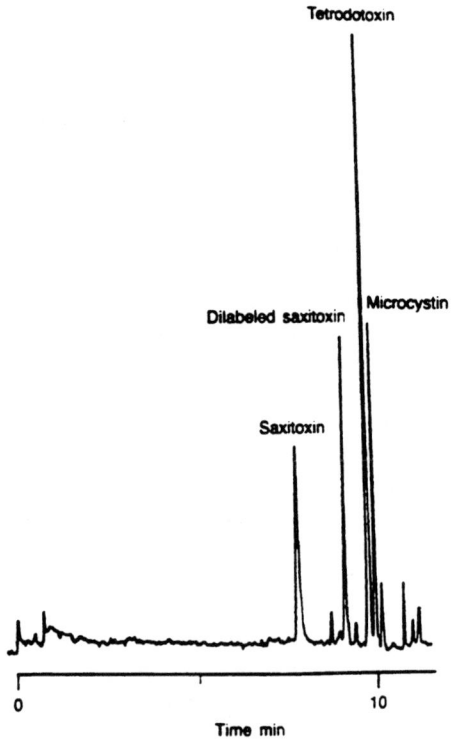

Figure 3.49 Capillary electropherogram of a mixture of fluroescamine-labeled toxins. Carrier electrolyte: 2 mM phosphate buffer, pH 8.7. Separation voltage +40 kV. The capillary dimensions were 100 μm and 100 cm. The sample was introduced by electromigration at +9 kV, 30 sec. The peaks correspond to ~100 pg of each toxin. The He–Cd laser was operated at 325 nm for excitation. The emission was recorded through a 400 nm cutoff filter. Reproduced with permission.[78]

shellfish have been connected to so-called "red tides," observed with increasing frequency along the east coast of United States. A sensitive and selective method of analysis is thus of great importance for clinical laboratories.

The problems with the known fluorescence labeling reagents prompted Novotny and co-workers to try to synthesize a new reagent that would be free of the problems discussed. They prepared a new compound, 3-(4-carboxybenzoyl)-2-quinoline carboxaldehyde (CBQA), which forms fluorescent isoindole derivatives with amino acids. They were able to form uniform and well-defined derivatives with all important amino acids with the exception of lysine, which produced multiple peaks due to the presence of two primary amino groups. This work is also one of the few examples in that

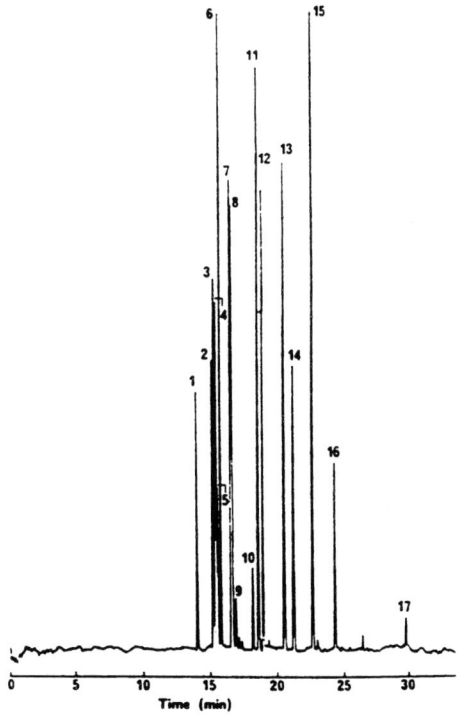

Figure 3.50 Eletropherogram of CBQA tagged amino acids obtained by hydrolysis of 1.9 pg lysozyme. Concentration of hydrolysate: 9.37×10^{-4} μg/μL. Capillary dimensions: 50 μm, 97 × 67 cm. Sample was introduced by electromigration with 9 sec sampling time at +5 kV. 1, Arg; 2, Trp; 3, Tyr; 4, His; 5, Met; 6, Ile; 7, Gln; 8, Asn; 9, Thr; 10, Phe; 11, Leu; 12, Val; 13, Ser; 14, Ala; 15, Gly; 16, Glu; 17, Asp. Carrier electrolyte: 0.05 M 2-[(N)-[tris (hydroxymethyl)methyl]amino]-ethanesulfonic acid, 50 mM SDS, pH = 7.02. Separation voltage: +25 kV. Reproduced with permission.[79]

156   CAPILLARY ELECTROPHORESIS OF SMALL MOLECULES AND IONS

particular research area with results documented not only on artificial mixtures of standards but on real samples as well.

Even though most of the work in fluorescence tagging has been carried out in the precolumn mode, experiments with postcolumn derivatization have also been performed. For example, Nickerson and Jorgenson[80] evaluated OPA derivatization of amino acids, peptides, and proteins in the postcolumn mode. The postcolumn mode avoids multiple peaks for identical an-

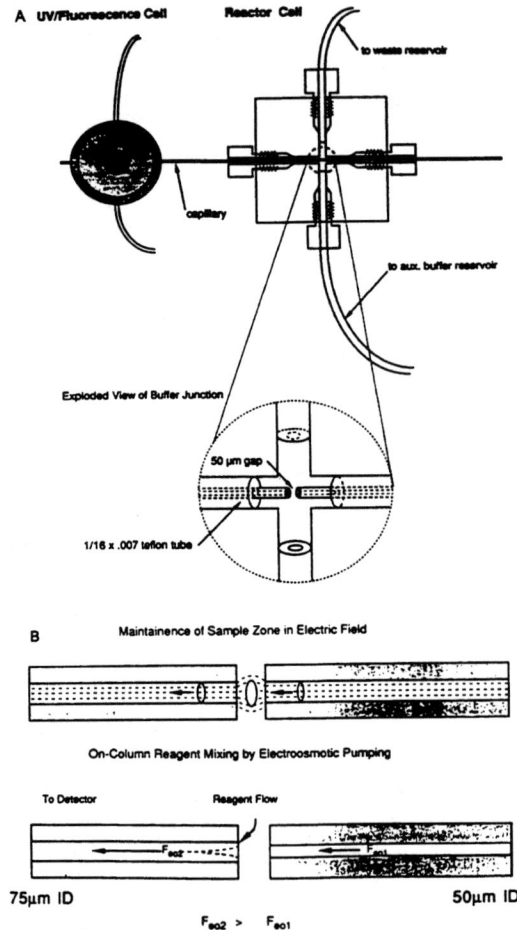

Figure 3.51   (A) Expanded view of a postcolumn reactor. (B) Principle of function of gap junction reactor. Top: analyte zone dispersion is minimized because of the containment by the electric field bridging the small 10–50 μm gap. Bottom: a secondary flow containing the fluorescent reagent is drawn into the capillary on the right-hand side by virtue of the difference in the electroosmotic flow in different diameters of capillary tubing (F_{eo1} and F_{eo2}). Reproduced with permission.[81]

alytes and also some separation problems encountered with the precolumn mode. The analytes are first separated in their original state, and the tagging step occurs after the separation. Even if there are multiple primary amino groups in a single molecule, resulting in several different fluorescing products, there will still be a single peak attributable to the original compound. It is clear, however, that not all precolumn labeling reagents can be employed for a postcolumn reaction. Some of the derivatization reactions may be too slow for the requirements of the postcolumn mode. Also, for postcolumn derivatization, the reagent must be nonfluorescent, forming a fluorescent product only on reacting with an analyte. The relative merits of pre- and postcolumn derivatization were investigated in a recent report.[81] The authors also employed a new design for a postcolumn reactor.

In all previously published reports,[80,82–84] the postcolumn reagent was added either by hydrostatic pressure or by a pump. In contrast to that conventional approach, the postcolumn reactor depicted in Figure 3.51 uses differential electroosmotic flow to introduce the reagent. The orifices of two capillaries with different internal diameters are positioned within a distance of ~ 50 μm from each other. This gap junction is immersed in the solution of the derivatization reagent. The electroosmotic flow is greater in the 75 μm capillary than at the 50 μm capillary. The flow imbalance between the two capillaries is made up by the reagent. In this way a sufficient volume of the labeling compound is drawn in to achieve derivatization. The authors conclude that their approach, characterized mainly by diffusional processes, yields less peak spreading than the reported conventional postcolumn reactors, where the mixing occurs by adding a laminar component to the electroosmotic flow. All comparisons between the two modes of derivatization in the report discussed[81] indicate a better sensitivity of detection in the precolumn mode. Also, four different reagents were evaluated in the precolumn mode versus only two with a postcolumn reaction. This is due to the greater ease of use and better sensitivity in the precolumn mode relative to the postcolumn derivatization. So far, the postcolumn reaction was shown useful only for those cases where the separation problems and multiple peak formation lower the utility of the precolumn method.

3.3.3.2 Indirect Fluorescence Detection

Considering all the difficulties encountered in the detection of fluorescence-labeled compounds and given the fact that the technique is still essentially in a research stage, it is logical to ask: Are there any other possible modes of operation not requiring derivatization of analytes and still employing the great potential of fluorescence detection?

A promising new method of fluorescence detection was recently described by Garner and Yeung.[85] The method can be used for UV absorbing analytes for which it is possible to find fluorescing compounds with excitation spectra closely matching UV absorption spectra of the analytes. The magnitude of fluorescence signal enhancement for such a pair of compounds is illustrated in two electropherograms in Figure 3.52.

Any electronic transition that allows absorption of UV light also makes possible the release of absorbed energy in the form of fluorescence. Whether or not we describe a compound as fluorescing depends solely on the relative portion of nonradiative and radiative release of energy. The first thing to note in Figure 3.52 is that, while the sensitivity scale was kept constant, the concentrations corresponding to the three peaks were changed. The Figure 3.52a shows results of detection by fluorescence for two compounds that are normally considered nonabsorbing—cresol red and orange G—and for one compound considered strongly fluorescing—fluorescein. No fluorescing compound is present in the electrolyte, and the fluorescence yield is over-

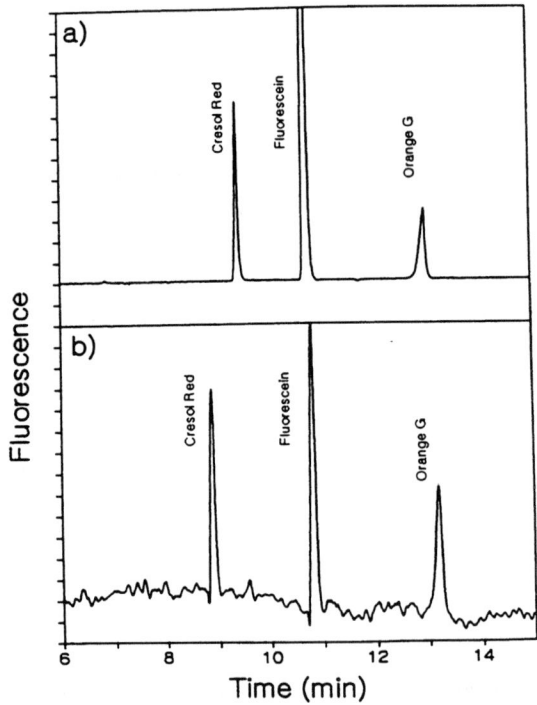

Figure 3.52 Comparison of fluorescence detection results with and without the addition of fluorescein to the carrier electrolyte. (a) Carrier electrolyte: 10 mM phosphate, pH 9. (b) 10 mM phosphate with 0.01 mM fluroescein, pH 9. Concentration of analytes: (a) 0.2 mM cresol red, 0.4 mM orange G, and 0.3 μM fluorescein; (b) 1 μM cresol red, 2 μM orange G, and 0.1 μM fluorescein. The injection for both electropherograms was made for 1 sec at 30 kV. Capillary (fused silica) dimensions were 21 μm and 60 × 70 cm. The separation voltage was adjusted at +30 kV. An argon laser produced an excitation wavelength of 488 nm. Reproduced with permission.[85]

whelmingly bigger for fluorescein than for the other two compounds. The concentrations of cresol red and orange G in the electropherogram in Figure 3.52b are two orders of magnitude below those in Figure 3.52a. The concentration of fluorescein in the sample is reduced only slightly, from 3×10^{-7} to 1×10^{-7} M, but a low concentration of fluorescein is now also present in the carrier electrolyte. Since the concentrations of the two "nonfluorescing" compounds have been lowered by a factor of 200 and the peak heights remained essentially unchanged, it is possible to conclude that the sensitivity of detection has been substantially enhanced by the addition of fluorescein to the carrier electrolyte.

A possible scheme explaining the events leading to the energy transfer between a merely UV absorbing compound A and the fluorescing compound F is given in Equations (3.19)–(3.20). The asterisks identify the respective excited states of the two compounds, and $h\upsilon$ stands for a quantum of light energy.

$$A + h\upsilon = A^* \tag{3.19}$$

$$A^* + F = F^* + A \tag{3.20}$$

$$F^* = F + h\upsilon \tag{3.21}$$

In the first step, the analyte molecule absorbs a quantum of light energy and is transformed to its excited state. Instead of a return to the ground state by one of the usual processes (e.g., dislocation, vibration, rotation, or self-quenching), the excitation of the analyte is transferred to the fluorophore F enhancing its fluorescence, Equation (3.20). The fluorophore then releases the transferred energy in form of fluorescence, Equation (3.21). The difficulty of the technique is less in finding a suitable match of fluorophores and chromophores than in problems related to the current stage of instrumentation used for fluorescence detection. The existing detectors are optimized for measurements of relatively large signals against a very low background signal. The technique produces, however, a relatively small signal against a background of considerable intensity. This, as we shall see next, is a problem that the technique of fluorescence energy transfer shares with another important detection method, laser-induced indirect fluorescence, but not with the direct detection of laser-induced fluorescence discussed on the examples of amino acids and marine toxins. The difficulty is connected to the inherent instability of lasers. Equation (3.15) is interpreted differently for indirect fluorescence than for indirect UV. The dynamic reserve DR is predetermined by the intensity stability of lasers, which is typically at one part per hundred or one part per thousand for unstabilized and stabilized lasers, respectively.[64] So that for the stabilized lasers we can write:

$$c_{lim} = c_m/10^3 \tag{3.22}$$

Choosing a concentration of the fluorescence background-providing ion in the carrier electrolyte as 5 mM, we obtain for c_{lim} the value of 5×10^{-6} M.

Considering the high cost of lasers and related modules required for their integration into a CE system, the micromolar detection limits calculated for laser-induced indirect fluorescence are disappointing. Even more so in comparison with the calculated [Equation (3.17)] and experimentally verified (Figure 3.41) values for indirect photometric detection obtained with a standard and relatively inexpensive CE system. A certain degree of improvement can be accomplished by minimizing the concentration of the fluorescence background-providing ion c_m. Gross and Yeung[86] have reported detection limits in the submicromolar range with a 250 μM concentration of salicylic acid. Salicylic acid is used to generate the background fluorescence.

The authors did not specify the stability of the laser source utilized to obtain the electropherogram in Figure 3.53. Applying Equation (3.22) for stabilized light source and the concentration of the salicylate, we arrive at $c_{lim} = 2.5 \times 10^{-7}$ M. This is in agreement with the observed detection limits, taking into account only the high-frequency noise and assuming that the low-

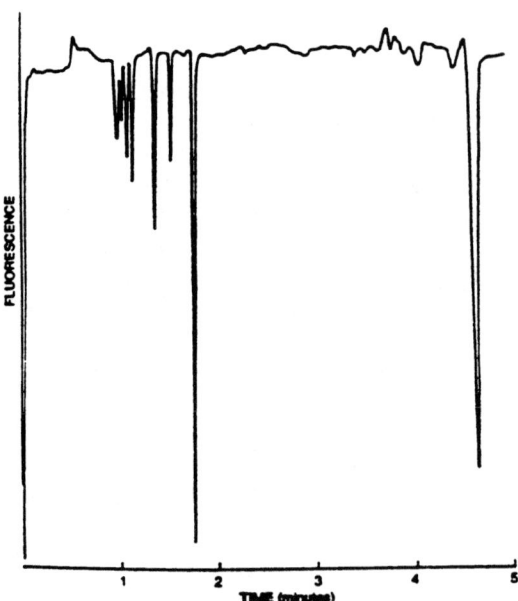

Figure 3.53 Laser-induced indirect fluorescence detection in conjunction with a counterelectroosmotic separation of inorganic anions. Migration times, peak identities, and concentration in the original sample: 53 sec, chloride, 19 μM; 54.6 sec, nitrate, 9.0 μM; 56 sec, perchlorate, 25.0 μM; 64.6 sec, permanganate, 12.0 μM; 76.4 sec, dichromate, 8 μM; 88.6 sec, iodate, 12 μM; 102.6 sec, phosphate, 54 μM. Carrier electrolyte: 250 μM salicylate, pH 4.0. C18-coated fused silica capillary, dimensions 14 μm, 51.2 × 33.1 cm. Electromigrative injection for 0.7 sec at +30 kV. Separation voltage: +30 kV. An argon laser was used for excitation at 331 nm. Reproduced with permission.[86]

frequency noise of the baseline is caused by thermal effects or other factors not covered by the calculation. The practical value of Equation (3.15) for predicting the sensitivity of indirect fluorescence detection is thus demonstrated in a similar way as in Section 3.3.2 for indirect photometric detection. The laser-induced indirect fluorescence detection was also reported for inorganic cations. (See Figure 3.54.)

As in the case of anionic separations, the sensitivity of laser-induced fluorescence for cations is at best comparable with that achieved by indirect photometric detection (see, for example, Figure 2.37). Indirect laser-induced fluorescence detection had been evaluated for a comparatively large number of analytes, mainly by Yeung and co-workers. The compounds for which a successful indirect fluorescence detection has been reported include amines,[87] alkylamines,[87] alkanolamines,[87] alkylammonium salts,[87] nucleo-

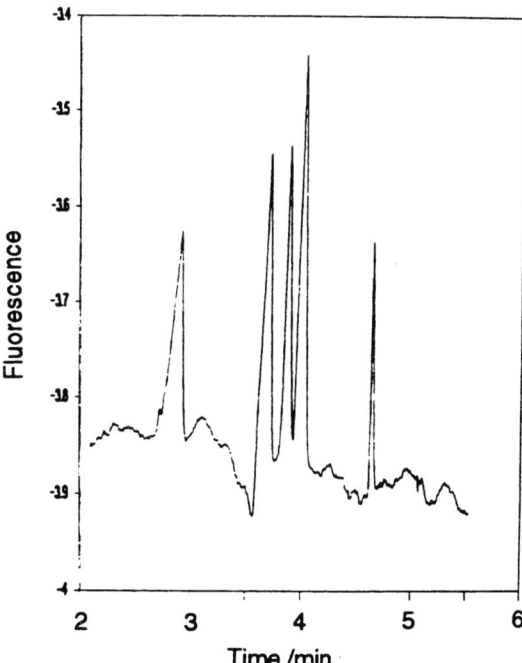

Figure 3.54 Indirect fluorometric detection of cations in conjunction with a coelectroosmotic separation by capillary electrophoresis. Migration times, peak identities, and concentration in the original sample: 173 sec, potassium, 40 μN; 221 sec, calcium, 20 μN; 223 sec, sodium, 20 μN; 240 sec, magnesium, 20 μN; 279 sec, lithium, 10 μN. Carrier electrolyte: 380 μM quinine sulfate and 580 μN sulfuric acid. Fused silica capillary dimensions: 18 μm, 82.3 × 70.7 cm. Sample was introduced by electromigration for 1 sec at +10 kV. The separation voltage was +40 kV. Reproduced with permission.[87]

tides,[86,88] nucleosides,[88] proteins,[88] dansylated amino acids,[88,89] underivatized (native) amino acids,[89] proteins,[88] and tryptic digests.[90]

In an important contribution, Amankwa and Kuhr[91] investigated the use of indirect detection with laser-induced fluorescence in conjunction with MECC (micellar electrokinetic capillary chromatography, see Section 2.4). They were able to detect alcohols and phenols (including alkylphenols and chlorophenols). A wide range of detectable analytes confirms the universal character of this mode of indirect detection. The authors also discuss the dynamic reserve (DR) of their instrument, reporting the value as 100 and 800–1000 for an unstabilized and stabilized laser source, respectively. Their assessment is in good agreement with the value in Reference 64 that is used to derive the Equation (3.22). For further details on indirect fluorescence, the reader is referred to the excellent review by Yeung and Kuhr.[92]

3.3.3.3 Conductivity Detection

There have been numerous reports of the use of conductivity detection in capillary electrophoresis. Conductivity measurements appear to be more easily adaptable to miniaturization than, for example, most optical techniques. In the form of so-called potential gradient measurement, conductivity detection has been used extensively in isotachophoresis. Since isotachophoresis is performed in a constant-current mode, a potential measurement is synonymous with the measurement of conductivity [$U = I/G$, where G is the conductivity (Siemens or ohms^{-1})]. The term *potential gradient* describes the placement of the measuring electrodes in the longitudinal direction of the separation capillary, as shown in Figure 3.55. The measured potential then indicates the potential gradient in the direction of migrating zones of analytes.

The isotachophoretic instrument in Figure 3.55 bears a strong resemblance to modern CE instrumentation. Note, however, that the capillary is formed by a PTFE foil–covered groove in a polymeric block. The fused silica capillaries used today for capillary electrophoresis are unsuitable for isotachophoresis because of the relatively high rates of the electroosmotic flow. The diagram illustrates clearly the longitudinal arrangement of the two electrodes. For isotachophoresis, the two electrolyte containers are filled with the two different electrolytes necessary for developing the separation. See Figure 3.10C,D and Section 3.1 for an additional explanation of the functional principle of isotachophoresis. Early pioneers[1] in capillary electrophoresis utilized equipment such as the one in Figure 3.55 to generate impressive separations of low-molecular-weight ionic compounds.

Before the start of the separation in Figure 3.56, both sides of the instrument depicted in Figure 3.55 are filled with an identical solution of carrier electrolyte. After a sample injection through septum 11, the separation is developed at a constant current of 20 μA. It is interesting to note that the detection limits in Figure 3.56 are in the 10^{-5} M range, a sensitivity of de-

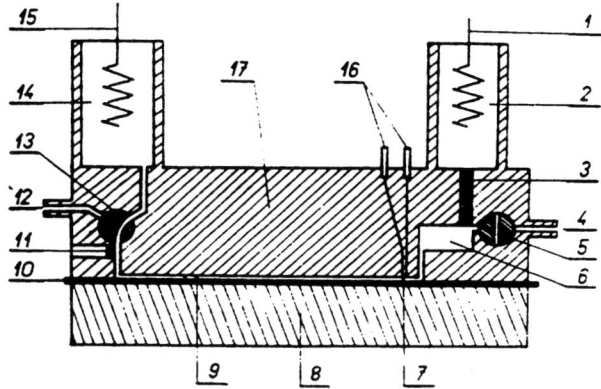

Figure 3.55 Cross section of an isotachophoretic instrument. 1 and 15: electrodes for application of the separation potential; 2 and 14: electrolyte reservoirs; 3: glass frit; 4 and 12: filling and draining tube connections; 5 and 13: PTFE stopcocks; 6: buffer reservoir; 7: pair of platinum electrodes for potential gradient measurement; 8: thermostatted metallic plate; 9: capillary groove; 10: PTFE foil; 11: septum; 16: connectors to the conductivity meter; 17: monolithic body of a polymeric material. Reproduced with permission.[93]

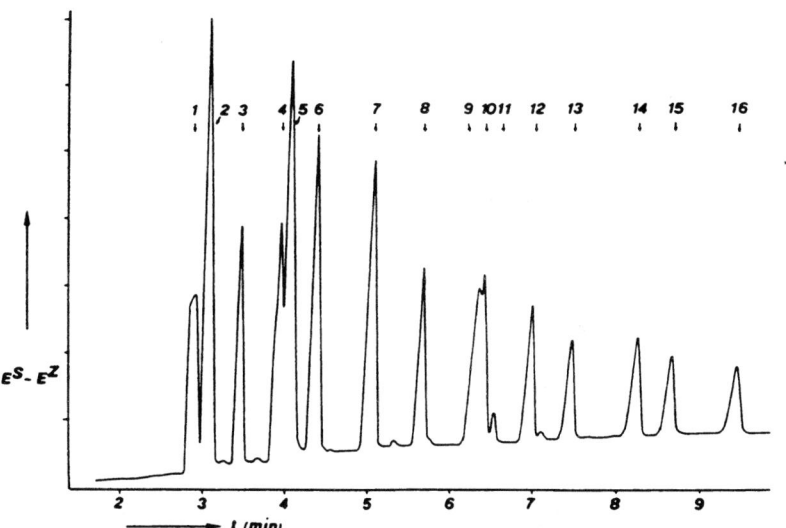

Figure 3.56 Separation of anions by capillary electrophoresis from 1979. E^S-E^Z indicates the detected potential gradient. Carrier electrolyte: 0.01 M 2-(N-morpholino) ethanesulfonic acid(MES)-histidine, 0.1% hydroxyethylcellulose (pH 6.05), separation current: 20 μA, capillary internal diameter: 0.2 mm. Peak identities: 1, chloride; 2, sulfate; 3, chlorate; 4, malonate; 5, chromate; 6, 3,5-pyrazoledicarboxylate; 7, adipate; 8, acetate; 9, propionate; 10, β-chloropropionate; 11, unknown; 12, benzoate; 13, napthalene-2-monosulfonate; 14, glutamate; 15, enanthate; 16, benzyl-DL-aspartate. The injection volume of 0.7 μl contained 17.5 picomole of each analyte. Reproduced with permission.[1]

tection for anions that has not been improved by any other report utilizing conductivity detection of anions between 1979 and today. Contributions from more recent work[94,95] consist mainly in new ways of adapting conductivity detectors to the narrow diameter of fused silica capillaries and in updating the electronic circuitry. In 1986, Bocek and co-workers[94] described a conductivity detection cell for connection to the end of a capillary (see Figure 3.57).

Zare[95] and co-workers utilized a computer-controlled CO_2 laser to drill two diametrically opposite holes into 50 or 75 μm I.D. fused silica capillaries (see Figure 3.58). These 40 μm I.D. holes were then used for a placement of 25 μm platinum electrodes exactly opposite to each other. The electrodes that were exactly positioned under a microscope were than glued in place by MW 1000 poly(ethyleneglycol) heated to liquidity and solidified by cooling. The resulting conductivity cell is the only example of on-column conductivity measurement on fused silica capillaries reported to date.

The on-column detector depicted also made possible the only improvement recorded so far in the sensitivity of conductivity detection. The detection limit for lithium in the electropherogram in Figure 3.59 is claimed to be in the 10^{-7} M range. Note that the same electrolyte as utilized in 1979 for anions is applied to the separation of cations in this example published in 1987.

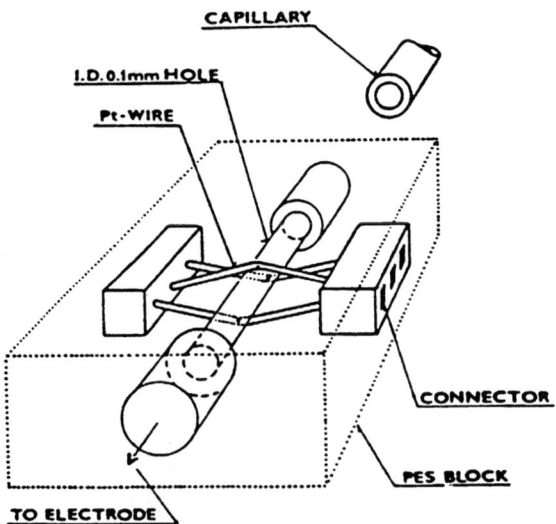

Figure 3.57 Drawing of a conductivity cell for capillary electrophoresis. The capillary I.D. and the cylindrical hole through the cell have identical dimensions. PES-polyester resin. Dimensions of the detection block were 50 × 50 × 15 mm. Reproduced with permission.[94]

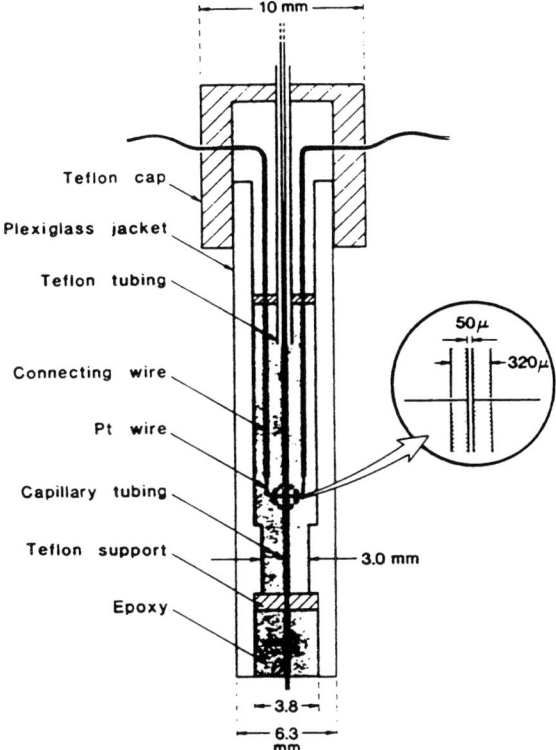

Figure 3.58 Diagram of an on-column conductivity cell for fused silica capillaries used in capillary electrophoresis. Reproduced with permission.[95]

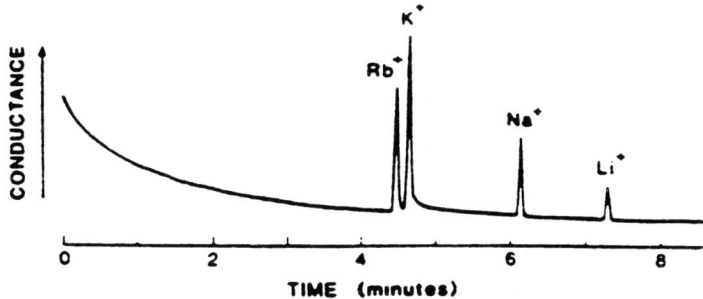

Figure 3.59 Electropherogram of a mixture of four cations detected by an on-column conductivity cell. The concentration of all four cations is 2×10^{-5} M. Sample introduction was carried out by electromigration for 5 sec at +5 kV. Capillary I.D.: 75 μm, length: 60 cm. Carrier electrolyte: 20 mM MES/histidine at pH 6. The separation voltage was +15 kV. Reproduced with permission.[95]

The same authors also reported successful experiments with an end-column conductivity cell,[96,97] in which a sensing microelectrode is placed at the outlet of the fused silica capillary. This arrangement is described as avoiding the electrical interferences caused by applied high voltage during a CE separation, while the separation efficiency is decreased only by about 25% in comparison with the on-column detector. The end-column design is derived from a prevailing approach in the amperometric detection and is discussed in more detail in the following section. With the exception of the report on cations by Zare et al.,[95] remaining publications[98–100] between ~ 1987 and 1992 do not reveal any further improvements in the sensitivity of conductivity detection beyond the levels already reached in 1979 by adaptation of isotachophoretic instruments.

The main problem in applying conductivity detection to CE is a necessity to satisfy two contradicting requirements. As explained in Sections 2.5–2.7, one of the key requirements in capillary electrophoresis is to use a carrier electrolyte coion that is very close in its electrophoretic mobility to that of the analyte ions. Only then is it possible to avoid excessive peak spreading observed either as tailing or fronting of analyte peaks. On the other hand, sensitive conductivity detection requires one to maximize the difference between the conductivity of the separated analyte zones and the conductivity of the eluent.[101] The same requirement also has to be applied to the optimization of conductivity detection in capillary electrophoresis:

$$\Delta G = G_{\text{analyte}} - G_{\text{carrier electrolyte}} \tag{3.23}$$

where $\Delta G[\mu S]$ is the magnitude of the conductivity signal, and G_{analyte} and $G_{\text{carrier electrolyte}}$ are the respective conductivities $[\mu S]$ of the analyte and carrier electrolyte zones. A contradiction arises from the fact that any attempt at maximizing ΔG has to result in the use of coions in the carrier electrolyte having an equivalent ionic conductivity and therefore also an electrophoretic mobility significantly different from those of the analyte ions. The differences in electrophoretic mobilities lead in turn to increasing asymmetry of the detected peaks. Any sensitivity gain according to Equation (3.23) is thus counteracted by decreases in peak heights due to peak broadening caused by tailing or fronting. The resolution of closely comigrating peaks is also strongly affected. Let us consider briefly some typical approaches to ΔG optimization in ion chromatographic conductivity detection and evaluate the results of similar hypothetical and real experiments in capillary electrophoresis.

One of the ion chromatographic methodologies optimizes ΔG in detection of anions by choosing anions of the lowest possible ionic equivalent conductivity for a mobile phase. Typical examples are benzoate or complex borate–gluconate anions. Ionic equivalent conductivities of both eluent anions are very different from those of the most frequently analyzed inorganic anions (e.g., bromide, chloride, sulfate, nitrite, nitrate, fluoride, phosphate); see Section 2.2. This does not affect the ion chromatographic peaks in any way, but the corresponding large difference in electrophoretic mobilities would cause an unacceptably high peak asymmetry for peaks detected in capillary electrophoresis.

In another ion chromatographic approach, the conductivity of the eluent surrounding and permeating the analyte zones is suppressed following the separation of the sample components on an ion exchange column. The principle of this methodology is in the chemical reaction leading to suppression of conductivity in eluents.

$$HCO_3^- + H^+ \rightarrow H_2CO_3 \qquad (3.24)$$

$$B(OH)_4^- + H^+ \rightarrow H_3BO_3 + H_2O \qquad (3.25)$$

$$OH^- + H^+ \rightarrow H_2O \qquad (3.26)$$

All eluents, three of which are shown in chemical reactions (3.24–3.26), used for chemically suppressed conductivity detection of anions are applied as an alkali salt of a weak acid. Their conversion to the conjugate weak acid is carried out by a transmembrane exchange of alkali cations for hydronium ions.[101] In CE, the utilization of carbonate and borate in a carrier electrolyte would result in fronting peaks for bromide, chloride, sulfate, nitrite, and nitrate. Carrier electrolytes containing hydroxide as a main electrolyte coion would generate strong tailing of peaks increasing in the series bromide, chloride, sulfate, nitrite, nitrate, fluoride, phosphate, carboxylates, and sulfonates. Chemical suppression has recently been reported to be feasible in capillary electrophoresis.[102] It is possible that suppressed conductivity detection will soon be able to match the number of applications and sensitivity levels achieved by indirect UV detection in capillary electrophoresis and play a similarly important role in capillary electrophoresis as it does in ion chromatography. Despite the obstacles discussed above, it would be premature to write off suppressed conductivity as unsuitable for CE. A possibility exists, for example, to avoid excessive peak asymmetries stemming from the analyte/electrolyte equivalent conductivity mismatch (see Section 2.7.3) by utilizing capillaries of smaller internal diameters (20 μm or less).

To conclude our comparisons with ion chromatography, let us consider indirect conductivity detection, which is used frequently in ion chromatographic detection of cations (Figure 3.60). For that application, indirect conductivity detection can give better results than chemically suppressed conductivity detection.[103]

The magnitude of the detection signal is optimized in a different way than in the preceding two examples. Highly conductive hydronium ions are chosen as main eluent ions, and the analyte zones are then detected as decreases in the background conductivity. If we were to choose hydronium ions as carrier electrolyte coions for separation and detection of cations, we would experience a strong asymmetry of analyte peaks due to tailing. An increase of ΔG due to the maximized value of the differential in Equation (3.23) ($G_{analyte} << G_{carrier\ electrolyte}$) would thus again be counteracted by the concomitant

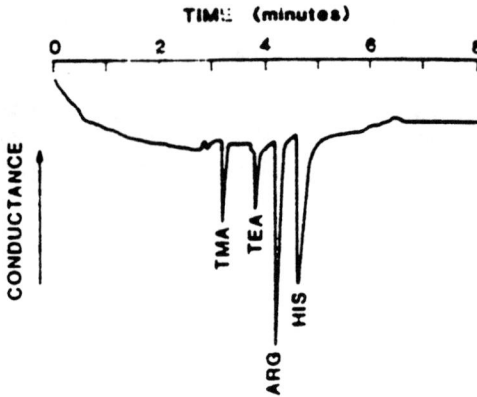

Figure 3.60 Indirect conductivity detection in capillary electrophoresis. The sample contained tetramethylammoinium ion, triethylamine at 8×10^{-5} M and arginine, histidine at 4×10^{-5} M. Hydrostatic sample introduction was carried out for 30 sec at 10 cm. Capillary I.D.: 50 μm, length: 51 cm. Carrier electrolyte: 5 mM potassium acetate at pH 5.4. The separation voltage was +15 kV. Reproduced with permission.[95]

deformations of peak shapes. Even if indirect conductivity detection is difficult in capillary electrophoresis in its most sensitive form given by Equation (3.23), it is possible in principle, and was successfully demonstrated for intermediate ΔG values.

In conclusion, the difficulty in applications of conductivity detection to capillary electrophoresis consists not only in the usual problems and challenges related to miniaturization, but also in conflicting requirements imposed by optimization of separation efficiency on the one hand and by the requirements necessary for sensitivity optimization on the other hand. The majority of reports to date describe solely the various aspects of the design of conductivity detectors for capillary electrophoresis. Most applications demonstrate only intermediate levels of sensitivity. More work needs to be done in optimizing carrier electrolyte chemistries and other analytical conditions for achieving a highly sensitive detection without any side effects on the efficiency of capillary electrophoretic separations. Optimized conductivity detection has great potential to become a universal detection method of choice, once a solution is found resolving the central problem of conflicting optimization requirements.

3.3.3.4 Amperometric Detection

One of the parameters optimized in the applications of amperometric detection to liquid chromatography is the coulombic efficiency, or the percentage of an electroactive analyte undergoing electrode reaction inside the detection cell. A typical coulombic efficiency in conventional amperometric cells at usual flow rates (thin layer cells, 0.5 to 1 ml/min[103]) does not exceed 10%. Attempts have been ongoing to miniaturize the chromatographic apparatus and the cell geometry to increase the coulombic efficiency. Whereas in conductivity detection miniaturization does not merely complicate the measurement, in amperometry, miniaturization is desirable for improvement of sensitivity due to higher coulombic efficiencies. However, it must also be mentioned that a specific problem exists in capillary electrophoresis, making the application of amperometry challenging: the requirement to perform electrochemical electrode reactions in the presence of high-voltage electric fields. While the potentials for detection are applied in the hundreds of millivolts range, and detection currents generated by electrode reactions are only in femtoamperes or picoamperes, the separations are driven by electrical currents in the microampere range. A strong need thus exists effectively to separate, to insulate the detection and electrophoresis circuitries. Wallingford and Ewing[104] were the first to design a functioning apparatus for amperometric detection in capillary electrophoresis (see Figure 3.61).

The key element of the design illustrated in Figure 3.61 is a conductive connection in the middle portion of the capillary, making it possible to sep-

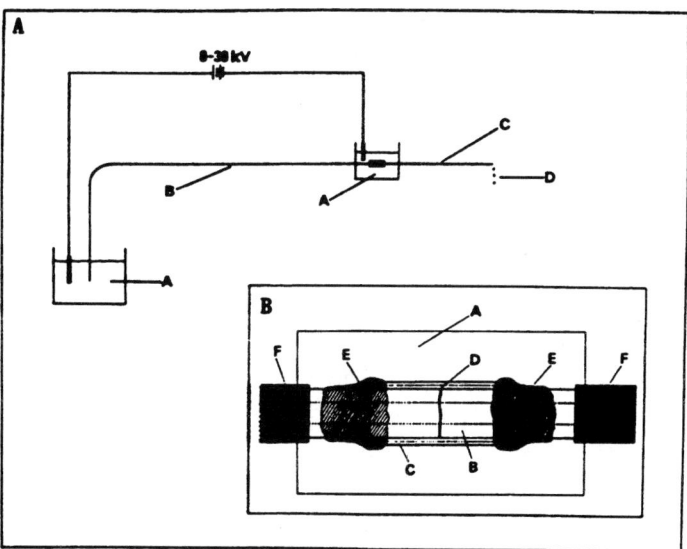

Figure 3.61 Separation of a CE system from the detection part of the fused silica capillary. (A) Schematic diagram of the system. A: Carrier electrolyte reservoirs, B: separation capillary, C: detection part of the capillary, D: carrier electrolyte dripping from the open end of the detection capillary. (B) Detailed diagram of the conductive joint. A: microscopic slide, B: fused silica capillary, C: porous glass capillary, D: crack, E: epoxy, F: polyimide coating. Reproduced with permission.[104]

arate the high voltage required for electrophoresis from the weak signals measured at the surface of a carbon fiber indicator electrode. To prepare a conductive connection from the inside to the outside of the capillary, while maintaining the integrity of the column of carrier electrolyte transporting the sample components toward the detection end of the capillary, a miniature crack is prepared and covered by a short length of closely fitting porous tubing acting as a diaphragm. Such a conductive joint is placed into a vessel containing an electrode at ground potential, and a volume of carrier electrolyte. An electric field applied between the conductive joint and the high-voltage end of the capillary drives the separations, but beyond the conductive joint the potential is effectively at the ground level. The electroosmotic flow continues to force electrolyte through the tube, transporting the separated sample components from the conductive joint to the detection end of the capillary. The high-voltage circuit controlled by the voltage source is thus effectively separated from the low-voltage circuit of the potentiostat and measuring electrodes (see Figure 3.62).

In Figure 3.62 is shown a miniature version of a standard three-electrode amperometric system. The detection potential is maintained between the working electrode and the reference electrode. Detection current is measured between the working electrode and the auxiliary electrode. The three-electrode system effectively precludes shifts of reference electrode potentials resulting from a passage of a large amount of electric charge. The working electrode is a carbon fiber (10 μm diameter) fitted into a drawn out end of a glass capillary and held in place by a drop of epoxy. After curing of the epoxy, the rest of the capillary was filled with mercury for a conductive

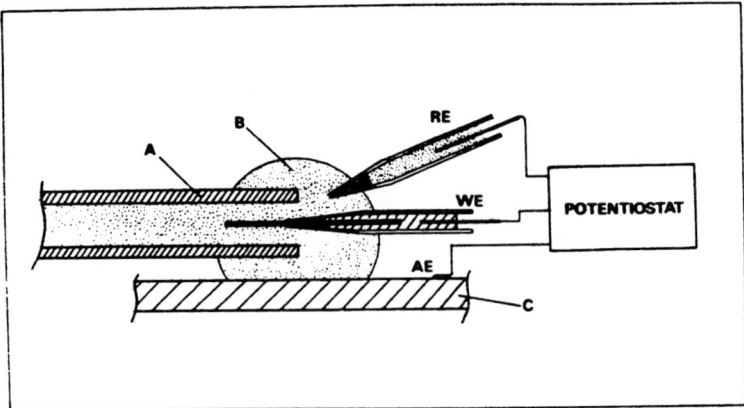

Figure 3.62 Schematic of an amperometric detector attached to the detection end of the CE system shown in Figure 3.61. A: Fused silica capillary, B: drop of carrier electrolyte, C: stainless steel plate, RE: reference electrode, WE: working electrode, AE: auxiliary electrode. Reproduced with permission.[104]

connection between the carbon fiber and the wire connection to the potentiostat. The saturated calomel reference electrode was improvised with the help of a pipette tip containing the reference electrode chemistry and connected to a potentiostat.

After the initial feasibility study[104] of their design of an amperometric detector, Ewing and co-workers reported a steadily improved sensitivity of detection.[105-108] The improvement was mainly due to smaller internal diameters of detection capillaries, also necessitating a smaller diameter of the carbon fiber electrode. The diameter of the carbon fiber utilized in experiments was reduced by an electrochemical etching procedure. In the process of adapting their detector to smaller capillary diameters, the authors had to abandon the three-electrode system in favor of a simpler two-electrode arrangement. As a result of a number of modifications of their original system, Ewing and co-workers were finally able to demonstrate[108] detection limits in the submicromolar range (10^{-8} M) for biogenic amines (Figure 3.63). In the same report, they also showed calibration plots over four orders of magnitude, with correlation coefficients higher than 0.99. Ewing et al. determined about 40% coulombic efficiency in 5 μm I.D. capillaries. Due to a very small sample volume (ca. 20 picoliters), absolute amounts of analytes that can be detected are in the femtomole (10^{-15}) range, with the significant exception of serotonin, which is detectable in the attomole (10^{-18}) range.

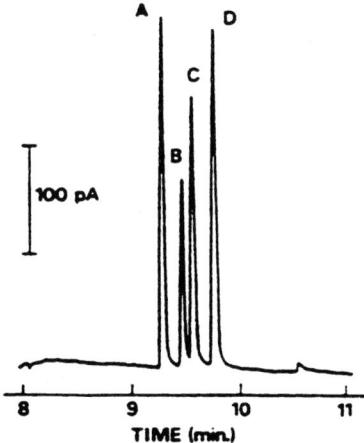

Figure 3.63 Amperometric detection of catecholamines and serotonin. Peak identities and concentrations: A, 2×10^{-5} M serotonin; B, 5×10^{-5} M dopamine; C, 5.5×10^{-5} M norepinephrine; D, 6×10^{-5} M epinephrine. Carrier electrolyte: 0.025 M MES in 20% isopropanol, pH 5.5. Sample is introduced by electromigration for 1 sec at +30 kV. The separation potential was +30 kV. The dimensions of the separation capillary were 69 cm and 9 μm. The detection capillary was ~1 cm long and 9 μm in internal diameter. The detection potential was +0.7 V versus saturated calomel electrode. Reproduced with permission.[108]

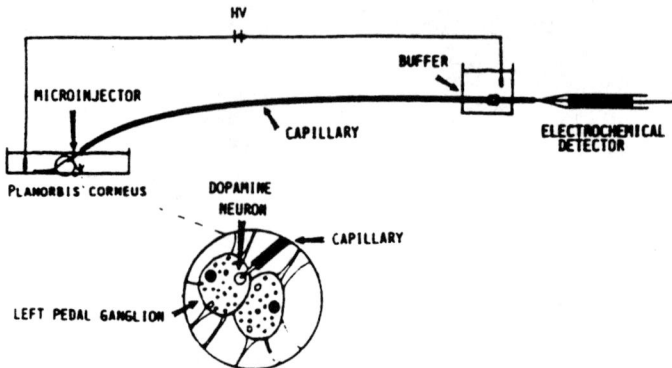

Figure 3.64 Experimental system for removal, electrophoretic separation, and amperometric detection of cytoplasmic samples from a neuron of *Planorbis corneus* (pond snail). The expanded view shows the relative sizes of neurons and the sampling end of the fused silica capillary inserted in the snail brain. Reproduced with permission.[109]

Practical relevance of the achieved mass sensitivity of amperometric detection was demonstrated in monitoring serotonin in cytoplasma of a single nerve cell.

For effective sampling the high-voltage end of the 5 μm I.D. capillary had to be etched to a much smaller size under a microscope. The neuron sampled in Figure 3.64 is one of the largest neurons available for experiments. It is about 200 μm in diameter, with a calculated cell volume of ~ 4 nanoliters. Electromigrative sampling was carried out for 2 sec at 100 V, resulting in a calculated sample volume of 36 picoliters or ~ 0.9% of the total cell volume. For sampling of mammalian neurons with diameters of only ~ 30 μm, sampling volumes in the femtoliter range will be required.

Amperometric detection of biogenic amines after liquid chromatographic separations, introduced in 1975 by Kissinger et al.,[110] led to a widespread use of that mode of detection in HPLC. Widespread use of amperometry with capillary electrophoresis may result from the development of a corresponding system for a routine analysis of single mammalian brain cells.

3.3.3.5 Mass Spectrometry and Capillary Electrophoresis

The potential rewards of the coupling of capillary electrophoresis and mass spectrometry have motivated many researchers to work in that field. In terms of the theory developed by Giddings,[111] the coupling of capillary electrophoresis and mass spectrometry represents a true two-dimensional separation system. In Giddings' theory, the separation power of a system in one dimension is described as a peak capacity $n(n = 0.5N^{1/2})$. The peak capacity is a number of concentration pulses (peaks) that can be resolved by a given separation tool. Reverse-phase chromatography with a separation efficiency of $N = 3000$ yields, for example, a peak capacity of 27. A capillary electrophoretic system generates a peak capacity of ~ 500 for $N = 1{,}000{,}000$.

Mass spectroscopy as a monodimensional separation tool exhibits peak capacities in excess of 1000, if the resolution of one unit of the mass-to-charge ratio is attained. A two-dimensional separation is advantageous because the peak capacity of such a system is a product—not a sum—of the respective monodimensional peak capacities, if the separation modes are completely orthogonal. A two-dimensional system consisting of capillary electrophoresis ($n = 500$) and a mass spectrometer ($n = 1000$) would thus exhibit a peak capacity of more than 500,000. This is a resolving power that currently only a coupled capillary GC/MS can approach. Similar resolving power is also needed for samples that cannot be processed by gas chromatography, such as for analytes that can be separated only in a liquid medium.

It has long been the desire of analytical chemists to accomplish a complete analysis of challenging mixtures such as biological fluids or industrial intermediary products such as coal tars. Also, if all concentration levels need to be determined, even some routine samples, such as, for example, river water, may still be outside an average laboratory's capability for conducting a complete assay.

Since capillary electrophoresis has evolved more rapidly than liquid chromatography, which also has the potential of achieving efficiencies of 10^6 theoretical plates per separation, it is expected that a CE/MS coupling may become more important in the future than the LC/MS combination currently preferred for nonvolatile samples. CE/MS was first reported in 1987 by Olivares, Nguyen, Yonker, and Smith,[112] who demonstrated submicromolar detection limits for alkylammonium compounds with the help of a simple electrospray ionization (ESI) interface. In subsequent reports,[113-120] however, Smith and co-workers used electrospray ionization in conjunction with a sheathing liquid. The sheathing liquid represents an elegant solution to the problem of compatibility of carrier electrolytes with mass spectroscopy and is reported to be useful not only for electrospray but also for fast atom bombardment (FAB) ionization as well.[121] The sheath flow overcomes problems arising from the need to evaporate aqueous and salt containing liquids tracelessly. In the sheath flow method, the liquid encountered at the electrospray interface is mostly volatile and independent of changes in the chemistry or flow rate of a carrier electrolyte. This is accomplished by ten to hundredfold higher flow rates of sheath liquid in comparison with the flow rates of a carrier electrolyte. Typically, the sheath liquid is composed of acetonitrile, methanol, acetone, or isopropanol.[115] (See Figure 3.65.)

The proper operation of electrospray ionization requires an uninterrupted electrical contact with the carrier electrolyte emerging from the detection end of the CE capillary. Such electrical contact is made through a layer of sheathing liquid interfacing with the stainless steel capillary, which in turn is fitted in the copper plate held at the ionization voltage (ESI voltage). This

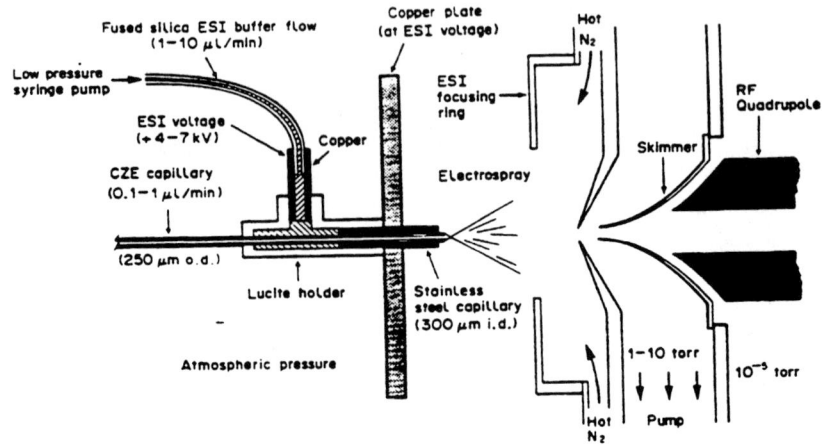

Figure 3.65 Schematic illustration of an electrospray ionization (ESI) interface utilizing sheath flow. The term ESI buffer is used for sheathing liquid. Reproduced with permission.[119]

is remotely analogous to the use of sheath flow in optical detection, where the sheathing liquid facilitates the entry of light into the sample by minimizing the reflection and refraction losses (Section 3.3.3.1). The ESI voltage is usually in the range of $+3-5$ kV for cations and -5 kV for anions, making it possible in some instruments to use the ESI voltage as the lower end of the circuit. The separation potential of, for example, $+20$ kV is then adjusted by using $+23$ kV for the separation part of the capillary and $+3$ kV at the ESI end. Another advantage of the apparatus in Figure 3.65 is that, in contrast with the earlier arrangements, fused silica capillaries can be used without any preparation whatsoever. Ions created by electrospray ionization enter the mass spectrograph through a focusing ring and a curtain of hot nitrogen. The hot nitrogen gas aids the desolvation of ionized analytes and minimizes solvent cluster formation later on under vacuum. The flight of ionized particles is further enhanced by a potential gradient between the ESI focusing ring (focusing lens) and the nozzle on the other side of the nitrogen curtain.

The focusing ring is typically held at $+350$ to 1000 V and the nozzle at $+200$ V. The skimmer is then at ground potential. The total distance between the fused silica capillary and the skimmer is ~ 1.5 cm. Behind the skimmer, ions enter the focusing region with the RF focusing quadrupole, which is held at a vacuum of 10^{-4} torr. The analysis quadrupole behind the ion lens is held at 10^{-5} torr. (See Figure 3.66.)

Electrical currents resulting from analyte ions reaching the conversion dynode at the back of the instrument are amplified in the electron multiplier

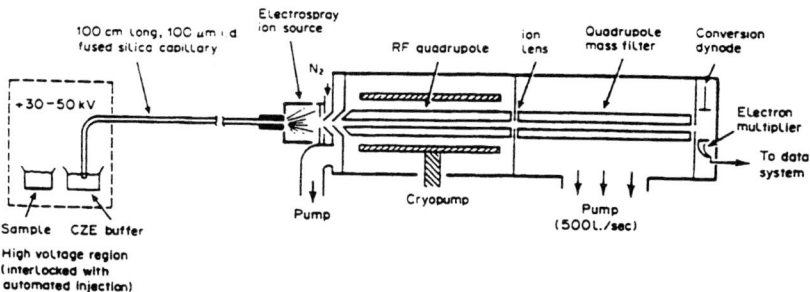

Figure 3.66 Schematic illustration of a complete CE/MS system. Details of the electrospray ion source, which include a coaxial sheath flow of a suitable liquid, are shown in Figure 3.65. Reproduced with permission.[119]

and after a conversion to proportional potentials acquired by suitable data systems.

The complete process of acquiring a mass spectrum consisting of the ionization and travel of ionized analytes through the spectrometer, followed by data processing steps, may require more than 1 sec with currently available instruments. This is too long for CE separations with peak widths of only a fraction of a second. In order to acquire complete mass information, the researchers in the field of CE/MS coupling have frequently adapted a procedure termed reduced elution speed (RES), based on decreasing the separation voltage during the time intervals when a zone of interest reaches the ESI end of the instrument. The width of peak zones can thus be extended up to several seconds to facilitate data acquisition by a mass spectrograph.

Alternatively, the analyst may choose to give up more of the information usually available and collect concentration signals only at one or several preselected mass-to-charge ratios (m/Z). This enables not only the CE instruments to be operated at optimum voltages, but it also increases the detection limits attainable by MS instruments. In fact, most detection limits calculated from the full mass spectra can be improved by at least an order of magnitude in the selected ion monitoring (SIM) mode or by choosing a narrower scanning range. To a first approximation, the relative sensitivity is inversely proportional to the range of masses scanned in a given application.[122] In many situations, the choice of the SIM mode may be dictated by a limited data storage capacity of a laboratory computer.

The SIM mode can be compared to the operation of a photodiode array UV spectrometer at a few preselected wavelengths rather than over an uninterrupted region of the UV spectrum. However, the resolving power and universal character of mass spectroscopy makes positive identification of unknown compounds much easier than in UV. Consider the following example of an incomplete separation of a mixture of phosphonium salts.

The total ion current recording in the upper trace of Figure 3.67 is a plot of total ionic current recorded at the sensing dynode independently of the mass–to–charge ratio (m/Z); see Figure 3.66. The same type of recording is also called a reconstructed total ion chromatogram or electropherogram to indicate that it can be generated by putting together the current versus migration time recordings for single values of m/Z.

The total ion current recording only shows an incomplete separation yielding two peaks where four are expected. Single-ion electropherograms in the four traces below the total current separation show convincingly the ability of mass spectroscopy not only to provide an additional separating dimension, but also to make possible an unequivocal identification of sample

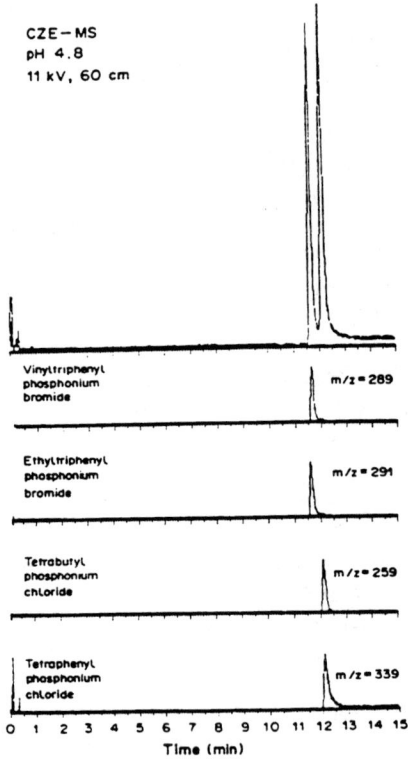

Figure 3.67 Total ion current (upper trace) and SIM electropherograms of a mixture of four phosphonium salts. Electromigration introduction of ~1 picomole of each phosphonium salt (ca. 10^{-6} M sample solution). Carrier electrolyte: 50 mM phthalate, pH 4.8, also containing ~0.1 mM potassium chloride. Separation voltage: +11 kV across a 60 cm × 100 μm fused silica capillary. Sheathing liquid: pH 8.9, 8.7% water in 2-propanol also containing 0.02 mM HCl and 0.1 M ammonium acetate. Reproduced with permission.[119]

components by their mass–to–charge ratio. The combination of mass spectroscopy and capillary electrophoresis offers uniquely powerful separations. Its full potential will be realized, however, only after solutions are found to a number of outstanding design and data processing problems. A significant reduction of the cost will also be necessary for CE/MS to become a truly routine technique.

3.3.3.6 Miscellaneous Experimental Detection Techniques

The steadily increasing range of detectors for capillary electrophoresis is not identical to the list of detection techniques for HPLC. While some detection methods requiring relatively large amounts of analyte (e.g., IR, NMR) may never be adapted to capillary electrophoresis, new approaches, unknown in HPLC (e.g., capillary vibration, thermooptical absorbance, etc.) are keeping the list of methods relatively long. Some of the techniques that are only experimental today may become widely utilized in the future. Other methods will always remain somewhat of a curiosity. CE detection techniques of predominantly experimental character are listed in Table 3.4.

3.4 Instrumentation and System Performance

The elements of a capillary electrophoresis system are briefly described in Section 2.1. This chapter requires familiarity with the functions of single modules of a CE system acquired not only from Section 2.1, but also from Sections 2.3 (Capillaries and Electroosmotic Flow), 2.6 (The Problem of Joule Heat Generation), 3.1 (Sample Introduction), and 3.2 (Selection and Handling of Capillaries).

Before the introduction of the first commercial systems in 1989, most researchers used various versions of improvised capillary electropherographs. The design of the system shown in Figure 3.68 had been in use for many years in the laboratory of one of the authors and represents an example of an optimum "home-made" design.

A regulated power supply (A) forms a platform for a wooden box (B) containing a high-voltage electrode (C) immersed in a volume of carrier electrolyte (D). The high-voltage electrode is connected by a cable (E) to the high-voltage circuitry of the power supply. Also located inside the wooden box is a high-voltage relay with corresponding circuitry connected to a depressable contact between the lid and the wall (not shown in Figure 3.68). The relay interrupts the high voltage whenever the wall switch is not depressed, as, for example, when the lid is not closed. This safety feature makes it impossible for operator to touch the high-voltage parts accidentally while the voltage is on. A fused silica capillary (F) is also immersed in the carrier electrolyte inside the box and protrudes through a slit between the lid and the side wall toward a UV detector (G) placed on an adjustable lab-

Table 3.4. EXPERIMENTAL DETECTION TECHNIQUES IN CAPILLARY ELECTROPHORESIS

Detection	Det. Limit (M)[a]	Typical Analytes	Comments	Ref. No.
Radioisotope	10^{-12}	^{32}P labeled compounds adenosine triphosphate	Detects β, γ emitters on column	123
Laser-induced capillary vibration	10^{-7}	Nonlabeled amino acids	Argon lasers on column	124
Laser-based refractive index	10^{-6}	Underivatized carbohydrates carbonic anhydrase	Position-sensing diode on column	125, 126
Thermooptical absorbance	10^{-7}	Derivatized amino acids	Pump laser: Ar probe laser: He–Ne on column	127, 128
Ion mobility	10^{-3}	Tetraalkylammonium salts	Electrospray ionization	129
Potentiometry	10^{-6}	Alkali, alkaline earth cations	Potentiometric microelectrode	130

[a] The estimates of detection limits are given as a multiple (2–3x) of noise.

INSTRUMENTATION FOR CAPILLARY ELECTROPHORESIS 179

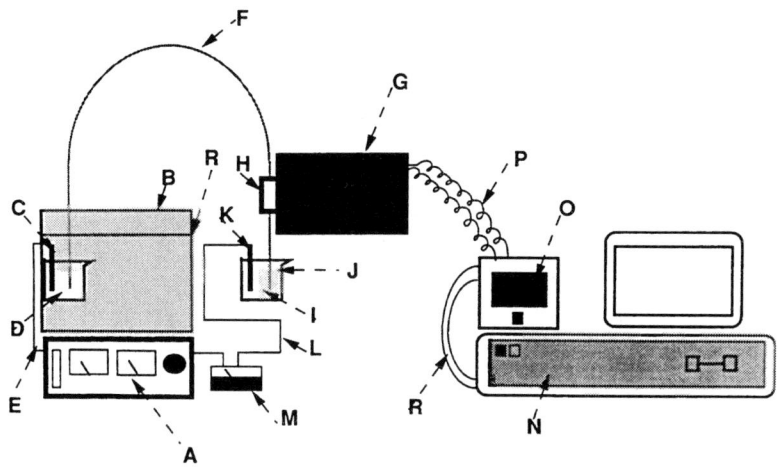

Figure 3.68 Home-made CE System, ~1987. Descriptive lettering is explained in the text.

oratory jack. The fused silica capillary passes through the detector cell (H), into the beaker (I), with another portion of carrier electrolyte. The liquid level inside the beaker at the detection side is adjusted at the same level (J) as in the beaker inside the wooden box to avoid distortions of separated peak zones due to a hydrodynamic flow. The ground electrode (K) is contacted by a wire connection leading to the ground inside the voltage power supply. The ground lead is connected to microammeter (M). To the right of the detector is placed a laboratory computer (N) for data acquisition and data processing. The box with a dark field on its front panel is an anolog–to–digital interface (O). Leads (P) conducting analog signals from the detector are attached to its input contacts in the rear. Another pair of leads (R) feeds the digitized detector signal from the interface to the computer.

To carry out a hydrostatic injection with the home-made system in Figure 3.68, the operator has to remove the capillary (F) from the electrolyte (D) and place the capillary into a sample solution in a suitable sample vial (not shown in Figure 3.68). The sample vial with the immersed end of fused silica capillary has to be held above the level of electrolyte (D). In order to reproduce the height easily, the sample vial was usually positioned at the upper edge (S) of the lower part of the wooden box (B). For the procedure to work properly, the capillary had to be full of carrier electrolyte. The operator maintained the configuration for a period of time that had to be determined as accurately as possible. Typical sampling times were measured in seconds (max. 30 sec), and reproducible sampling thus required considerable effort and experience on the part of the operator. The best precision of peak areas obtainable with the home-made instrument was never under 10% RSD. During the sampling procedure the voltage was switched off and the high-voltage electrode remained in the electrolyte (D).

Sampling was interrupted by removing the capillary from the elevated sample vial and placing it back into the electrolyte beaker inside the wooden box. Following that, the operator closed the lid, depressing the safety switch in the process. A smooth and highly regular execution of the next step was crucial for the reproducibility of migration times. With the main switch on the left side of the power supply in "on" position, the operator had to adjust the separation voltage with the help of a dial on the right side of the front panel. The voltage adjustment required about 2 to 3 sec and had to be reproduced within $\sim$ 0.5 to 1 sec. This requirement stemmed from the fact that peaks could be as close together as 1–2 sec. Reproducibility of migration times was never better than $\sim$ 5%. In our experience, the home-made instrument was quite suitable for qualitative work, optimizing the selectivity of separations, for example, but inadequate for quantitative analyses. Interestingly, this was also the prevailing opinion of the majority of researchers in the time prior to the introduction of commercial CE systems. In some minds, the poor precision of results obtained with home-made instruments even led to serious doubts regarding the general suitability of capillary electrophoresis for use in quantitative analysis.

In all commercial instruments, the parts in contact with the high voltage are placed in an enclosure behind a door. The operator can access high-voltage parts only by interrupting the high-voltage circuit between the electrodes and the power supply. This is accomplished by a variety of means, usually involving a contact switch between the door leading to the inside compartment on one side and the body of the instrument on the other. Figure 3.69 offers a view of the inside compartment of a commercial capillary electropherograph. The vials filled with carrier electrolyte alternate with sample

INSTRUMENTATION FOR CAPILLARY ELECTROPHORESIS 181

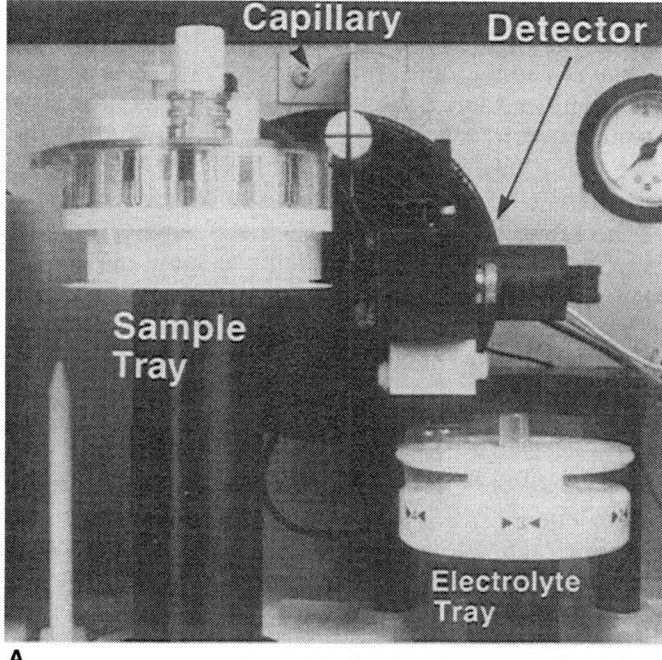

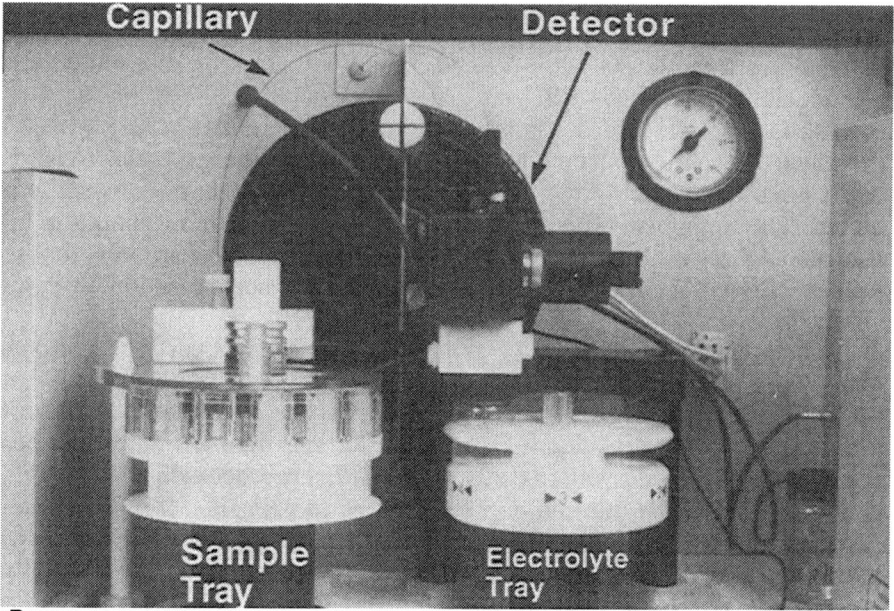

Figure 3.69 Inside compartment of a commercial capillary electrophoresis instrument. A: Sample tray is in an elevated position for hydrostatic sampling. B: Sample tray is in a position for separation.

vials in the carousel on the left side of Figure 3.69A. During a separation and in times between runs, the capillary is placed in one of the electrolyte vials. A change of capillary position between the vials is achieved by lifting the end of the capillary above the upper side of the tray, by turning the tray to a new position and by lowering the end of the capillary to the vial in that position. The arm lifting the capillary is discernible as a white quadrangle resting on the rear half of the sample carousel. All movements of the sample tray and of the lifting arm are fully automated. From the sample tray, the capillary reaches through the detector cell located in the upper right half of the compartment into an electrolyte container placed in another carousel. The carousel on the detector side of the instrument is operated manually. The thickly insulated high-voltage cable is connected on one side to the high-voltage electrode attached to the lifting arm and on the other to the high-voltage supply inside the rectangular box located close to the left wall of the compartment.

Hydrostatic sampling is initiated by lifting the sample carousel to the position shown in Figure 3.69A. The carousel is held in the elevated position for a time preprogrammed by the operator. At the end of sampling period, the carousel is brought down again, to its position shown in Figure 3.69B. Lifting and lowering of the carousel is carried out by a smoothly functioning mechanism driven by a programmable stepper motor and in consequence with a precision that cannot be matched by human operators. The typical precision of directly measured peak areas lies between 1–2% RSD. With normalization procedures eliminating the effects of sample conductivity, temperature fluctuations, etc. (see Section 3.4.5), it is possible to achieve reproducibility below 1% RSD.

With the sample carousel in the lower position, and the capillary back in the electrolyte, the instrument begins to ramp up the potential toward a value programmed for separation. Again, no action of human hands is required. The timing is software controlled with a precision measured in milliseconds. The resulting precision of directly measured migration times is usually below 1% RSD and can be improved by normalization to values below 0.1%.

Commercial instruments have changed the outlook of early practitioners of capillary electrophoresis. The introduction of well-designed and fully automated instruments has eliminated any reason for doubt about the quantitative ability of the technique. Good precision of migration times and peak areas, however, is derived not only from the microprocessor-controlled execution of the sample introduction and voltage adjustments. It relies on other contributions as well. One of the additional functions of modern CE instruments contributing to better precision is temperature control. Another emerging methodology enabling better precision of results is the programmed change of current and voltage during separations. Also important are, among other things, the proper choice of data acquisition rates and various normalization procedures. Good precision of results obtained by capillary electrophoresis is thus based on the overall performance of a well-designed system.

3.4.1 Effects of Temperature Fluctuations

In an illustrative experiment,[131] the home-made system discussed was modified to enable controlled changes of ambient temperature in the immediate vicinity of the wall of the 75 μm I.D. fused silica capillary.

Temperature changes inside the water jacket surrounding the capillary were imposed by changing the temperature of the thermostatted water bath. The water was circulated between the jacket and water bath by a single-piston chromatographic pump at a rate of 3 ml/min. The 5 mM chromate, pH 8, also containing an alkylammonium salt (Waters OFM BT Anion) for the reversal of electroosmotic flow, was used as a carrier electrolyte. A standard mixture containing seven inorganic anions identified in Figure 3.71 was introduced into the capillary as described in Figure 3.68. The sample introduction time was 30 sec. Indirect UV detection was carried out at 254 nm with a time constant of 0.1 sec.

In the following discussion, it is important to keep in mind that the temperatures indicated in Figure 3.71 represent only the ambient temperature. The actual temperature inside the capillary was not measured during the experiment.

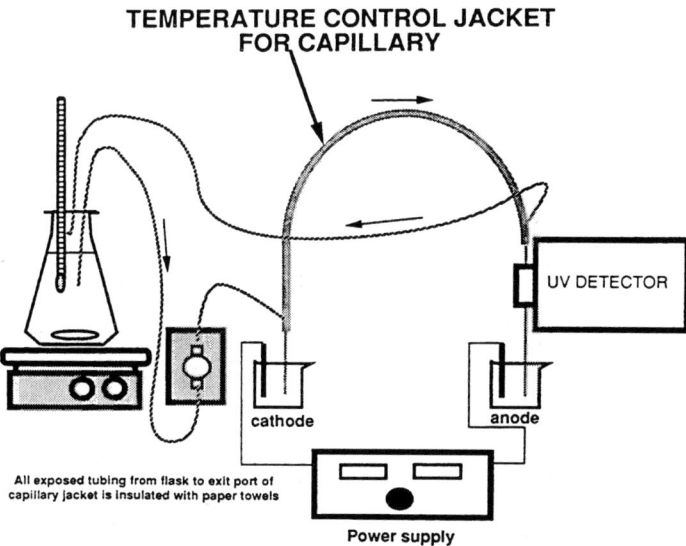

Figure 3.70 CE system from Figure 3.68 modified for experiments with controlled changes of temperature. Courtesy W.R. Jones, Waters Chromatography Division of Millipore.

184 CAPILLARY ELECTROPHORESIS OF SMALL MOLECULES AND IONS

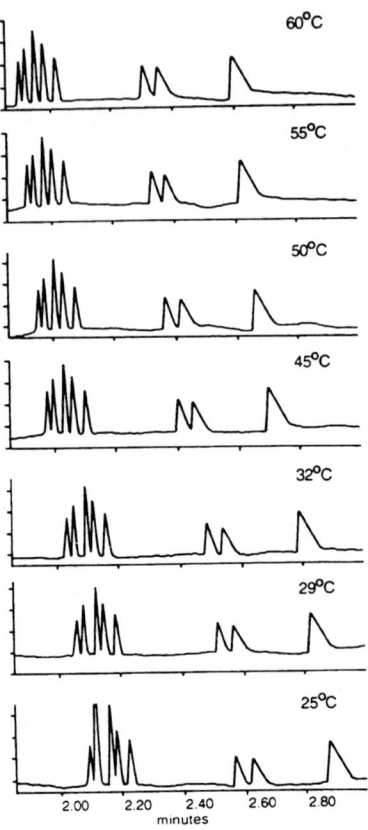

Figure 3.71 Effect of temperature on electropherograms of seven inorganic anions. Identical detection signal and migration time scales are used in all six recordings. Peak identities from left to right in the electropherograms: bromide, 1 ppm; chloride, 1 ppm (chloride contamination at 25°C); sulfate, 2 ppm; nitrite, 2 ppm; nitrate, 2 ppm; phosphate, 4 ppm; fluoride, 1 ppm; and carbonate, 2 ppm. Courtesy W.R. Jones, Waters Chromatography Division of Millipore.

Inspection of six electropherograms allows two conclusions: First, there is a considerable change in migration times, even with a single temperature step, approximately 2% between 25 and 29°C and 3% between 25 and 32°C. Second, the change in temperature appears to be influencing all migration times in a similar way. The sequence of migration remains unchanged over the entire temperature range, even though the quality of resolution of some peaks suffers in going from 25 to 60°C. Both observations are in accord with Equation (2.2), relating migration times to electrophoretic mobilities, and with Equation (2.6), expressing electrophoretic mobility as a function of viscosity. The primary effect of temperature is through the change of viscosity.

Viscosity variations affect all ions in the same way. Temperature-induced changes of viscosity are frequently approximated as 2% per degree. Interestingly, the observed shift of migration times is only ~ 0.5% per degree, allowing one to conclude that the actual electrolyte temperature is changing less steeply than the temperature of the water medium.

Temperature fluctuations between 25 and 32°C are frequently observed in laboratories without closely regulated air conditioning. A change of migration times by more than ~ 3% leads to misidentification of closely migrating peaks, and is thus not acceptable for routine use. There are at least two different ways of eliminating the effect of temperature on migration time. The first approach consists of an application of normalization procedures, and is described in Section 3.4.5. The second approach involves thermostatting the separation capillaries. Various forms of temperature control in commercial capillary electrophoresis systems are discussed next.

The electrical current required to drive the electromigration of ions in capillaries can cause a considerable increase in the temperature of carrier electrolytes. In Section 2.6, we discussed the reasons why adequate heat dissipation is only possible in capillaries with small internal diameters (100 μm or less). Even with such narrow capillaries, however, the temperature of a carrier electrolyte during a run is seldom identical to the ambient temperature. In an instrument relying only on natural heat convection between the capillary wall and stagnant air, the difference between the ambient temperature and the average temperature in the capillary during run can be as much as 45°C.[132] In our experiment with the capillary in a water jacket, the temperature difference must have been much smaller. Otherwise, boiling would have occurred during the three separations in the range between 50 and 60°C. Temperature shifts affect not only the precision of peak areas and migration times, but they can also degrade thermally labile compounds. Karger et al.[133] have described a heat-induced formation of a second form of myoglobin during a CE separation. The conversion was due to a change in the oxidation state of heme iron. The same report also mentioned thermally initiated conformational transitions resulting in asymmetric peaks and other atypical phenomena for α-lactalbumin.

The objective of temperature control in CE systems is thus not only to stabilize temperature fluctuations, but also to achieve the largest possible reduction of the differential between the carrier electrolyte and the ambient temperature. Temperature control can be accomplished by one of the three instrument designs illustrated in Figure 3.72.

A schematic diagram of temperature control utilizing rapid circulation of air inside an enclosed compartment, containing not only the capillary, but also the carrier electrolyte vessels and the detector cell, is shown in Figure 3.72A. This approach permits a large number of system elements to be included in the controlled environment. The sample vials are exposed to the same temperature as the carrier electrolyte outside the capillary. The operator can easily adapt capillaries from different sources for use with this in-

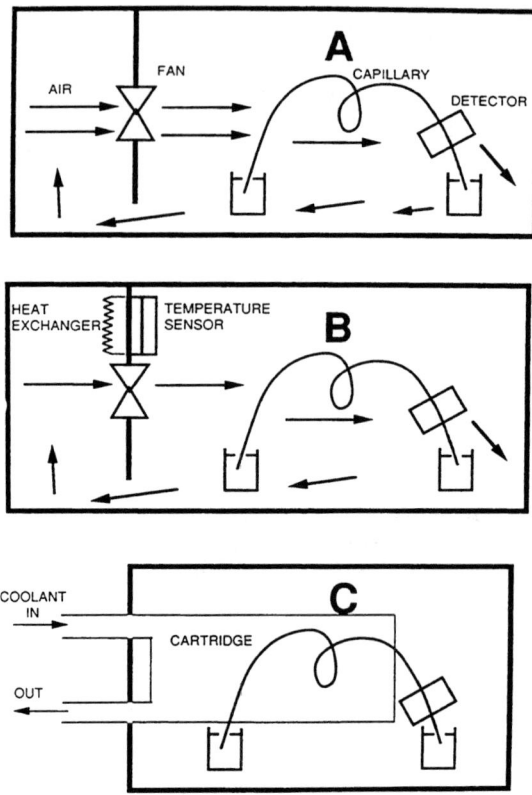

Figure 3.72 Three types of temperature control in commercial CE systems. A: Cooling by air circulation. B: Thermostatting of circulating air in thermally insulated enclosure. C: Portion of a capillary is immersed in liquid medium that is kept at a constant temperature.

strument. The design does not rely on the use of prefabricated cartridges encapsulating the capillaries. By minimizing air exchange with the surrounding environment, the instrument exhibits sufficient thermal capacity so that it does not follow rapid fluctuations of ambient temperature caused, for example, by drafts. It does, however, adjust slowly to gradual, lasting changes of ambient temperature. While the heat dissipation and temperature stabilization is fully sufficient for the majority of situations, in laboratory environments experiencing large temperature shifts, it will be necessary to utilize one of the normalization procedures described in Section 3.4.5. Figure 3.72B depicts a temperature control system relying on thermostatting of air, circulating in an enclosed and thermally insulated compartment.

In this type of instrument, the rate of air circulation is slightly lower than in the previous approach. A temperature sensor is installed inside the compartment. It provides a signal to the temperature module proportional to the temperature at a given time. The temperature control module evaluates a differential between the actual temperature value and a preprogrammed, constant operating temperature. The resulting differential is translated into a heating or cooling current for the thermal electric element. The thermal electric element is in conductive contact with the circulated air through a heat exchanger and is thus capable of heating or cooling it in response to the temperature fluctuations sensed by the temperature sensor. Thermostatting of air does not require the use of prefabricated cartridges and allows as great a freedom in the choice of capillaries as the design based on air cooling. Some manufacturers,[134] however, use special cartridge holders even in air thermostatted instruments for easier alignment of optics and do not include the sample tray in a temperature-controlled environment.

An instrument utilizing a liquid medium for the temperature control of the capillary is shown in Figure 3.72C. The capillary is an integral part of a prefabricated cartridge filled with a suitable liquid (chlorofluorohydrocarbons). The liquid medium is recirculated at a steady rate through a liquid thermostat maintaining a preset temperature. Since many liquids have approximately ten times better heat exchange coefficients than gasses (see Table 2.6), this mode of temperature control could in principle lead to the lowest temperature rise of carrier electrolytes during separation. A disadvantage, however, lies in the difficulty to include the samples and electrolyte containers in the temperature-controlled environment. This approach may also not be suitable for frequent changes between capillaries from different sources. On the other hand, integration of capillaries in cartridges facilitates the alignment of the optical path for optical detection modes.

Another approach to efficient temperature control[135] consists of placing capillaries in close contact with thermostatted metallic blocks. The metallic blocks have detection openings and optical elements with which a capillary is permanently aligned. As in the previous method, integration of optical elements in the temperature control module makes the proper alignment of the detector cell easier. On the other hand, the relative complexity of the modules probably discourages a frequent change of capillaries.

The efficiency of heat dissipation and attainable precision of migration times of CE systems utilizing three different temperature control techniques shown in Figure 3.72 was evaluated by a pharmaceutical laboratory.[136] A test mixture consisting of codeine, caffeine, and butalbital was separated in a 75 μm I.D. fused silica capillary. The components are named in the same sequence as the corresponding peaks in Figure 3.73. The carrier electrolyte was a 50 mM sodium phosphate solution at pH 7.0. The efficiency of heat dissipation was evaluated by monitoring the shifts of migration times for caffeine at several different field strengths. For easier interpretation the mi-

188 CAPILLARY ELECTROPHORESIS OF SMALL MOLECULES AND IONS

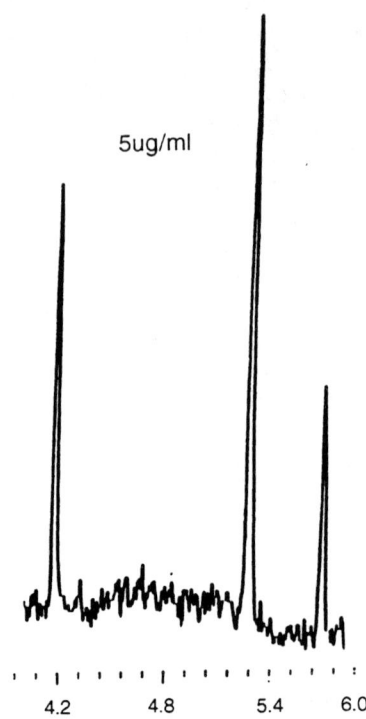

Figure 3.73 Electropherogram of a pharmaceutical text mixture generated by a system utilizing temperature control by air cooling (see Figure 3.72A). The injected concentration (5 μg/ml) is near the detection limit for each of the components. The electromigrative sample introduction was carried out for 10 sec at 50 V/cm. The separation voltage was 300 V/cm. See text for additional explanation. Reproduced with permission.[136]

gration time was converted to electrophoretic mobility according to Equation (2.2). Results for the three systems in question are presented along with those from a home-made system relying only on natural heat convection in air.

The following conclusions can be drawn from the results shown in Figure 3.74: First, none of the evaluated systems could prevent a temperature increase over ambient. Second, while the temperature increase in the home-made system was excessive, the heat dissipation properties of the three control systems were comparable. The larger temperature increase in the air thermostatted instrument, relative to the other two systems, is explained by the fact that a 27 cm long section of the capillary remained outside the stream of thermostatted air.

Also included in the same report[136] was an evaluation of the reproducibility of migration times and peak areas obtained with three temperature-con-

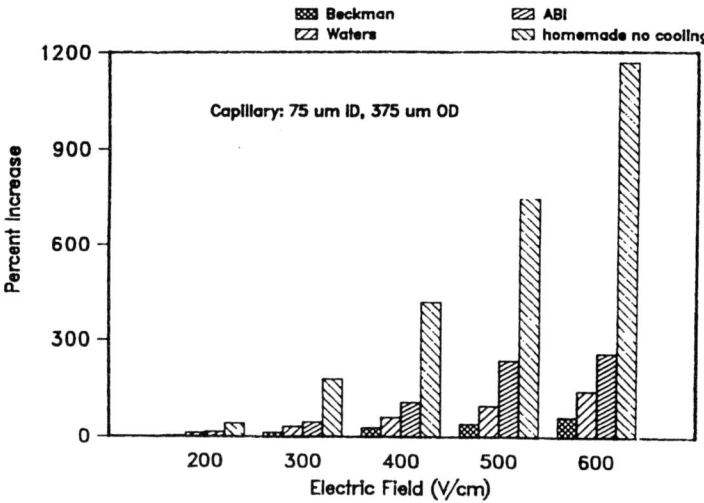

Figure 3.74 Percent increase in electrophoretic mobility of caffeine as a function of applied electric field for three different types of temperature control. Beckman: liquid medium temperature control; Waters: temperature control through a rapid circulation of air; ABI: thermostatting of air. Adapted from Reference.[136]

trolled CE systems. The reproducibility of directly measured (not normalized) migration times after sample introduction by electromigration was in the range of 0.09 to 1.5% RSD for caffeine. The precision of directly reported (not normalized) peak areas for caffeine fluctuated between 0.8 and 6% RSD, depending on instrument and concentration (0.005 to 1.0 mg/ml). Two of the three instruments gave a slightly higher imprecision ($\sim$ 9% RSD) for one of the concentrations (0.005 and 0.01 mg/ml, respectively). The authors of the evaluation stated that in their opinion, the pressure differential sample introduction methods were too different for the three instruments in question to merit comparison. Even the precision data for the electromigration, quoted previously, are more likely to be influenced by instrument features other than the efficiency of temperature control. However, from the discussed data, it is safe to conclude that the temperature control features included in the three evaluated commercial instruments and illustrated in Figure 3.72 can bring about a genuine improvement in overall analytical conditions in comparison with most home-made instruments.

3.4.2 Effect of Different Sample Conductivity

The problem of quantitation following electromigrative introduction of samples with different conductivities is recognized and discussed in the literature.[5,8] A similar problem as with electromigration, however, is also observed for certain samples introduced by a pressure differential (i.e., hydrostatic, pressure, and vacuum). For an illustration, let us consider the five samples summarized in Table 3.5.

Table 3.5 contains five samples with identical chloride and sulfate levels in the presence of fluoride. The fluoride concentration increases from sample to sample up to a ratio of 1:75 (ppm Cl to ppm F). If we analyze the five samples with the help of capillary electrophoresis, we obtain the series of electropherograms in Figure 3.75. The shifts of migration times in Figure 3.75 are significant and would make reliable peak recognition impossible. On the other hand, the migration time shifts appear to be regular and dependent on the concentration of fluoride. This regularity suggests the conclusion that the observed changes may be caused by changes in sample conductivity, since as the fluoride concentration increases so does the conductivity of the sample. Sample conductivities (G) are listed along with the concentrations in Table 3.5.

We should briefly turn our attention to the fact that the highest sample conductivity in Table 3.5 actually exceeds the conductivity of the carrier electrolyte, which is at 1131 μS/cm.[137] This observation helps to explain some additional differences between the electropherogram corresponding to the sample with the highest conductivity and the electropherograms of the lower-conductive samples in Figure 3.75. Additional features are, first, a new peak preceding the chloride and, second, an increase in the size of the peak for sulfate. An explanation of these effects lies in electrostacking.[138] However, unlike the effect described in Section 3.1.1, the electrostacking responsible for the artifacts observed with sample of highest conductivity occurs from the carrier electrolyte and not in the sample zone. Because of the higher electric field (lower conductivity) in the carrier electrolyte, in comparison with the sample, a small amount of sulfate (sulfuric acid is used for pH adjustment) and bromide (bromide is the anionic component of Waters OFM BT) is preconcentrated at the boundary between the sample and the electrolyte inside the carrier electrolyte container at the high-voltage end of the capillary. Once inside the sample zone, additional amounts of

Table 3.5. FIVE DIFFERENT SAMPLES FOR EVALUATION OF INFLUENCE OF SAMPLE CONDUCTIVITY ON MIGRATION TIMES. ADAPTED FROM REFERENCE 137

Sample	Chloride [ppm]	Sulfate [ppm]	Fluoride [ppm]	G [μS/cm]
1	4	4	4	55
2	4	4	30	203
3	4	4	60	356
4	4	4	100	574
5	4	4	300	1519

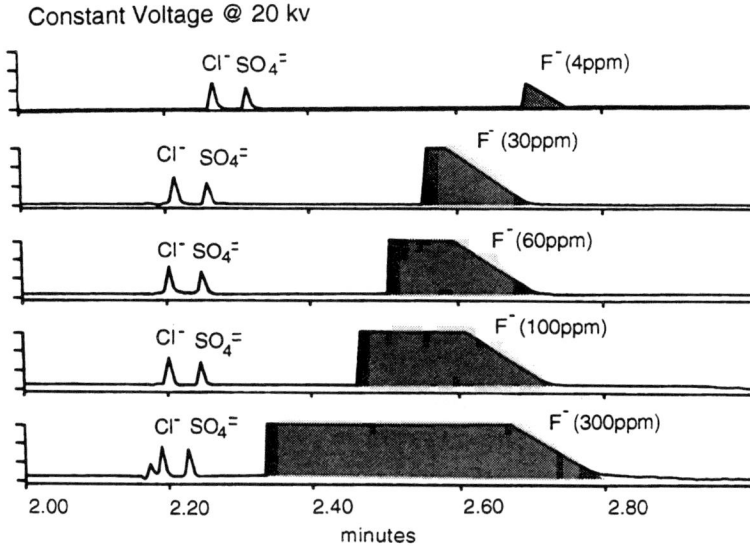

Figure 3.75 Constant-potential mode. Shifts of migration times due to changes in the concentration of a major component. Carrier electrolyte: 5 mM sodium chromate, 0.5 mM OFM BT (Waters electroosmotic flow modifier), pH 8.0. Sampling: hydrostatic for 30 sec at 9.8 cm. Capillary: 75 μm × 52 × 60 cm. Separation voltage: −20 kV. Detection: indirect at 254 nm, time constant 0.1 sec, 20 Hz data acquisition. Concentrations are as indicated in Table 3.5. Adapted with permission from Reference 137.

bromide and chloride are subjected to the same separation as the original sample components. This phenomenon (reverse electrostacking) represents yet another reason (see Section 3.1) why the ionic strengths of samples should always be adjusted at lower levels than those of carrier electrolytes.

Let us turn back to our original problem of shifting migration times and the preliminary explanation that these shifts could be caused by changes in sample conductivity. According to Fuchs,[139] the capillary used for electrophoretic separations shown in Figure 3.75 can be treated as a serial arrangement of two electrical resistances: R_s, the resistance of the sample zone, and R_{el}, the resistance of the carrier electrolyte column inside the capillary. Once migration begins, several things happen. The analyte ions leave the sample zone and are replaced with background electrolyte ions. This means that the resistance of the sample zone changes to a certain extent because of differences in the conductivity of background and analyte ions. Also, diffusion will blur the edges of the sample zone. Still, the sample zone will maintain its identity as a region of conductivity that is different from the conductivity of the rest of the capillary. This region will migrate with the electroosmotic flow rate, behind the analytes of interest. As a result, the figures calculated in the following and listed in Tables 3.6 and 3.7 are illustrative rather than exact values, except at the start of the separation. At the start of a separation, the total resistance of capillary R_{tot} can thus be calculated from the resistance of two main zones inside the capillary:

$$R_{tot} = R_s + R_{el} \tag{3.27}$$

The length of a sample zone required in further calculations of R_s and R_{el} can be obtained from Equation (3.2) for sampling rates (ΔS), using $\Delta h = 9.8$ cm and $L_t = 60$ cm. The resulting sampling rate of 2.82×10^{-2} cm/sec is converted to the length of the sampling zone by $L_s = \Delta S \times$ sampling time $= 2.82 \times 10^{-2} \times 30 = 0.842$ cm. The two serial resistances can be calculated from $L_s = 0.842$, $L_t = 60-0.842 = 59.158$ cm and from the specific conductances G [μS/cm] listed in Table 3.5.

$$R_s = L_s/(G\pi r^2) = 0.842/(G\pi\, 0.00375^2) \tag{3.28a}$$

$$R_{el} = L_{el}/(G\pi r^2) = 59.158/(G\pi\, 0.00375^2) \tag{3.28b}$$

Now we can calculate all values of resistances in the CE system, as shown in Table 3.6. Using Ohm's Law, we can calculate the voltage drop for each of the two zones from the total separation voltage of 20,000 V. The total voltage P_{tot} is divided into two partial voltages (P_s and P_{el}) in the same ratio as that of the partial resistance R_s or R_{el} to the total resistance R_{tot}. To consider the influence of different conductivities of zones on the electrophoretic migration rates, it is useful to compare the respective field strengths ($E_s = P_s/L_s$ and $E_{el} = P_{el}/L_{el}$) in Table 3.7.

As seen in Table 3.7 and considering Equation (2.2), the migration time shifts in Figure 3.73 are actually a result of two contradicting trends. The field in a sample zone decreases with increasing conductivity of sample. This effect alone would lead to longer migration times. However, the sample zone is much shorter than the section of capillary that is filled with carrier electrolyte, and the direction of the change of field strength inside the electrolyte is thus decisive. The field strength inside the carrier electrolyte increases in going from lower to higher conductive samples, causing higher mobilities

Table 3.6. RESISTANCES OF ZONES IN THE CAPILLARY DURING THE SEPARATIONS SHOWN IN FIGURE 3.75. REPRODUCED WITH PERMISSION[139]

Sample	R_s [$\Omega \times 10^7$]	R_{el} [$\Omega \times 10^9$]	R_{tot} [$\Omega \times 10^9$]
1	34.7	1.18	1.53
2	9.39	1.18	1.27
3	5.37	1.18	1.23
4	1.33	1.18	1.21
5	1.26	1.18	1.195

INSTRUMENTATION FOR CAPILLARY ELECTROPHORESIS

Table 3.7. ELECTRIC FIELD DISTRIBUTION IN THE CAPILLARY DURING THE SEPARATIONS SHOWN IN FIGURE 3.75. REPRODUCED WITH PERMISSION[139]

Sample	E_s [V/cm]	E_{el} [V/cm]
1	5390	260
2	1756	313
3	1037	323
4	654	329
5	250	335

and shorter run times. Contrary to initial expectations, running the five samples from Table 3.5 in a constant-current mode instead of the constant-potential mode only changes the direction of the problem. Similar calculations as for the constant-potential mode indicate an unchanging field strength in the electrolyte and a decreasing field in the sample zone for the constant-current mode.[139] The migration times are thus getting longer in going from the low-conductive sample 1 to more conductive samples 2–5. This direction of change was confirmed experimentally[137] (see Figure 3.76).

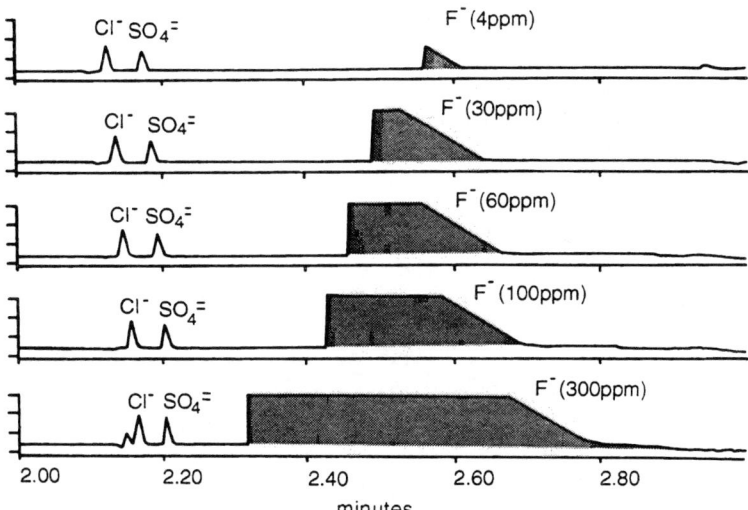

Figure 3.76 Constant-current mode. Shifts of migration times due to changes in the concentration of a major component. Carrier electrolyte: 5 mM sodium chromate, 0.5 mM OFM BT (Waters electroosmotic flow modifier), pH 8.0. Sampling: hydrostatic for 30 sec at 9.8 cm. Capillary: 75 μm × 52 × 60 cm. Separation current: 20 μA. Detection: indirect at 254 nm, time constant 0.1 sec, 20 Hz data acquisition. Concentrations are as indicated in Table 3.5. Adapted with permission from Reference 137.

Note the peak preceding chloride and increased peak height of sulfate in the 300 ppm F electropherogram. Reverse electrostacking, which is responsible for these artifacts, was explained in connection with Figure 3.75.

As discussed in the following section, changing migration times also mean, in the majority of cases, changing peak areas. For this reason, it is important to use the proper normalization techniques described in Section 3.4.5 for reliable quantitation of samples with fluctuating levels of conductivity.

3.4.3 Effect of Migration Rates on Peak Areas

In the preceding two sections, we discussed how temperature and sample conductivity can influence migration times in capillary electrophoresis. In this section, we analyze the influence of changing migration times on peak areas. For the sake of simplicity, we assume that the electrophoretic zone has a constant width and constant concentration of analyte. This is a reasonable assumption since zone spreading by longitudinal diffusion is only minimal.

Let us now evaluate the magnitude of change of peak areas for the same amount of analyte, recorded at different migration rates. In the majority of cases, these shifts in migration times occur by four different mechanisms. First, by changing electric field strengths (e.g., different separation voltage, different sample conductivity, different capillary diameter, different temperature, etc.); second, by viscosity changes (e.g., temperature, organic solvent composition); third, by changes in electroosmotic flow (e.g., pH, viscosity, changing number of charges on capillary walls); fourth, from changing electrophoretic mobilities of analytes due to shifts of dissociation equilibria, ion pairing, wall interactions, etc.

The expected change in response for observing an object moving at different speeds can be illustrated by the simple experiment in Figure 3.77.

In Figure 3.77, a narrow object is moved behind a wall at several different speeds between 0.1 and 10 cm/sec. The object can be observed through a slit in the wall that is narrower than the observed object. The observer's eye, the slit in the wall, and the moving object are oriented in the same plane. If any part of the object is touched by the observer's line of vision, the source of light positioned farther away from the wall is obscured. The tiny speck of light in the middle of the wall turns dark.

Let us assume that an observer is equipped with an electronic timer and is able to record the intervals of darkness in the wall slit. The length of such intervals can be calculated from the length of the object and the speed with which it is moving; 10 to 0.1 sec for 0.1 to 10 cm/sec and a 1 cm long object. If we now replace the observer's eye by a photodiode, an electronic signal processing module and a chart recorder running at a constant speed (10 cm/min or 0.16 cm/sec) and at a constant sensitivity (light in the window, pen not touching paper; window dark, pen on the paper), we can record lines of

INSTRUMENTATION FOR CAPILLARY ELECTROPHORESIS 195

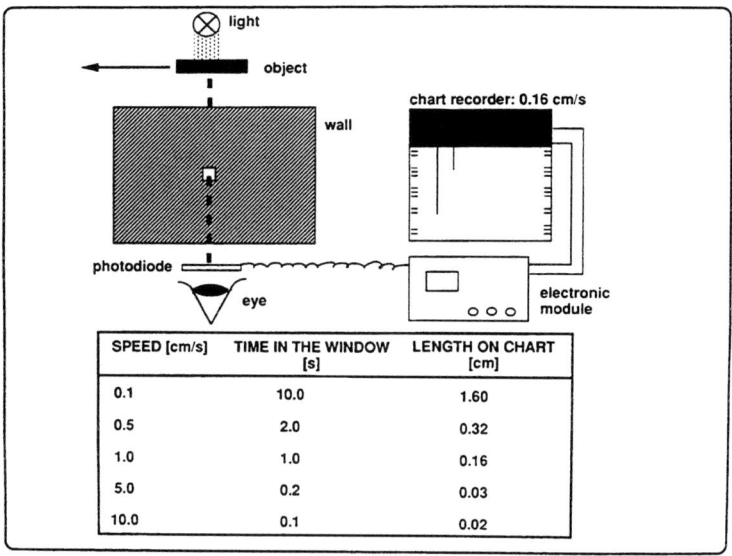

Figure 3.77 Moving object viewed through a narrow window.

different length dependent on different velocities of the object. Notice how strongly the recording of the object varies and how different it is from the actual length of the object under observation (1.6 to 0.016 cm vs. 1.0 cm). The similarity between the simple experiment in Figure 3.77 and the time dimension in electropherograms is now obvious. The faster a zone of analyte migrates past a detector, the narrower the peak and vice versa. Since we have set out to discuss the influence of migration rates on recorded peak areas, and not only on peak widths, we have to add an additional dimension to our experiment.

In Figure 3.78, we have a translucent capillary filled by a carrier electrolyte that absorbs only small portions of light passing through it on its way to the photodiode. The photodiode is connected to an electronic module transposing the full light intensity at the photodiode and the resulting current produced as a signal of 0.0 mV. Whenever the light passage between the light source and the photodiode is interrupted, for example, by the passage of an opaque zone, the photodiode stops producing a current, and as a consequence the electronic module generates a signal of 10 mV. The changes of signal are recorded by a chart recorder running at a constant chart speed (10 cm/min) and with a constant sensitivity of 10 mV per 10 cm.

Migration rates of the opaque zone in Figure 3.78 are made similar to migration times of the electropherogram in Figure 2.10. The first peaks in Figure 2.10 appear around 1.5 min. The eighteenth peak migrates at ~ 2 min. Peaks 27 and 28 are detected at ~ 2.5 min. The length of the opaque zone is

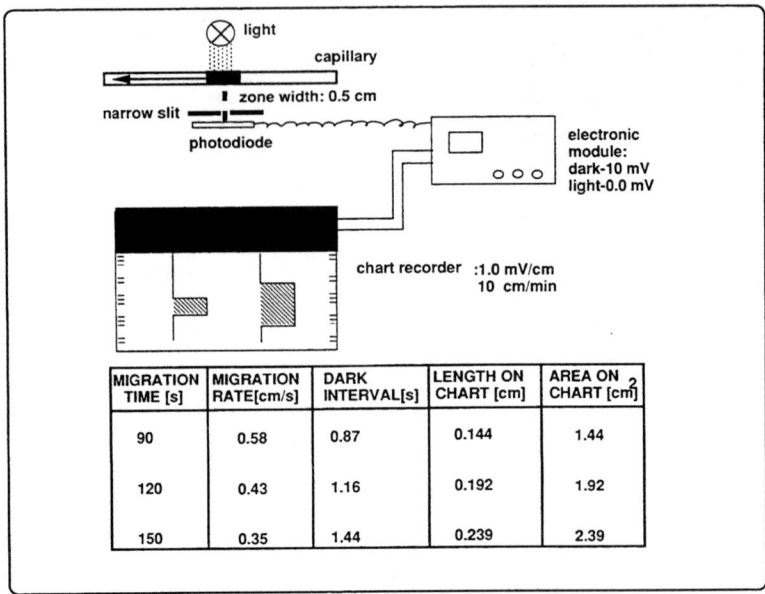

Figure 3.78 Opaque zone moving past an illuminated photodiode.

derived from the base width of ~ 1.0 sec at a migration rate of 0.5 cm/sec. These parameters are matched closely by the first six peaks in Figure 2.10. Conditions of the hypothetical experiment under consideration are thus much closer to reality than in the preceding experiment.

As the migration time changes, the opaque zone of analyte in Figure 3.78 exhibits a different time of residence between the light source and photodiode similarly as the object in the linear experiment in Figure 3.77. However, the recording is now two dimensional, and different lengths of the zone on the chart paper are now also recorded as different areas of a rectangular signal pulse. A detector signal of constant amplitude is recorded over varying lengths of chart paper.

Obviously, real peaks in capillary electrophoresis can be approximated more closely by Gaussian curves than by rectangular profiles. A Gaussian distribution curve shows the effects of longitudinal diffusion. Analyte zones are broadened not only in the migrational direction, but in the opposite direction as well (Section 2.5).

Consequently, the problem of recording a constant signal over different lengths of time changes into recording the same concentration distribution over different lengths of time. A zone of analyte of unchanging longitudinal distribution of concentrations produces the same values of maximal or minimal deflections of the detector signal, independently of how fast it moves. What changes, as in the previous example, is only the zone width. In terms of parameters characterizing the Gaussian curves, we can speak of a constant distribution height, h, and changing variances, σ.

For the magnitude of noise between 4 and 5% of total peak height, the base width of a peak corresponds to $\sim 5\sigma$. For our peaks in Figure 2.10, which are 1 sec wide and migrate at 0.5 cm/sec, the peak areas can be calculated as the term A in Figure 3.79. The maximum width on the chart, representing 5σ, is divided by five to obtain the value of σ for the calculation of A. The sensitivity of chart recorder is, as in the previous case, 10 cm per full deflection (h). Again, we can observe that areas of concentration pulses, this time in the shape of a Gausian distribution, change with migration rates. Luckily, the change of areas is strictly linear. The ratio of two peak areas for an identical zone is the same as the ratio of corresponding migration times or migration rates. This is of great importance for the elimination of imprecision by mathematical normalization procedures (Section 3.4.5).

At first glimpse, changes of peak areas with migration times seem to make universal calibration in capillary electrophoresis with indirect UV detection difficult or even impossible. As described in more detail in Section 3.3.2, the universal calibration in indirect detection is made possible by a replacement of equimolar amounts of UV background-providing coion in the carrier electrolyte, by an equimolar amount of analyte.

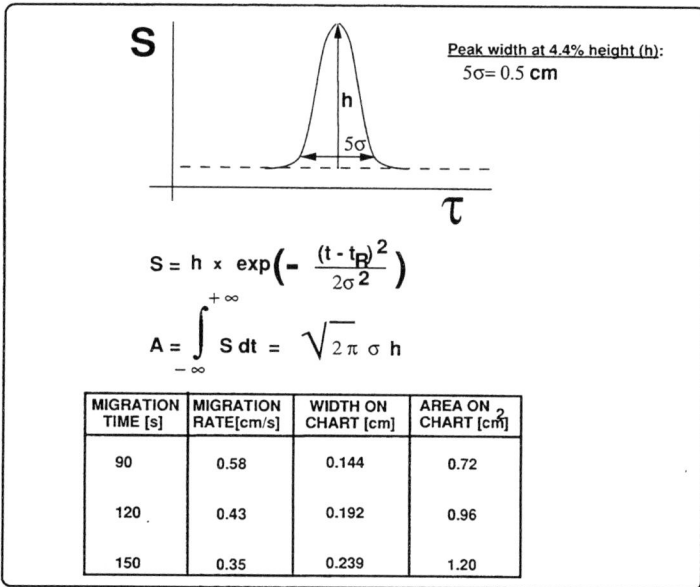

Figure 3.79 Different areas under a Gaussian distribution curve characterized by a constant height and changing values of variance.

In column chromatography, for which the universal calibration was first developed,[62] peaks move through a detector at a constant speed. Consequently, a zone of identical depletion of background ions yields identical peak areas, independently of the retention time. It thus becomes possible to calibrate an entire separation with just one standard of known concentration. Hence the name universal calibration. The concentrations of all other components can then be calculated as equimolar to the decrease in background ion concentration measured as a peak area at any retention time. In capillary electrophoresis, on the other hand, two zones having an identical distribution of the depleted background ion, but reaching the detector at different migration times, will yield different peak areas because of their different migration rates. In reality this problem is much less severe. We have already observed in our simple experiments that a change of peak areas is a linear function of migration times. As discussed in Section 5.3, in some cases, the trend in recoveries of electromigrative sample introduction compensates for changes in peak areas; less mobile ions have longer migration times, resulting in larger peak areas, but less mobile ions also give lower recoveries in electromigrative sample introduction, a phenomenon yielding smaller peak areas. Under certain conditions, both dependencies cancel each other out.

3.4.4 Effect of Data Acquisition Rates

Chart recorders, if used at an adequate chart speed for capillary electrophoresis, 10–20 cm/min, produce excessive amounts of paper. Another disadvantage shared with simple integrators without a data storage capacity is the inability to display only a selected range of the separation or to repeat integration of the same separation with several different sets of integration parameters. For these and other reasons, most operators prefer to use computers for acquisition and processing of CE data.

Data acquisition by computers is fundamentally different from the analog representation of an electropherogram obtained by a chart recorder. Analog-to-digital (A/D) conversion is required for computers. The original analog format is lost, and the subsequent processing is carried out with a digital version of an electropherogram. The operator has to have an understanding of the performance characteristics of an A/D converter in his system in order to be able to avoid the possibility of distortion of recordings by digitalization.

There are two basic types of A/D converters.[140] The first type, based on voltage–to–frequency conversion (VFC), is used mainly in single-channel integrators. The strength of VFC-equipped integrators is its very high resolution of detector signal. The VFC instruments were specifically designed to measure low millivolt outputs from LC and GC detectors. A disadvantage of VFC instruments is that they are relatively slow, having been designed before the advent of the highly efficient techniques such as capillary GC and capillary electrophoresis. The output frequency cannot be generally controlled by the operator and is determined by the VFC instruments as a multiple of the frequency of the analog input signal from the detector.

The more modern, second type, the dual slope, integrating A/D converters are capable of sampling at a much higher rate than the VFC instruments. The frequency of the output is usually adjustable by the operator.

The general performance of A/D converters has to be judged by their resolution in two dimensions. Resolution along the ordinate, or resolution of the analog detector signal indicating the magnitude of the concentration pulse in the detector cell, is specified by the number of bits. A 12-bit A/D converter is capable of resolving a detector input of 1 V in 4096 (2^{12}) data points. This resolution is of course adequate for an analog signal between 100 and 1000 mV, but may not be sufficient for smaller detector signals. A peak height of 1 mV would be described only by four values along the detector signal coordinate, if the total number of points of a 12-bit converter were assigned to 1 V input. Based on his knowledge of the number of bits in his A/D converter, the operator has to change the sensitivity setting on his detector to assign a larger voltage to the same magnitude of concentration change, if a small signal is to be resolved properly. The 1 V outputs in some detectors are decoupled from sensitivity settings on the front panel, however, making such adjustments impossible. An adequate number of bits (e.g., 16 bits or better) has to be available for the latter type of detectors for resolution of the smallest signals.

The resolution along the time coordinate is determined by the sampling frequency of an A/D converter. In order to preserve the integrity of an analog signal, any operator is naturally inclined to choose the highest possible frequency, which is 50 or 60 points per second in most instruments. This approach creates a big demand on data storage space in a laboratory computer. In order to avoid a frequent and laborious downloading of acquired data, for example, from a hard disk to diskettes, to make space for new measurements, most workers will resort to using lower sampling rates. Based on the formula derived by Rossi,[141] the appropriate sampling rates can be estimated for anticipated widths of peaks.

$$\%E = 33.3/(n^2 ht) \tag{3.29}$$

where $\%E$ is the estimated percent error in the measurement of a Gaussian peak area, n is the number of data points describing the peak along the time axis, h is the height of a Gaussian curve (see Figure 3.79), and t is the time interval of the peak.

For a Gaussian peak of unit height and 1 sec wide, we can for example try first a sampling rate of 5 points/sec; the calculated error generated by sampling is 1.33%. If we increase the sampling rate to 10 points/sec, the error in the measured area, caused by sampling, is going to be only 0.33%.

In a strict sense, Equation (3.29) is admissible only for use with Gaussian peaks. Only under optimum conditions can peaks in capillary electrophoresis be expected to be Gaussian. A useful model of the so-called exponentially modified Gaussian function (EMG) was developed for mathematical modeling of distorted chromatographic peaks. In his report, Rossi[141] describes the use of software for numerical integration of EMG curves approximated by a given number of points to give percent error of the estimate for the true peak area. It is very likely that with the continuing improvements in computer hardware, the consideration regarding resolution will diminish in importance. With the existing hardware, however, the misadjustment of sampling rates or wrong choice of A/D converters may still lead to unacceptable levels of error.

3.4.5 Normalization Procedures in Capillary Electrophoresis

In the relatively short time since the introduction of commercial capillary electropherographs, numerous workers have found that the quantitation in CE can be improved by a variety of normalization procedures. Correspondingly, a number of reports has been published recommending various types of normalization of migration times and peak areas in capillary electrophoresis. A summary of the procedures published to date is given in Table 3.8.

Table 3.8. VARIOUS APPROACHES TO NORMALIZATION OF CE DATA DESCRIBED IN RECENT LITERATURE

Reference	Summary
Zare et al., *J. Chromatogr.* **480**, 95 (1989)	Correction of peak widths for different rates of migration, "temporal" peak widths converted into "spatial" peak widths
Zare et al., *Anal. Chem.* **61**, 766 (1989)	Peak areas for a constant amount of sample change with migration time
Grossman et al., *ABL*, February 1990	Normalization of migration time by calculation of electrophoretic mobilities. Normalization of peak areas by division with migration times
Karger et al., *Anal. Chem.* **63**, 1346 (1991)	Correction of mobilities by a viscosity coefficient
Guiochon et al., *Anal. Chem.* **63**, 1154 (1991)	Concentrations of analytes are calculated by a method that normalizes for different mobilities of ions
Everaerts et al., *J. Chromatogr.* **537**, 407 (1991)	Normalization procedure for obtaining "absolute" mobilities from migration times
Everaerts et al., *J. Chromatogr.* **549**, 345 (1991)	Variation of peak areas with changing migration times is not linear. New way of correcting for different migration rates is proposed
Yeung et al., *Anal. Chem.* **63**, 2842 (1991)	Proposes and evaluates the use of migration indices based on Walden Rule

As explained with the help of practical examples in preceding chapters, fluctuating experimental parameters can cause variations in both migration times and peak areas. Within a narrow range of migration times, we can expect peak areas to be a linear function of migration rates (see Section 3.4.3). In the majority of cases, peak areas can thus be normalized, simply by division with corresponding times of migration, as suggested in 1990 by Grossman and co-workers.[142] For substantial migration time fluctuations, the more complex, theoretically sound correction developed by Ackermans, Everaerts, and Beckers[143] has to be employed. Literature references discussing normalization of migration times have evolved over time from preliminary suggestions to the first attempts to provide a comprehensive solution to the problem of migration time fluctuations. Initially, the calculations of effective mobilities[142,144] [μ_{ion}, Eq. (2.3)] were suggested as a best approach to the normalization of migration data. The effective mobilities were shown to improve precision of migration times (from ca 0.6 to ca. 0.1% RSD) for ten consecutive counterelectroosmotic CE runs of a peptide standard.[142]

However, as pointed out by Lee and Yeung,[145] even the effective electrophoretic mobility is affected by changes in viscosity of a carrier electrolyte [Equation (2.6)]. Moreover, such changes of viscosity due to temperature changes (2% viscosity change per °C) are the most frequent cause of imprecision of migration times. In the same article, Lee and Yeung also show that a calculation of ratio of mobilities or migration times of two peaks from the same separation is an effective way to eliminate the influence of viscosity on the migration data. Let us, for example, create a ratio of effective mobilities, for an analyte and reference peak of monovalent species ($z_i = 1$) from a same electropherogram. If we use Equation (2.6), we obtain:

$$\mu_{ion(1)}/\mu_{ion(reference)} = r_{ion(reference)}/r_{ion(1)} \tag{3.30}$$

If, for the sake of simplicity, we also assume that the term μ_{EOF} in Equation (2.3) is negligible, we can extend the validity of Equation (3.30) to observed migration times.

$$t_{ion(reference)}/t_{ion(1)} = \mu_{ion(1)}/\mu_{ion(reference)} = r_{ion(reference)}/r_{ion(1)} \tag{3.31}$$

where the migration times for analyte and reference peaks correspond to values obtained at zero electroosmotic flow. Neither Equation (3.31) nor Equation (3.30) includes any viscosity-dependent term. Even for those cases, where the magnitude of electroosmotic flow cannot be considered negligible, numerical calculations show that a normalization of migration times, by using a reference peak, effectively eliminates any influence of changed viscosity. Consider, for example, the temperature influence study from Figure 3.71. The uncorrected migration times differ by approximately 10% across the range of evaluated temperatures.

The separation system employed to generate data in Table 3.9 exhibits a significant contribution of electroosmotic flow to the apparent mobility of the analyte anions. Compare, for example, the results of curve fitting in

Section 2.2, Figure 2.11. Although the assumption used to derive Equation (3.31) does not hold for data in Table 3.9, the calculation of retention time ratios proves to be an effective method of elimination of temperature-induced variance even in that case. The bromide peak was used as a reference, and the data in Table 3.10 are ratios of respective migration times and the migration time of bromide at the same temperature.

Note that Table 3.10 also contains the values of percent relative standard deviation for each of the anions. The normalized migration data fluctuate by less than 0.5% for the investigated ambient temperatures between 25 and 60°C. In Section 5.3, we discuss another successful application of migration time ratios. In that case, the migration time ratio improves fluctuations of migration times of trace anions with varying excess of boric acid. Normalization by migration time ratios can thus be expected to improve the precision of results for a great many situations. An additional attraction of that approach is that it can be performed automatically by many integrators as a part of their internal standard routine. It is important to keep in mind, however, that the migration time ratios may not be able to eliminate the influence of temperature and viscosity under all circumstances. This type of approximation may fail, for example, in counterelectroosmotic separations, where the changes of viscosity influence the electroosmotic flow and electrophoretic mobility in mutually opposing ways. In an attempt to provide a more universal method of normalization than the migration time ratios, Lee and Yeung[145] developed a migration index (MI) that takes into account mutually independent changes of electrophoretic and electroosmotic mobilities.

$$MI = L_t/1 \, _0\!\int^{t_m} i \, dt = k\eta/\varepsilon[\zeta_c + (2\zeta_a/3) f(\kappa\alpha)] \quad (3.32)$$

where L_t is the total length of capillary, t_m the migration time, i the current, k the specific conductivity of the electrolyte, η the viscosity of the electrolyte, and ε the dielectric constant of the carrier electrolyte. ζ_c and ζ_a are the zeta potentials of the capillary wall and analyte particles, respectively. The mobility index is derived from an alternative form of Equation (2.6):

$$\mu_{ion} = (2\varepsilon\zeta_a/3\eta) f(\kappa\alpha)] \quad (3.33)$$

Table 3.9. MIGRATION TIMES [MIN] OF ANIONS AND TEMPERATURES FROM FIGURE 3.71. ADAPTED FROM REFERENCE 131

Temp. [°C]	Br^-	Cl^-	SO_4^{2-}	NO_2^-	NO_3^-	F^-	HPO_4^{2-}	HCO_3^-
60	1.984	1.913	1.944	1.974	2.015	2.305	2.353	2.595
55	1.918	1.938	1.970	1.998	2.040	2.330	2.373	2.618
50	1.951	1.971	2.003	2.031	2.073	2.368	2.418	2.660
45	1.976	1.997	2.033	2.058	2.100	2.403	2.448	2.700
32	2.031	2.053	2.094	2.117	2.158	2.485	2.538	2.793
29	2.057	2.080	2.121	2.143	2.184	2.515	2.568	2.823
25	2.096	2.116	2.162	2.184	2.224	2.568	2.623	2.880

Table 3.10. NORMALIZATION BY MIGRATION TIME RATIOS. THE DATA CORRESPOND TO THE SERIES OF ELECTROPHEROGRAMS FROM FIGURE 3.71. ADAPTED FROM REFERENCE 131.

Temp. [°C]	Br^-	Cl^-	SO_4^{2-}	NO_2^-	NO_3^-	F^-	HPO_4^{2-}	HCO_3^-
60	1.000	1.010	1.026	1.042	1.064	1.217	1.242	1.370
55	1.000	1.010	1.027	1.042	1.064	1.215	1.237	1.365
50	1.000	1.010	1.027	1.041	1.063	1.214	1.239	1.363
45	1.000	1.011	1.029	1.042	1.063	1.216	1.239	1.366
32	1.000	1.011	1.031	1.042	1.063	1.224	1.250	1.375
29	1.000	1.011	1.031	1.042	1.063	1.223	1.248	1.372
25	1.000	1.010	1.031	1.042	1.061	1.225	1.251	1.374
%RSD		0.053	0.213	0.036	0.101	0.383	0.468	0.342

Equation (3.33) includes as a parameter a function of κ (the reciprocal of the analyte double layer thickness) and α (a parameter defining the size of a nonspherical particle) instead of the simple z_i/r_i ratio used in Equation (2.6). The more complex expression for mobility, Equation (3.33), was developed for larger molecules, such as peptides and proteins. Such large molecules do not yield a satisfactory correlation in terms of Equation (2.6). See Section 2.2 for a description of the use of Equation (2.6) for a correlation of simple inorganic anions.

The independence of electrophoretic and electroosmotic parameters in the mobility index by Lee and Yeung is expressed by two different zeta potential terms. In that form, Equation (3.32) is capable of characterizing a mutually opposing as well as a single directional influence of analytical conditions on observed mobilities of analytes. A practitioner does not have to know the value of all terms in Equation (3.32) to be able to use the migration index. One possibility of obtaining the MI is to integrate the current from the start of the separation ($t=0$) to the time of migration of analyte in question ($t=t_m$) and to divide the resulting charge (As) by L_t. Another possibility makes use of the fact that MI represents the slope of the i versus the migration rate plot. The value of the slope can be determined from a plot of migration rates (L_d/t_m) against current values at several different separation potentials.

An interesting aspect of MI is the deliberate use of the Walden Rule. According to the Walden Rule the product $k\eta$ can be expected to remain constant for small variations in temperature. The viscosity term is thus not eliminated as in the ratio of migration times, but compensated for by an inclusion of a conductivity term in the expression for MI. Lee and Yeung[145] report an improvement of precision data for eleven consecutive electropherograms with the help of the migration index. The original variance of ~ 0.7% RSD could be decreased to 0.05% RSD by a normalization with MI. Migration indices were used not only to improve the precision of migration data within a series of runs, but also to enable comparison of two sets of data obtained under different analytical conditions. The authors[145] reported a successful outcome of evaluations of MI for programmed potential runs, for comparisons of data from capillaries of different lengths, different zeta potentials, and different internal diameters. As anticipated, the ratios of migration times were as useful for normalization as migration indices for data from consecutive runs, programmed potential runs, and for comparison of capillaries of different lengths. On the other hand, normalization by migration times ratios failed for data from capillaries with different zeta potentials and from capillaries with different internal diameters. For the two latter cases, even the normalization by MI did not perform satisfactorily. Consequently, for such difficult comparisons, Lee and Yeung introduced an adjusted migration index (AMI)

$$AMI = [(MI)_{eo}MI]/[(MI)_{eo} - MI] \tag{3.34}$$

Where $(MI)_{eo}$ denotes the MI of a neutral marker representing the electroosmotic mobility. The AMI indices made possible a comparison of migration data not only from capillaries of different zeta potential, but also from those of different internal diameter. Reflecting on the discussion in the preceding paragraphs, we can conclude that the first approach to normalization of CE data should probably involve the migration time ratios and division of peak areas by the corresponding migration time. The suitability of the simplistic version of normalization should be verified by an experiment and if necessary replaced by Lee and Yeung's migration indices and by the method for the normalization of peak areas proposed by Ackermans, Everaerts, and Beckers.[143]

Additional alternatives for normalization of CE data are listed in Table 3.8.

References

1. F. E. P. Mikkers, F. M. Everaerts, and T. P. E. M. Verheggen, *J. Chromatogr.* **169**, 11 (1979).
2. S. E. Morig, J. C. Colburn, P. D. Grossman, and H. H. Lauer, *LC GC* **8**, 34 (1990).
3. D. S. Burgi and R. L. Chien, *J. Microcol. Sep.* **3**, 199 (1991).
4. R. L. Chien, *Anal. Chem.* **63**, 2866 (1991).
5. D. J. Rose and J. W. Jorgenson, *Anal. Chem.* **60**, 624 (1988).
6. K. Otsuka and S. Terabe, *J. Chromatogr.* **480**, 91 (1989).
7. E. V. Dose and G. Guichon, *Anal. Chem.* **64**, 123 (1992).
8. X. Huang, M. J. Gordon, and R. N. Zare, *Anal. Chem.* **60**, 377 (1988).
9. X. Huang, T. K. Pang, M. J. Gordon, and R. N. Zare, *Anal. Chem.* **59**, 2747 (1987).
10. J. Tehrani, R. Macomber, and L. Day, *HRC* **14**, 10 (1991).

11. T. Tsuda and R. N. Zare, *J. Chromatogr.* **559**, 103 (1991).
12. T. Tsuda, T. Mizuno, and J. Akiyama, *Anal. Chem.* **59**, 799 (1987).
13. M. M. Bushey and J. W. Jorgenson, *Anal. Chem.* **62**, 978 (1990).
14. J. C. Giddings, *HRC* **10**, 319 (1987).
15. M. Deml, F. Foret, and P. Bocek, *J. Chromatogr.* **320**, 159 (1985).
16. H. F. Yin, S. R. Motsch, J. A. Lux, and G. Schomburg, *HRC* **14**, 282 (1991).
17. R. W. Wallingford and A. G. Ewing, *Anal. Chem.* **60**, 1972 (1988).
18. C. A. Monnig and J. W. Jorgenson, *Anal. Chem.* **63**, 802 (1991).
19. P. Jandik, P. R. Haddad, and P. Sturrock, *CRC Crit. Rev. Anal. Chem.* **20**, 13 (1988).
20. R. L. Chien and D. Burghi, *J. Chromatogr.* **559**, 141 (1991).
21. R. L. Chien and D. Burghi, *J. Chromatogr.* **559**, 153 (1991).
22. F. M. Everaerts, J. L. Beckers, and Th. P. E. M. Verheggen, *Isotachophoresis*, Elsevier, Amsterdam (1976).
23. P. Bocek, *Top. Curr. Chem.* **95**, 131 (1981).
24. F. Kohlrausch, *Ann. Phys. Chem. N. F.* **62**, 209 (1897).
25. P. Jandik and W. R. Jones, *J. Chromatogr.* **546**, 431 (1991).
26. V. Dolnik, M. Deml, and P. Bocek, *J. Chromatogr.* **320**, 89 (1985).
27. CS ITP Analyzer, VVZ PJT, Spisska Nova Ves, Czechoslovakia (1988).
28. D. Kaniansky and J. Marak, *J. Chromatogr.* **498**, 191 (1990).
29. V. Dolnik, K. A. Cobb, and M. Novotny, *J. Microcol. Sep.* **2**, 299 (1990).
30. F. Foret, V. Sustacek, and P. Bocek, *J. Microcol. Sep.* **2**, 127 (1990).
31. W. R. Jones and P. Jandik, *Am. Lab.* **22**, 51 (1990).
32. M. Merion, R. H. Aebersold, and M. Fuchs, Poster Presentation, HPCE '91, San Diego (1991).
33. M. Swartz and M. Merion, Poster Presentation, HPCE '92, Amsterdam (1992).
34. S. Hjerten, *Chromatogr. Rev.* **9**, 122 (1967).
35. P. Gebauer, M. Deml, P. Bocek, and J. Janak, *J. Chromatogr.* **267**, 455 (1983).
36. M. Jansson, A. Emmer, and J. Roeraade, *HRC* **12**, 797 (1989).
37. T. Tsuda, J. V. Sweedler, and R. N. Zare, *Anal. Chem.* **62**, 2149 (1990).
38. J. P. Chervet, United States Patent 5,057,216, October 15, 1991.
39. G. B. Gordon, United States Patent 5,061,361, October 29, 1991.
40. J. A. Lux, U. Haeusig, and G. Schomburg, *HRC* **13**, 374 (1990).
41. Technical Information, Polymicro Technologies, Phoenix, AZ (1990).
42. A. E. Bruno, E. Gassmann, N. Pericles, and K. Anton, *Anal. Chem.* **61**, 876 (1989).
43. R. M. McCormick and R. J. Zagursky, *Anal. Chem.* **63**, 750 (1991).
44. J. S. Fritz, D. T. Gjerde, and C. Pohlandt, *Ion Chromatography*, A. Huethig Verlag, Heidelberg (1982).

45. D. A. Skoog and D. M. West, *Principles of Instrumental Analysis*, Holt, Reinhart and Winston, New York (1971).
46. M. Fuchs, P. Timmoney, and M. Merion, Poster No. PM-24, HPCE '91, San Diego, CA (1991).
47. Y. Walbroehl and J. W. Jorgenson, *J. Chromatogr.* **315**, 135 (1984).
48. F. A. Jenkins and H. E. White, *Fundamentals of Optics*, Fourth Edition, McGraw-Hill Book Company, New York (1976).
49. C. W. Allen, *Astrophysical Quantities*, Chapter 6, 2nd ed. The Athlone Press, London (1963).
50. P. Wernly and W. Thorman, *Anal. Chem.* **63**, 2878 (1991).
51. K. Oka, K. Oshima, T. Mizuguchi, and M. Funato, Poster Presentation, HPCE '90, San Francisco (1990).
52. G. J. M. Bruin, G. Stegeman, A. C. Van Asten, X. Xu, J. C. Kraak, and H. Poppe, *J. Chromatogr.* **559**, 163 (1991).
53. C. Burgess and A. Knowles, eds., *Standards in Absorbance Spectroscopy*, Chapman and Hall, London, 1981.
54. S. E. Moring, J. C. Colburn, P. D. Grossman, and H. H. Lauer, *LC GC* **8**, 34 (1990).
55. T. Wang, R. A. Hartwick, and P. B. Champlin, *J. Chromatogr.* **462**, 147 (1989).
56. T. Wang, J. H. Aiken, C. W. Huie, and R. A. Hartwick, *Anal. Chem.* **63**, 1372 (1991).
57. B. A. Bidlingmeyer, S. N. Deming, W. P. Price, B. Sachok, and M. Petrusek, *J. Chromatogr.* **186**, 419 (1979).
58. H. Small and T. E. Miller, *Anal. Chem.* **54**, 462 (1982).
59. F. Foret, M. Deml, V. Kahle, and P. Bocek, *Electrophoresis* **7**, 430 (1986).
60. S. Hjerten, K. Elenbring, F. Kilar, J. L. Liao, A. J. C. Chen, C. Siebert, and M. D. Zhu, *J. Chromatogr.* **403**, 47 (1987).
61. S. G. Shirazi and G. Guichon, *Anal. Chem.* **62**, 923 (1990).
62. E. S. Yeung, *Acc. Chem. Res.* **22**, 125 (1989).
63. M. W. F. Nielen, *J. Chromatogr.* **588**, 321 (1991).
64. M. Fuchs, Millipore Corporation, personal communication.
65. R. P. W. Scott, *Liquid Chromatographic Detectors* (*J. Chromatogr. Library*, Vol. 33), Elsevier, Amsterdam, 2nd ed. (1986), pp. 12–14.
66. W. G. Kuhr, *Anal. Chem.* **62**, 403R (1990).
67. Z. Deyl and R. Struzinsky, *J. Chromatogr.* **569**, 163 (1991).
68. S. Wu and N. J. Dovichi, *J. Chromatogr.* **480**, 141 (1989).
69. M. V. Novotny, K. A. Cobb, and J. Liu, *Electrophoresis* **11**, 735 (1990).
70. J. W. Jorgenson and K. D. Lukacs, *Anal. Chem.* **53**, 1298 (1981).
71. E. J. Guthrie and J. W. Jorgenson, *Anal. Chem.* **56**, 483 (1984).
72. J. S. Green and J. W. Jorgenson, *J. Chromatogr.* **352**, 337 (1986).
73. Y. Kurosu and M. Saito, *HRC* **14**, 186 (1991).

74. Y. F. Cheng and N. J. Dovichi, *Science* **242**, 562 (1988).
75. Y. F. Cheng, S. Wu, D. Y. Chen, and N. J. Dovichi, *Anal. Chem.* **62**, 496 (1990).
76. B. Nickerson and J. W. Jorgenson, *HRC* **11**, 878 (1988).
77. [86] in Reference 67.
78. B. W. Wright, G. A. Ross, and R. D. Smith, *J. Microcol. Sep.* **1**, 85 (1989).
79. J. Liu, Y. Z. Hsieh, D. Wiesler, and M. Novotny, *Anal. Chem.* **63**, 408 (1991).
80. B. Nickerson and J. W. Jorgenson, *J. Chromatogr.* **480**, 157 (1989).
81. M. Albin, R. Weinberger, E. Sapp, and S. Moring, *Anal. Chem.* **63**, 417 (1991).
82. D. J. Rose and J. W. Jorgenson, *J. Chromatogr.* **447**, 117 (1988).
83. S. L. Pentoney, X. Huang, D. S. Burgi, and R. N. Zare, *Anal. Chem.* **60**, 2625 (1988).
84. T. Tsuda, Y. Kobayashi, A. Hori, T. Matsumoto, and O. Suzuki, *J. Chromatogr.* **456**, 375 (1988).
85. T. W. Garner and E. S. Yeung, *Anal. Chem.* **62**, 2193 (1990).
86. L. Gross and E. S. Yeung, *J. Chromatogr.* **480**, 169 (1989).
87. L. Gross and E. S. Yeung, *Anal. Chem.* **62**, 427 (1990).
88. W. G. Kuhr and E. S. Yeung, *Anal. Chem.* **60**, 2642 (1988).
89. W. G. Kuhr and E. S. Yeung, *Anal. Chem.* **60**, 1832 (1988).
90. B. L. Hogan and E. S. Yeung, *J. Chromatogr. Sci.* **28**, 15 (1990).
91. L. A. Amankwa and W. G. Kuhr, *Anal. Chem.* **63**, 1733 (1991).
92. E. S. Yeung and W. G. Kuhr, *Anal. Chem.* **63**, 275A (1991).
93. P. Bocek, M. Deml, and J. Janak, *J. Chromatogr.* **106**, 283 (1975).
94. F. Foret, M. Deml, V. Kahle, and P. Bocek, *Electrophoresis* **7**, 430 (1986).
95. X. Huang, T. K. J. Pang, M. J. Gordon, and R. N. Zare, *Anal. Chem.* **59**, 2747 (1987).
96. X. Huang, R. N. Zare, S. Sloss, and A. G. Ewing, *Anal. Chem.* **63**, 189 (1991).
97. X. Huang and R. Zare, *Anal. Chem.* **63**, 2193 (1991).
98. J. L. Beckers, T. P. E. M. Verheggen, and F. M. Everaerts, *J. Chromatogr.* **452**, 591 (1988).
99. X. Huang, J. A. Luckey, M. J. Gordon, and R. N. Zare, *Anal. Chem.* **61**, 766 (1989).
100. X. Huang, M. J. Gordon, and R. N. Zare, *J. Chromatogr.* **480**, 285 (1989).
101. D. T. Gjerde and J. S. Fritz, *Ion Chromatography*, 2nd Edition, Huethig Verlag, New York (1987).
102. P. K. Dasgupta, Plenary Lecture, International Ion Chromatography Symposium, Denver, CO (1991).
103. P. Jandik, P. R. Haddad, and P. E. Sturrock, *CRC Critical Reviews in Analytical Chemistry* **20**, 1 (1988).
104. R. W. Wallingford and A. G. Ewing, *Anal. Chem.* **59**, 1762 (1987).
105. R. W. Wallingford and A. G. Ewing, *Anal. Chem.* **60**, 1972 (1988).
106. R. W. Wallingford and A. G. Ewing, *Anal. Chem.* **60**, 258 (1988).

107. A. W. Ewing, R. A. Wallingford, and T. M. Olefirowicz, *Anal. Chem.* **61,** 299A (1989).
108. R. W. Wallingford and A. G. Ewing, *Anal. Chem.* **61,** 98 (1989).
109. T. M. Olefirowicz and A. G. Ewing, *Anal. Chem.* **62,** 1872 (1990).
110. R. M. Riggin and P. T. Kissinger, *Anal. Chem.* **49,** 2109 (1977).
111. J. C. Giddings, *HRC,* **10,** 319 (1987).
112. J. A. Olivares, N. T. Nguyen, C. R. Yonker, and R. D. Smith, *Anal. Chem.* **59,** 1230 (1987).
113. R. D. Smith, J. A. Olivares, N. T. Nguyen, and H. R. Udseth, *Anal. Chem.* **60,** 436 (1988).
114. R. D. Smith, C. J. Barinaga, and H. R. Udseth, *Anal. Chem.* **60,** 1948 (1988).
115. R. D. Smith, J. A. Loo, C. J. Barinaga, C. G. Edmonds, and H. R. Udseth, *J. Chromatogr.* **480,** 211 (1989).
116. C. G. Edmonds, J. A. Loo, C. J. Barinaga, H. R. Udseth, and R. D. Smith, *J. Chromatogr.* **474,** 21 (1989).
117. J. A. Loo, H. R. Udseth, and R. D. Smith, *Analytical Biochemistry* **179,** 404 (1989).
118. J. A. Loo, H. K. Jones, H. R. Udseth, and R. D. Smith, *J. Microcol. Sep.* **1,** 223 (1989).
119. R. D. Smith, H. R. Udseth, J. A. Loo, B. Wright, and G. A. Ross, *Talanta* **36,** 161 (1989).
120. R. D. Smith, H. R. Udseth, C. J. Barinaga, and C. G. Edmonds, *J. Chromatogr.* **559,** 197 (1991).
121. M. A. Moseley, L. J. Deterding, K. B. Thomer, and J. W. Jorgensen, *J. Chromatogr.* **480,** 197 (1989).
122. R. M. Caprioli, W. T. Moore, M. Martin, B. B. DaGue, K. Wilson, and S. Moring, *J. Chromatogr.* **480,** 247 (1989).
123. S. L. Pentoney, R. N. Zare, and J. F. Quint, *Anal. Chem.* **61,** 1642 (1989).
124. J. Wu, T. Odake, T. Kitamori, and T. Sawada, *Anal. Chem.* **63,** 2216 (1991).
125. A. E. Bruno, B. Krattiger, F. Maystre, and H. M. Widmer, *Anal. Chem.* **63,** 2689 (1991).
126. C. Y. Chen, T. Demana, S. D. Huang, and M. D. Morris, *Anal. Chem.* **61,** 1590 (1989).
127. M. Yu and N. J. Dovichi, *Anal. Chem.* **61,** 37 (1989).
128. N. J. Dovichi, *Chemistry in Britain,* September (1988), p. 895.
129. R. W. Hallen, C. B. Shumate, W. F. Siems, T. Tsuda, and H. H. Hill, *J. Chromatogr.* **480,** 233 (1989).
130. C. Haber, I. Silvestri, S. Roosli, and W. Simon, *Chimia* **45,** 117 (1991).
131. W. R. Jones, Waters Chromatography Division of Millipore, personal communication, November 13, 1989.
132. S. Terabe, K. Otsuka, and T. Ando, *Anal. Chem.* **57,** 834 (1985).
133. R. S. Rush, A. S. Cohen, and B. L. Karger, *Anal. Chem.* **63,** 1346 (1991).
134. S. R. Weinberger and J. L. Mills, US Patent 5,066,382, November 19, 1991.

135. B. L. Karger, A. Paulus, and A. S. Cohen, US Patent 4,898,658, February 6, 1990.
136. G. Yowell, S. Fazio, and R. Vivilecchia, An Evaluation of Current Commercially Available Capillary Electrophoresis Instruments, Sandoz Research Institute, East Hanover, NJ (1990).
137. W. R. Jones and B. Rahn, Waters Chromatography Division of Millipore, personal communication, December 12, 1991.
138. W. R. Jones, unpublished data.
139. M. Fuchs, Waters Chromatography Division of Millipore, personal communication, March 12, 1992.
140. N. Dyson, *Chromatographic Integration Methods*, RSC Chromatography Monographs, T. Graham House, Cambridge (1990).
141. D. T. Rossi, *J. Chromatogr. Sci.* **26**, 101 (1988).
142. P. D. Grossman, H. H. Lauer, S. E. Moring, D. E. Mead, and M. F. Oldham, J. H. Nickel, J. R. P. Goudberg, A. Krever, D. H. Ransom, and J. C. Colburn, *ABL* **8**, 35 (1990).
143. M. T. Ackermans, F. M. Everaerts, and J. L. Beckers, *J. Chromatogr.* **549**, 345 (1991).
144. J. L. Beckers, F. M. Everaerts, and M. T. Ackermans, *J. Chromatogr.* **537**, 407 (1991).
145. T. T. Lee and E. S. Yeung, *Anal. Chem.* **63**, 2842 (1991).

CHAPTER

4

Selected Applications of Counterelectroosmotic Capillary Electrophoresis

As explained in Section 2.4, counterelectroosmotic separations belong to one of the three main separation modes in capillary electrophoresis (the remaining two are coelectroosmotic CE and separations in gel-filled capillaries).

In this chapter, selected applications are described in considerable detail, in order to illustrate the potential of counterelectroosmotic capillary electrophoresis for the analysis of small molecules and ions. Because of its high sensitivity, high reproducibility, and the simplicity of sample handling, counterelectroosmotic CE is highly useful for solving analytical problems in medicine, food chemistry, biotechnology, the pharmaceutical industry, and environmental chemistry. In addition to other important applications that are outside of the focus of this book (i.e., proteins, peptides, and nucleic acid fragments), counterelectroosmotic capillary electrophoresis is particularly useful for carbohydrates, due to its capability of rapid and sensitive analysis of underivatized and derivatized aldoses, ketoses, and uronic acids. Time-consuming derivatization steps are not always necessary and, unlike in GC, isomeric distributions are retained. CE separations of compounds that are relevant to pharmaceutical and biotechnological industries are also of growing interest. Flavonoids, for example, which are ubiquitous secondary plant metabolites possessing high pharmaceutical relevance, can be readily separated by counterelectroosmotic CE. Furthermore, counterelectroosmotic CE represents a simple, rapid, and economic alternative for the analysis of other phytochemically important substances like tannins and chiral compounds.

In a wide range of applications, counterelectroosmotic CE matches the ability of established separation techniques. In special cases, this technique has even shown itself to be superior to HPLC, due to its improved resolution and higher separation efficiency.[1,2]

4.1 Counterelectroosmotic CE of Carbohydrates

Capillary electrophoresis emerges as an alternative to current analytical techniques for carbohydrates (thin-layer chromatography,[3] gas chromatography,[4] and high-performance liquid chromatography[5-7]). The photometric detection of underivatized carbohydrates in CE (as in HPLC) is a difficult task, due to the low UV absorption of this class of compounds. Recently, however, considerable progress has been achieved by a variety of methods aimed at increasing the sensitivity of carbohydrate detection either by borate complexation or by derivatization with other suitable reagents. Promising results have also been obtained by indirect UV detection (see Section 4.1.2). General principles of this method, which utilizes a highly UV absorbing compound in the carrier electrolyte, are discussed in Section 3.3.2.

4.1.1 Direct Detection Techniques for Carbohydrates

As direct detection techniques, we understand such approaches to the analysis of carbohydrates utilizing a complexation or derivatization reaction (or both) improving the detection sensitivity for either one group or the entire class of carbohydrates. The complexation usually occurs inside the capillary following the sampling of carbohydrates or their derivatives into a CE carrier electrolyte containing the borate anion.

The derivatization reactions, on the other hand, are usually performed prior to the actual sampling. In both approaches, the quality of separation depends, with only a few exceptions, on the properties of reaction products (complex, derivative) and not on the properties of the original uncomplexed and underivatized carbohydrates.

4.1.1.1 CE of Carbohydrate–Borate Complexes

Under alkaline conditions, borate ions react with vicinal hydroxyl groups, transforming carbohydrates into negatively charged borate complexes exhibiting an increased UV absorption and capable of migrating in an electric field.[8-11]

Sugar–borate complexes have been also used for the separation of carbohydrates in chromatographic techniques, such as high-performance liquid chromatography and thin-layer electrophoresis.[11]

A temperature of 60°C is necessary for a sufficient separation of underivatized carbohydrate–borate complexes under the conditions of capillary

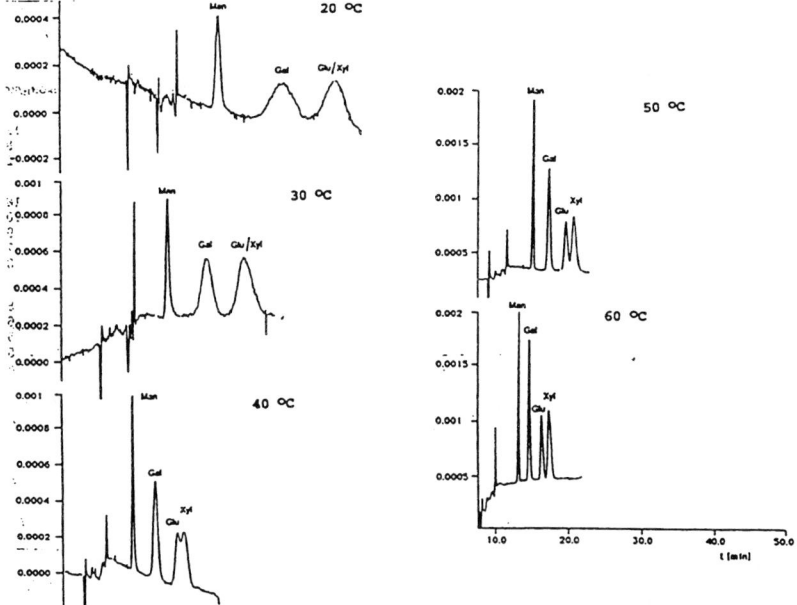

Figure 4.1 Effect of temperature on CE of complexed monosaccharides. Carrier electrolyte: 50 mM tetraborate, pH 9.3; Fused silica capillary, 94 cm, 75 μm. Voltage, +20 kV. Direct UV detection at 195 nm. Sample: mannose (Man), galactose (Gal), glucose (Glu), xylose (Xyl), 10 mM each, dissolved in water; temperature 20–60°C. Reproduced with permission.[8]

electrophoresis. Figure 4.1 shows the extent of the influence of temperature on selectivity of separation. The UV detection of borate complexes of carbohydrates is possible with detection limits in the nanomolar range.[8] The electrophoretic mobilities of disaccharides are lower than those of monosaccharides.

4.1.1.2 CE of Fluorescent Derivatives of Oligosaccharides

The high cost of lasers has thus far precluded the use of fluorescence detection in many laboratories. Section 3.3.3.1 discusses applications of fluorescent labeling to various classes of compounds. There are two carbohydrate specific reagents described in the literature.

Fluorogenic reagents such as 3-(4-carboxybenzoyl)-2-quinoline-carboxoaldehyde or 3-benzoyl-2-naphthaldehyde react with aldoses as well as with aldose polysaccharides. Reaction products can be separated by counterelectroosmotic CE and detected by laser-induced fluorescence.[12] A pretreated polyacrylamide gel capillary and an addition of borate to the buffer system are needed to separate sugar oligomers (see Figure 4.2).

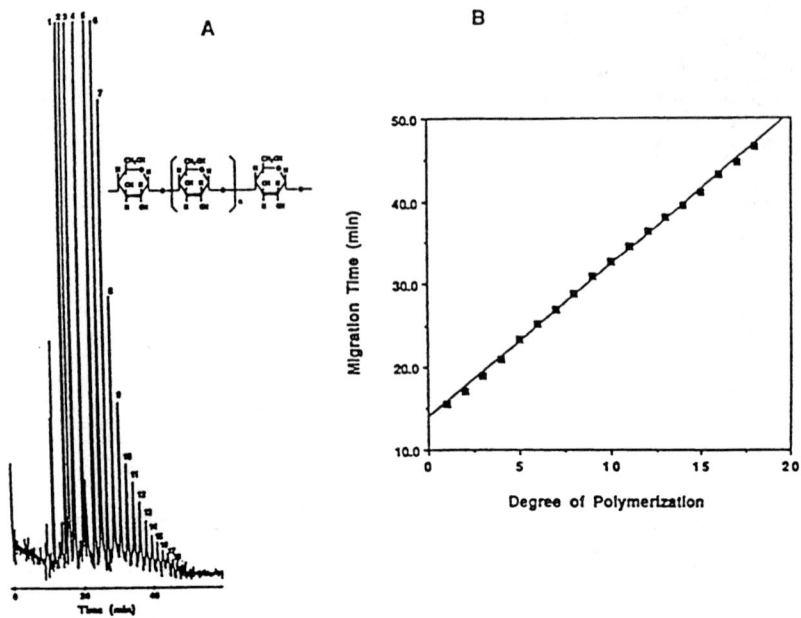

Figure 4.2 (A) Electrophoretic separation of partially hydrolyzed Dextrin 15 in a polyacrylamide gel-filled capillary. Capillary: 26 cm × 50 μm I.D., buffer: 0.1 M Tris–0.25 M borate–7 M urea (pH 8.33); applied field, 269 V/cm (20 μA). (B) Correlation between molecular masses of the oligomers from dextrin and their corresponding migration times. Reproduced with permission.[12]

4.1.1.3 Capillary Electrophoresis of Aminopyridine Derivatives of Carbohydrates

The derivatization of carbohydrates with aminopyridine improves their detectability in UV light. Successful coupling of 2-aminopyridine to carbohydrates by reductive amination requires the presence of a free aldehyde group and, for this reason, has been possible only with aldoses and uronic acids.

The separation of 13 N-2-pyridylglycamines of various reducing carbohydrates is shown in Figure 4.3.

The separation of carbohydrate derivatives occurs as a result of differences in the stability of their borate complexes, which in turn depends strongly on the number and configuration of hydroxyl groups as well as on the presence of substituents.[9] The separation mode is as illustrated in Figure 2.8B. Lower-mobility analytes have shorter migration times. The general migration sequence of borate complexes of aminopyridine derivatives is similar to that of borate complexes of *p*-aminobenzoic acid derivatives and is discussed in more detail in Section 4.1.1.4.

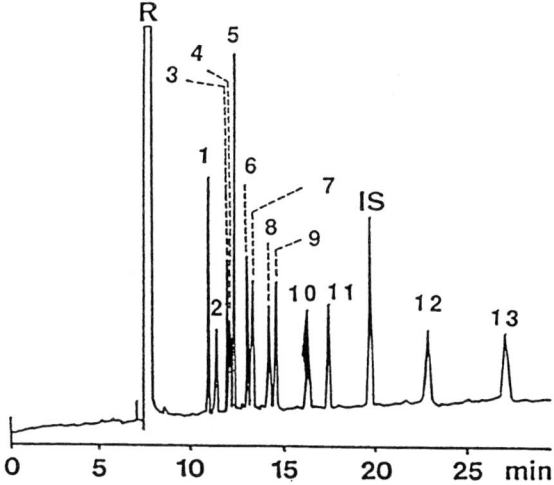

Figure 4.3 Capillary electrophoretic separation of derivatized carbohydrates. Derivatizing reagent: 2-aminopyridine. Fused silica capillary 72 × 50 cm, 50 μm I.D. Carrier electrolyte: 150 mM borate, pH 10.5. Voltage +20 kV. UV detection: 237 nm. Sampling time: 1 sec. Peak identification: R = reagent, IS = internal standard, (1) 2-deoxy-D-ribose, (2) maltotriose, (3) maltose, (4) I-rhamnose, (5) D-lyxose, (6) D-xylose, (7) D-ribose, (8) D-glucose, (9) L-arabinose, (10) D-fucose, (11) D-galactose, (12) D-glucuronic acid, (13) D-galacturonic acid. Reproduced with permission.[13]

Uronic acids are detected last because of their additional carboxyl group, which is completely dissociated, contributing strongly to the observed mobility under the described electrophoretic conditions. As a matter of fact, 2-aminopyridine derivatives of mannuronic acid, glucuronic acid, and galacturonic acid gave electrophoretic mobility values (μ_{ion}) similar to those calculated for underivatized uronic acids, which had been separated on the basis of differences in dissociation in a capillary filled with sorbate, pH 12.1 (see Section 4.1.2).

4.1.1.4 CE of p-Aminobenzoic Acid Derivatives of Carbohydrates

The use of *p*-aminobenzoic acid as a derivatization reagent permits an efficient derivatization of aldoses, ketoses, and uronic acids as well as their separation as borate complexes by means of counterelectroosmotic capillary electrophoresis. Furthermore, due to a pronounced difference in the chemical properties of reaction products, a number of carbohydrates could be separated that are not resolved upon derivatization with some other reagents (i.e., ethyl-p-aminobenzoate, Section 4.1.1.5).

Identity of p-Aminobenzoic Acid Derivatives of Carbohydrates

The identity of the reductive amination products of glucose and fructose, after their reaction with p-aminobenzoic acid, can be verified by ^{13}C NMR spectroscopy. The chemical shift of the C1 carbon of glucose became δ = 44.5 ppm after reductive amination, compared with the original value of δ = 98 ppm for the β-pyranose form of D-glucose. And the chemical shift of the carbon C2 of fructose became δ = 57.71 and 58.35 ppm, compared with the original value of δ = 97.74 ppm for the β-pyranose form of D-fructose (Figure 4.4). Furthermore, the double signal indicated the existence of diastereomers. This is corroborated by ^{1}H NMR spectrometry, which reveals that the two diastereomers of the p-aminobenzoic acid derivative of fructose are present in a ratio of approximately 1:1.7.

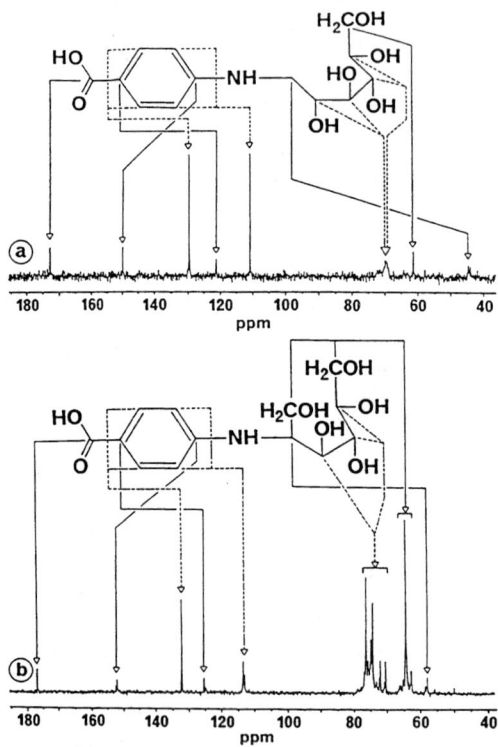

Figure 4.4 ^{13}C NMR spectra of (a) glucose and (b) fructose after reductive amination with p-aminobenzoic acid. Reproduced with permission.[13]

Optimization of Precolumn Derivatization

The effect of concentration of *p*-aminobenzoic acid on precolumn derivatization was examined by using selected carbohydrates (fructose, glucose, and glucuronic acid). As shown in Figure 4.5a, the peak areas observed at 285 nm (the wavelength for maximum absorption), adjusted to that of cinnamic acid as an internal standard, increased continuously with increasing concentration of *p*-aminobenzoic acid to give convex curves for glucose and glucuronic acid as well as a straight line for fructose.

With increasing concentration of acetic acid in the reaction mixture, slight linear increases in peak area were observed for all three of the monosaccharides (Figure 4.5b). Therefore, the concentrations of *p*-aminobenzoic acid and acetic acid were held constant at 7 and 10% (w/v), respectively, to

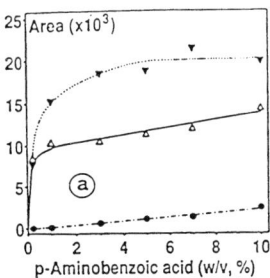

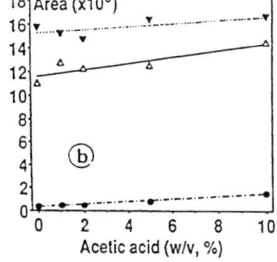
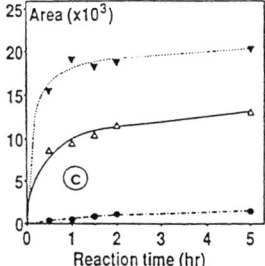

Figure 4.5 Effects of various factors on precolumn derivatization of selected monosacarides: ● fructose; Δ glucuronic acid; ▼ glucose. (a) Concentration of *p*-aminobenzoic acid. The concentrations of acetic acid and sodium cyanoborohydride were 10 and 1% (w/v), respectively. Reaction temperature, 50°C; reaction time, 2 h. (b) Concentration of acetic acid. The concentration of *p*-aminobenzoic acid and sodium cyanoborohydride were 7 and 1% (w/v), respectively. Reaction temperature, 50°C; reaction time: 2 h. (c) Time course of derivatization. The concentrations of *p*-aminobenzoic acid, acetic acid, and sodium cyanoborohydride were 7, 10, and 1% (w/v), respectively. Reaction temperature, 50°C. In all experiments the reaction products were analyzed by using 150 mM borate buffer, pH 10.0, at +28 kV. On-column UV detection was carried out at 285 nm. The concentration of carbohydrates in the sample solution was commonly 2 mg/ml. Reproduced with permission.[13]

guarantee optimum sensitivity for both ketoses as well as aldoses and also for uronic acids. However, since at the chosen concentrations p-aminobenzoic acid and acetic acid were found to increase the conductivity of the carrier electrolyte, affecting the resolution, the reaction mixtures were diluted routinely prior to their capillary electrophoretic analysis.

Another study of the derivatization method indicated that peak areas continued to increase for at least 5 h (Figure 4.5c), but 2 h were arbitrarily adopted for economy of time. With this reaction time, the adjusted peak areas were approximately 90% of those at 5 h, as far as glucose and glucuronic acid were concerned, and 75% for fructose.

The yields of derivatized monosaccharides were found to increase continuously with increasing reaction temperature. However, since high temperatures around the boiling point necessitated sealing the reaction mixture in an ampule to prevent rapid evaporation of solvent, a reaction temperature of 50°C was found to be more convenient.

Although the yields of p-aminobenzoic acid derivatives are not quantitative, they are reproducible enough to allow highly precise determination of aldoses, ketoses, and uronic acids. Another advantage is that the reaction solutions can be introduced into the capillary without any cleanup, since neither the underivatized portions of the monosaccharide samples nor the byproducts interfere with the CE separation. The only interference observed with this method was the masking of the rhamnose peak by a prominent peak due to the reagent.

Capillary Electrophoresis of p-Aminobenzoic Acid Derivatives of Carbohydrates

Figure 4.6 depicts an electropherogram of a reaction mixture of 13 sugars and sugar acids with concentrations ranging from 0.15 to 1.73 mg/ml.

At the selected pH value, the p-aminobenzoic acid derivatives are negatively charged, both due to their inherent charges as well as due to their complexation with tetrahydroxyborate, which, rather than boric acid, is complexed by polyols. Therefore, they will migrate away from the detector toward the anode. However, due to the large electroosmotic flow in the system, which is faster than electrophoretic migration, analytes are propelled together with the bulk solution toward the cathode, but at a much slower rate (see also Figure 2.8B).

Relationship between Electrophoretic Mobility of the p-Aminobenzoic Acid Derivatives and Carbohydrate Structure

The electrophoretic mobilities, μ_{ion}, given in Table 4.1 have been calculated by the following form of Equation (2.3):

$$\mu_{ion} = (L_t L_d / P)(1/t_m - 1/t_{eof}) \tag{4.1}$$

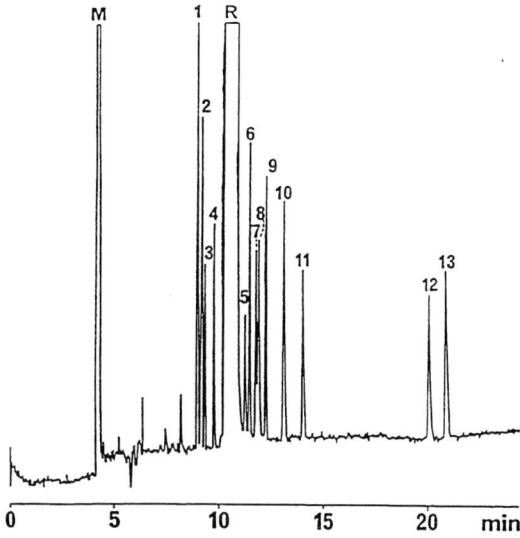

Figure 4.6 Separation of a mixture of carbohydrates (0.15–1.73 mg/ml) derivatized with p-aminobenzoic acid. Fused silica capillary 72 cm × 50 cm, 50 μm I.D.; carrier electrolyte: 150 mM borate, pH 10.0. Voltage: +28 kV. Current: 79 μA. UV detection: 285 nm. Temperature: 30°C. Injection by vacuum, 1 sec. Peak identification: M, mesityl oxide; R, reagent; 1, cellobiose; 2,2-deoxy-D-ribose; 3, melibiose; 4, lactose; 5, sorbose; 6, xylose; 7, glucose; 8, fructose; 9, arabinose; 10, fucose; 11, galactose; 12, glucuronic acid; 13, galacturonic acid. Reproduced with permission.[13]

where L_t is the total capillary length [cm], L_d the capillary length to the detector [cm], P the applied voltage [V], t_m the solute migration times [sec], and t_{eof} the migration time of the electroosmotic flow marker mesityl oxide [sec].

From Table 4.1, it is apparent that di- and trisaccharide derivatives are among the sugar derivatives detected first (note the similarity of the general migration sequence with that illustrated in Figure 4.3 for the aminopyridine derivatives). The di- and trisaccharide derivatives have lower mobilities than most of the monosaccharide derivatives because of their lower charge density and lower complex stability. Of the investigated disaccharides, maltose and cellobiose derivatives form the least negatively charged complexes with tetrahydroxyborate.

Melibiose and lactose derivatives, on the other hand, have relatively higher mobilities due to their galactose unit, which is present in its annular form and enables complexation at the cis-oriented hydroxyl groups of C3 and C4. Moreover, the β-linked disaccharides exhibit higher mobilities than their α-linked equivalents. Comparing the mobilities of monosaccharides de-

Table 4.1. ELECTROPHORETIC MOBILITIES [μ_{ion}, SEE EQ. (4.1)] and STRUCTURAL CHARACTERISTICS OF THE INVESTIGATED 4-AMINOBENZOATE DERIVATIVES OF CARBOHYDRATES*

No.	Carbohydrate	Mobility (10^{-4} cm^2 V^{-1} sec^{-1})	Config. of OH on C2-C3	Config. of OH on C3-C4	Config. of OH on C4-C5
1	Maltotriose	2.178	threo		
2	Maltose	2.619	threo		
3	Cellobiose	2.662	threo		
4	2-Deoxy-D-ribose	2.704		erythro	erythro
5	N-acetyl-D-galactosamine	2.717		erythro	threo
6	Melibiose	2.746	threo	threo	erythro
7	Lactose	2.849	threo		
8	2-Deoxy-D-galactose	2.912		erythro	threo
9	D-ribose	2.949	erythro	erythro	
10	D-lyxose	3.041	erythro	threo	
11	N-acetyl-D-glucosamine	3.049		threo	erythro
12	L-sorbose	3.121		threo	threo
13	D-xylose	3.159	threo	threo	
14	D-mannose	3.204	erythro	threo	erythro
15	D-glucose	3.204	threo	threo	erythro
16	D-fructose	3.218		threo	erythro
17	L-arabinose	3.274	threo	erythro	
18	D-fucose	3.393	threo	erythro	threo
19	D-galactose	3.509	threo	erythro	threo
20	D-mannuronic acid	3.835	erythro	threo	erythro
21	D-glucuronic acid	3.989	threo	threo	erythro
22	D-galacturonic acid	4.034	threo	erythro	threo

*150 mM borate, pH 10.0, 28 kV L_t = 72 cm, L_D = 50 cm.

rivatives, it is evident that hexoses generally migrate faster than pentoses as a result of the higher probability of the formation of borate complexes with an increase in the number of hydroxyl groups. Furthermore, complex stability depends strongly on the configuration of hydroxyl groups. This becomes apparent by relating values of electrophoretic mobility to the configuration of the three vicinal hydroxyl groups at carbons C2 to C4 (Table 4.1), and to a lesser extent at carbons C3 to C5. This reveals that the tendency to complex with borate increases in the following order: e-e (adjacent *erythro* pairs), e-t (an *erythro-diol* adjacent to a *threo*-diol), t-t (adjacent *threo* pairs), and t-e (a *threo*-diol adjacent to an *erythro*-diol). This sequence corroborates a previous finding that *threo*-diols are energetically more favorable than *erythro*-diols for complex formation.[14] Arabinose forms the strongest complexes with borate and, hence, exhibits the highest electrophoretic mobility of all the investigated aldopentoses. In turn, ribose gives the lowest value of μ_{ion} of the investigated aldopentoses due to its adjacent *erythro* pair. With regard to the investigated aldo-hexoses, galactose forms stronger complexes than glucose, which, in turn, should complex more strongly than mannose. But for unknown reasons, glucose and mannose give identical values of electrophoretic mobility. The effect of a substituent on electrophoretic mobility is apparent from the values of μ_{ion} calculated for N-acetylgalactosamine and N-acetylglucosamine, which are significantly lower than those obtained for galactose and glucose. This example also illustrates the great importance of a free hydroxyl group at C2. While galactose migrates faster than glucose because of its *threo-erythro* pair at carbons C2 to C4, N-acetylglucosamine has a higher electrophoretic mobility because of its *threo-erythro* pair at carbons C3 to C5. This configuration evidently exhibits a stronger tendency to complex with borate than the *erythro-threo* pair N-acetylgalactosamine. Finally, as with the aminopyridine derivatives (Section 4.1.1.3), derivatives of carbohydrates containing additional carboxyl functions, like mannuronic acid, glucuronic acid, and galacturonic acid, have higher mobilities similar to those caused by borate complexation only. Correspondingly, they are detected last.

Effect of pH on Electrophoretic Mobility

Figure 4.7 shows the relationship between electrophoretic mobilities of *p*-aminobenzoic acid derivatives of selected carbohydrates and the pH. The concentration of borate in the carrier electrolyte was kept constant at 150 mM during this study.

It is evident that with increasing pH both electrophoretic mobilities and selectivity are increased due to the facilitated formation of borate complexes.[15] The increase in mole fraction of the complexed form causes an increase in the migration rate of a derivative against the electroosmotic flow, resulting in a slow down of its migration. However, the magnitude of the retardation in relation to the pH increase differs among carbohydrate spe-

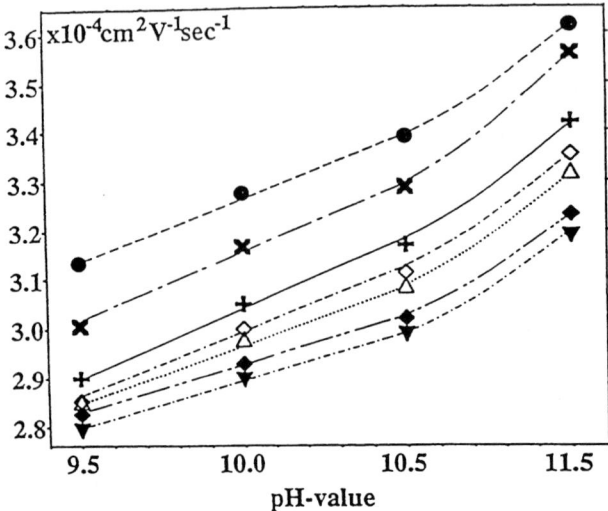

Figure 4.7 Effect of carrier electrolyte pH on the separation of p-aminobenzoic acid derivatives of selected carbohydrates. Fused silica capillary: 72 × 50 cm, 50 μm I.D. Carrier electrolyte: 150 mM borate; voltage: +28 kV. Temperature: 30°C. UV detection at 285 nm. ▼ sorbose, ♦ xylose, Δ glucose, ◊ fructose, + arabinose, × fucose, • galactose.

cies, as the pH dependence of complexation varies with the configuration of the hydroxyl groups. A pH value of 10.0 was chosen, because it offered sufficient resolution for all pairs of interest, for example, in the analysis of the component monosaccharides in plant polysaccharides, in particular glucose and arabinose, within a relatively short time of analysis.

Effect of Temperature on Resolution of Carbohydrates

It has been shown that in the presence of borate, resolution and efficiency were dramatically improved by performing the capillary electrophoretic separation of underivatized carbohydrates at elevated temperatures up to 60°C.[8] An opposite effect, however, was observed when saccharides derivatized with p-aminobenzoic acid were separated in borate-containing carrier electrolyte (Figure 4.8).

The possibilities for borate complexation are broad, due to the fact that a native saccharide may assume several different forms in aqueous solution. On one hand, an increase in temperature increases the reaction rate, allowing the reaction equilibrium to be reached faster and the peak shapes to become narrower. On the other hand, since monosaccharides are present only in their open-chain form, the temperature cannot exert such a dramatic effect on the reaction rates. Consequently, higher working temperatures lead

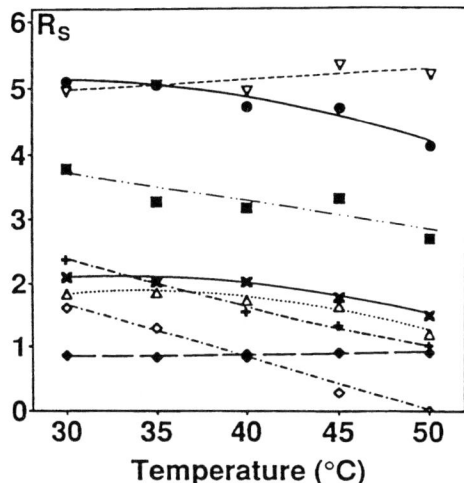

Figure 4.8 Effect of separation temperature on resolution. Fused silica capillary: 72 × 50 cm, 50 μm I.D. Carrier electrolyte: 150 mM borate. Voltage: +28 kV; pH 10.0. UV detection at 285 nm. ◆ glucose/fructose; ◇ cellobiose/2-deoxy-D-ribose; △ sorbose/xylose; × fructose/arabinose; + xylose/glucose; ■ arabinose/fucose; ▽ melibiose/lactose; • fucose/galactose.

to an increase in longitudinal diffusion of the sample compounds, causing broader peak shapes.

Reproducibility of Electrophoretic Mobilities

Table 4.1 lists electrophoretic mobilities of 22 carbohydrates, which were analyzed at a constant voltage of +28 kV in a fused silica capillary of 50 μm internal diameter and an effective length of 50 cm using 150 mM borate, pH 10.0, as the carrier electrolyte. For determining the reproducibility of electrophoretic mobilities, eighteen repeat runs were carried out for seven different carbohydrates. The mean coefficient of variation was 0.72%, ranging from 0.51 to 0.83% for galacturonic acid and xylose, respectively.

Resolution

In order to obtain a baseline resolution under the described conditions, the minimum difference in electrophoretic mobility between two carbohydrates has to be approximately 0.02×10^{-4} cm^2 V^{-1} s^{-1}. The high degree of resolution achieved is a result of the high separation efficiency common in CE (e.g., xylose: 1.04×10^5 theoretical plates). In comparison with the separation of underivatized sugar–borate complexes,[8] it can be noted that derivatization does not only improve sensitivity, but also selectivity and resolution of counterelectroosmotic CE separations. The resolution and separation efficiency of carbohydrate derivatives represent an improvement over the separations of native carbohydrates as well. This improvement is due to differences in ionization at extremely alkaline pH values of the carrier electrolyte.[16] However, an advantage of the latter technique is its capability to separate and to detect even aldonic acids and (1-2)-linked disaccharides such as saccharose.

Quantitation

As discussed in Section 3.4.5, quantitative information from capillary electrophoretic analyses usually requires an application of normalization procedures. In the quantitative evaluation of the CE separations of *p*-aminobenzoic acid derivatives, the normalization consisted in dividing the peak areas by migration time and relating it to that of cinnamic acid as an internal standard. The calibration curves for various aldoses and uronic acids exhibited higher than 0.960 correlation coefficients, over a concentration range of 0.1–10 mM. The Y intercepts were negligible. The range of good correlation (R^2 higher than 0.960) in calibration plots for ketoses was found to be in the concentration range of 0.7–90 mM.

The linear regression analysis yielded the following results for calibration plots: arabinose, $Y = 1.602X + 0.162$ ($R^2 = 0.998$); fucose, $Y = 1.428X + 0.087$ ($R^2 = 0.999$); galactose, $Y = 1.425X - 0.170$ ($R^2 = 0.989$); glucose, $Y = 1.628X - 0.0004$ ($R^2 = 0.991$); xylose, $Y = 1.668X - 0.01$ ($R^2 = 0.998$); fructose, $Y = 0.641X - 0.893$ ($R^2 = 0.998$); galacturonic acid, $Y = 0.965X - 0.091$ ($R^2 = 0.961$), and glucuronic acid, $Y = 1.146X + 0.184$ ($R^2 = 0.963$). The reproducibility of relative peak areas was determined by six measurements of constant concentrations of analytes. A mean coefficient of variation of 4.7% was obtained.

Sensitivity

The mass detection limits for glucose and fructose were found to be approximately 15 and 300 fmol, respectively, at a signal–to–noise ratio of 3. The concentration detection limits were about 4 and 80 μM, respectively, for glucose and fructose. The mass detection limit is similar to that of CE of reducing mono- and oligosaccharides tagged with fluorophores, which has been reported to lie in the upper attomole range.[17] The CE mass detection limits for glucose and fructose are more than four orders of magnitude better than those of carbohydrates separated and detected by high-performance anion-exchange chromatography utilizing triple-pulse amperometric detection on a gold-working electrode.[18] In terms of concentration sensitivity, CE with direct UV detection (i.e., 8-nl injection of glucose derivatized with *p*-aminobenzoic acid) matches the detection limits of pulsed amperometric detection (200-μl injection). Hence, the discussed CE method can be considered as being one of the most sensitive approaches available for the analysis of carbohydrate mixtures.

Analysis of Monosaccharides in Complex Matrices

The practical value of capillary electrophoretic separations of *p*-aminobenzoic derivatives of carbohydrates in borate-containing carrier electrolyte is confirmed by the analyses of monosaccharides in the polysaccharides extracted from *Radix althaeae* (Figure 4.9a) and by the analysis of carbohydrates after hydrolysis of rice-straw hemicellulose (Figure 4.9b).

Compared with the electropherogram of a standard mixture of carbohydrates, a significant reduction in the analytical run time can be noted due to the increase in the sample conductivity caused by the uronic acids, which amount to approximately 40% of all carbohydrates present in *Radix althaeae*. The influence of sample conductivity on results in capillary electrophoresis is discussed in Section 3.4.2.

4.1.1.5 CE of Ethyl-p-Aminobenzoate Derivatives of Carbohydrates

Until recently, the only reported use of ethyl-*p*-aminobenzoate had been for the precolumn derivatization of aldoses and their subsequent separation by high-performance liquid chromatography.[19,20] According to the most recent results,[21] however, ethyl-*p*-aminobenzoate permits an efficient derivatization of not only the aldoses, but also of ketoses. The precolumn derivatization of aldoses and ketoses by this reagent can then be followed by a CE separation in a borate-containing carrier electrolyte under general conditions that are very similar to those discussed in the preceding two chapters. Figure 4.10 illustrates the results of an optimization study of precolumn derivatization of carbohydrates by ethyl-*p*-aminobenzoate.

Based on the data in Figure 4.10, the optimum conditions for precolumn derivatization were defined by the following set of parameters. The concentration of ethyl-*p*-aminobenzoate and acetic acid in the reaction mixture is 10% (w/v) each. The concentration of sodium cyanoborohydride is 1% (w/v). The reaction time is 2 h and the reaction temperature, 50°C. The reaction mixtures obtained by the precolumn derivatization technique under discussion can be sampled directly for CE separations. No sample preparation step is required. This represents a considerable advantage over HPLC. In the latter technique, the excess of ethyl-*p*-aminobenzoate has to be removed by an extraction step prior to an injection. An optimized CE separation of ethyl-*p*-aminobenzoate derivatives of selected carbohydrates is shown in Figure 4.11.

Maximum resolution was observed for the majority of peak pairs around pH 10. Few of the pairs such glucose and arabinose or rhamnose and xylose exhibited a continuous improvement of resolution with increasing values of

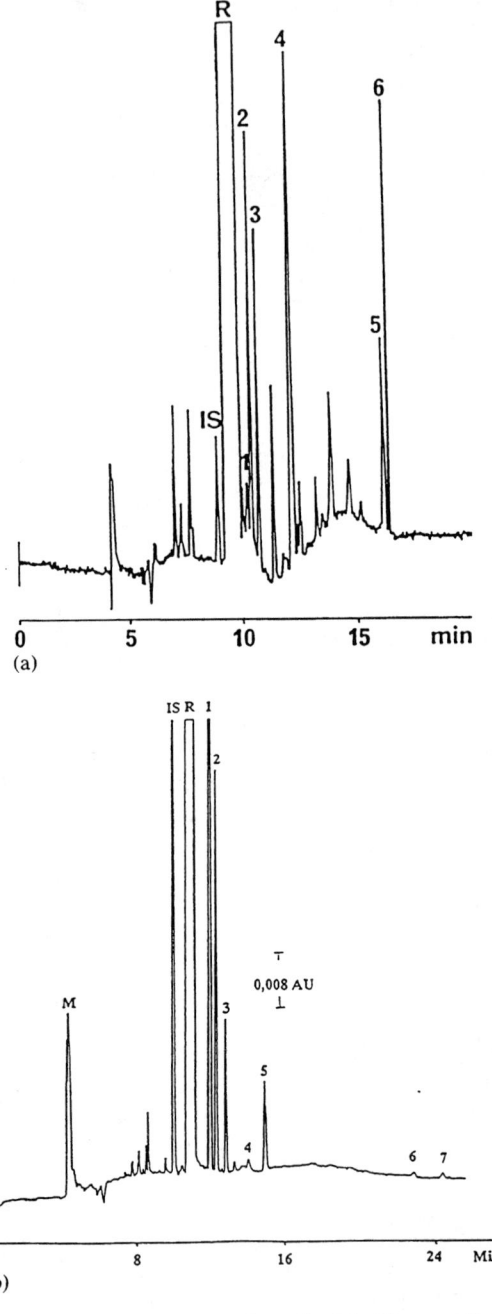

Figure 4.9 Analytical conditions were as described in Figure 4.8. Sample introduction by vacuum, 1 sec. (a) Analysis of monosaccharides contained in polysaccharides extracted from *Radix althaeae*. Peak identification: IS, internal standard (cinnamic acid); R, reagent; 1, xylose; 2, glucose; 3, arabinose; 4, galactose; 5, glucuronic acid; 6, galacturonic acid. (b) Analysis of carbohydrates after hydrolysis of rice straw hemicellulose.

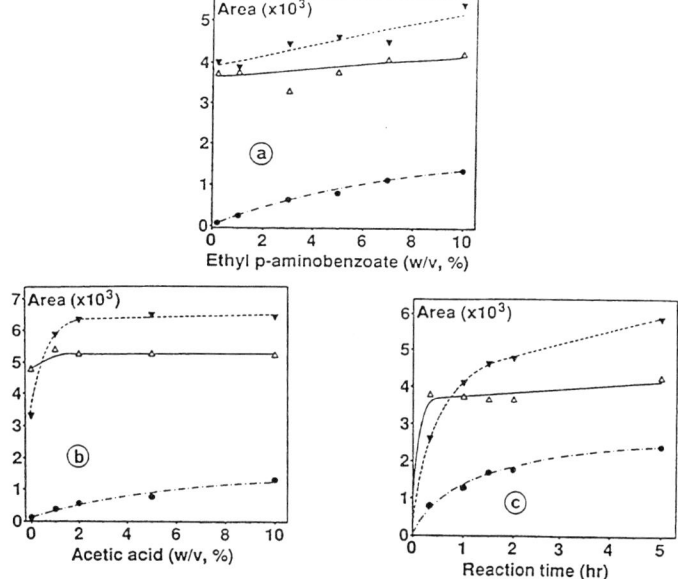

Figure 4.10 Effects of various factors on precolumn derivatization of selected monosaccharides by ethyl-p-aminobenzoate. (a) Concentration of ethyl p-aminobenzoate. The concentrations of acetic acid and sodium cyanoborohydride were 10 and 1% (w/v), respectively. Reaction temperature, 50°C; reaction time, 2 h. (b) Concentration of acetic acid. The concentrations of ethyl-p-aminobenzoate and sodium cyanoborohydride were 10 and 1% (w/v), respectively. Reaction temperature, 50°C; reaction time: 2 h. (c) Length of derivatization. The concentrations of ethyl p-amniobenzoate, acetic acid, and sodium cyanoborohydride were 10, 10, and 1% (w/v), respectively. Reaction temperature, 50°C. In all experiments the reaction products were detected after their CE separation at 305 nm (UV absorption maximum of all derivatives). The carrier electrolyte consisted of 175 mM borate, pH 10.5. Fused silica capillary dimensions were: 72 × 50 cm, 50 μm I.D. Separation voltage and temperature were +25 kV and 30°C, respectively. The concentration of monosaccharides in the sample solution was commonly 2 mg/ml. • Fructose; ▼ glucose; ∆ glucuronic acid. Reproduced with permission.[21]

pH. Since the formation of any given sugar-derivative–borate complex is facilitated under alkaline conditions, the molar fraction of the complex increases at higher pH values.[15] Such an increase enhances the electrophoretic mobilities (μ_{ion}, Figure 2.5) of derivatized saccharides, resulting in a reduction of their observed mobilities (μ_{obs}). However, the magnitude of such decreases in μ_{obs} with increasing pH differs among the derivatized carbohydrate species. An explanation for this is in the fact that the pH dependence of complexation varies with the configuration of hydroxyl groups. Under these conditions, a pH value of 10.5 offered a resolution greater than 1.2 for most of the important carbohydrates, in particular for glucose and arabinose.

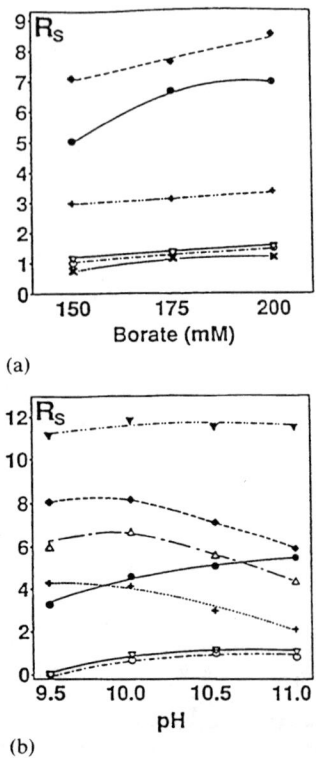

Figure 4.11 Effects of (a) pH and (b) borate concentration of carrier electrolyte on the separation of ethyl-*p*-aminobenzoate derivatives of selected carbohydrates. The separation conditions were as given in Figure 4.10, with the exception of Figure 4.11a for pH and Figure 4.11b for borate. (▼) 2-deoxy-D-ribose/xylose, (◆) 2-deoxy-D-ribose/rhamnose, (Δ) glucose/fucose, (+) fucose/galactose, (•) rhamnose/xylose, (∇) xylose/sorbose, (o) glucose/arabinose, and (x) glucose/mannose. Reproduced with permission.[21]

Figure 4.11b shows that an increasing borate concentration enhances the resolution. As indicated, this is a consequence of increased molar fractions of the sugar-derivative–borate complexes. However, another effect of increasing salt concentration is a decrease of the electroosmotic flow rate, due to a lower value of zeta potentials (see Section 2.3). A good value for the borate concentration appears to be 175 mM. This concentration of borate in the carrier electrolyte maximizes the resolution without causing excessively long run times.

Figure 4.12 shows a counterelectroosmotic electropherogram of a reaction mixture of 14 saccharides obtained under optimized conditions. This electropherogram illustrates a major advantage of derivatization by ethyl-*p*-aminobenzoic acid. The borate complexes of the derivatives are well sepa-

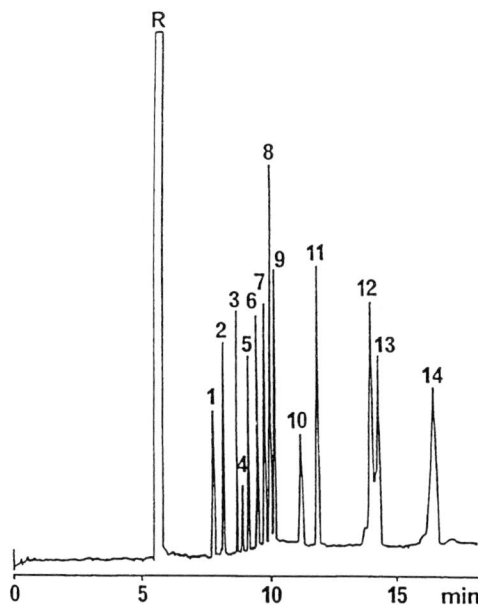

Figure 4.12 Separation of a mixture of mono- and oligosaccharides derivatized with ethyl-*p*-aminobenzoic acid. CE conditions were as indicated in Figure 4.11, including the 175 mM concentration of borate in the carrier electrolyte. Vacuum injection: 1.0 sec. Peak identification: R, reagent; 1, 2-deoxy-D-ribose; 2, maltotriose; 3, rhamnose; 4, cellobiose; 5, xylose; 6, ribose; 7, lactose; 8, glucose; 9, arabinose; 10, fucose; 11, galactose; 12, mannuronic acid; 13, glucuronic acid; 14, galacturonic acid.

rated from the large reagent peak. Unlike *p*-aminobenzoate (see Figure 4.6), ethyl-*p*-aminobenzoic acid has a very low value of electrophoretic mobility, and its rate of migration is determined largely by electroosmotic flow. Any masking of small analyte peaks, by the much larger reagent peak is thus precluded. From the techniques discussed so far, aminopyridine offered a similar advantage as ethyl-*p*-aminobenzoic acid by not interfering with the peaks due to the analytes (see Figure 4.3). However, we shall also recall that aminopyridine is useful for the derivatization of aldoses only and that it cannot be utilized for the detection of ketoses.

Table 4.2 contains values of electrophoretic mobilities (μ_{ion}) calculated from Figure 4.12 and from other electropherograms generated under the same conditions. The electrophoretic mobilities were calculated using Equation (4.1).

From the values in Table 4.2, it is evident that aldopentoses with cis-oriented hydroxyl groups at C3/C4 (arabinose and ribose) migrate faster than those with trans-oriented hydroxyl groups in the same position (lyxose and

Table 4.2. RELATIONSHIP BETWEEN ELECTROPHORETIC MOBILITIES [μ_{ion}, SEE EQ. (4.1)] OF ETHYL 4-AMINOBENZOATE DERIVATIVES AND CARBOHYDRATE STRUCTURES

Ethyl-p-Amino-Benzoate Derivatives	μ_{ion} (10^{-4} cm^2 V^{-1} sec^{-1})	Orientation of the Vicinal Hydroxyl Groups at C-3/C-4
Arabinose	1.917	cis
Lyxose	1.522	trans
Ribose	1.753	cis
Xylose	1.669	trans
Galactose	2.257	cis
Glucose	1.906	trans
Fucose	2.138	cis
Rhamnose	1.490	trans
Galacturonic acid	2.841	cis
Glucuronic acid	2.626	trans
N-acetylgalactosamine	1.373	cis
N-acetylglucosamine	1.833	trans

xylose). The former two compounds are known to undergo complexation with borate more easily than the latter two.

The same is true for aldohexoses and hexuronic acids. N-acetylhexosamines, on the other hand, showed a reverse effect regarding 3,4-disposition, presumably due to the contribution of the N-acetyl group at C2. These observations are in agreement with those made by Honda et al.[22] for N-2 pyridylglycamines.

From the borate complexes of ethyl-p-aminobenzoic acid derivatives of ketoses, the complex derived from fructose with its cis-oriented hydroxyl groups at C4/C5 was found to migrate faster than that of sorbose with its trans-oriented hydroxyl groups at C4/C5.

A comparison with high-performance liquid chromatography on an amine-bonded vinyl alcohol copolymer column[20] reveals that monosaccharides such as fucose and rhamnose or xylose and arabinose, which cannot be separated by chromatography, are well resolved by means of capillary electrophoresis. On the other hand, however, mannose and arabinose, which can be well resolved on the amine-bonded vinyl alcohol copolymer column, cannot be separated by capillary electrophoresis. Therefore, it can be concluded that high-performance liquid chromatography and the capillary electrophoretic methods presented here are not competing, but rather complementary. As in the preceding section, quantitative information from CE analyses of ethyl-p-aminobenzoate derivatives was normalized by dividing peak areas by the respective migration times and relating the resulting ratio to that of 2-

deoxy-D-ribose used as an internal standard. The calibration plots showed excellent linearity over a range of 0.5–20 mM.

The mass detection limit for glucose, by the discussed method, is approximately 7 fmol at a signal–to–noise ratio of 3. This mass detection limit is close to that of CE in conjunction with fluorogenic derivatization and fluorescence detection. The mass detection limits of mono- and oligosaccharides derivatized with fluorophores were reported to lie in the upper attomole range.[17]

The reader should note that these mass detection limits of CE/fluorescence and CE/UV detection are five magnitudes better than those for carbohydrates separated by HPLC using triple-pulse amperometric detection with a gold-working electrode.[18] The concentration sensitivity of the CE method with precolumn derivatization by ethyl-*p*-aminobenzoate is about 2 μM. This level of sensitivity is just about matched by the concentration detection limits attainable by pulsed amperometric detection with a 200-μl injection. It is interesting to point out that the CE methods utilizing only ~ 4 nl of a sample are shown here to give approximately the same concentration sensitivity as the HPLC technique with a 50,000 times larger sample volume. As explained in Section 3.3, the high sensitivity of CE is due to a combination of factors such as separation efficiency, electrostacking, etc.

4.1.1.6 Counterelectroosmotic CE of 4-Aminobenzonitrile Derivatives of Carbohydrates

In order to expand the capabilities of the CE approach to carbohydrate analysis further, a new precolumn derivatization method has been developed.[23,24] Figure 4.13a shows the reaction scheme of reductive amination of carbohydrates with 4-aminobenzonitrile in the presence of 10% acetic acid and 1% sodium cyanoborohydride in methanol.

Derivatization of carbohydrates by aminobenzonitrile enables a sensitive UV detection and provides new selectivity for separations of oligo- and monosaccharides. The large peak caused by the excess concentration of the precolumn reagent is well separated from all analytes of interest.

Whereas 2-aminobenzonitrile sugar derivatives have a UV-absorption maximum of approximately 213 nm, 4-aminobenzonitrile derivatives have their maximum at 285 nm.[23] Figure 4.13b shows an example of a separation 4-aminobenzonitrile derivatives of oligo- and monosaccharides.

The rate of precolumn derivatization by 4-aminobenzonitrile is determined by relatively slow reaction kinetics and can be controlled by temperature. At 90°C, the reaction gives a maximum yield within 15 min for aldoses. Ketose reactions are slower, but their yield is high enough to attain an excellent sensitivity after 15 min of derivatization (see Figures 4.14a and b).

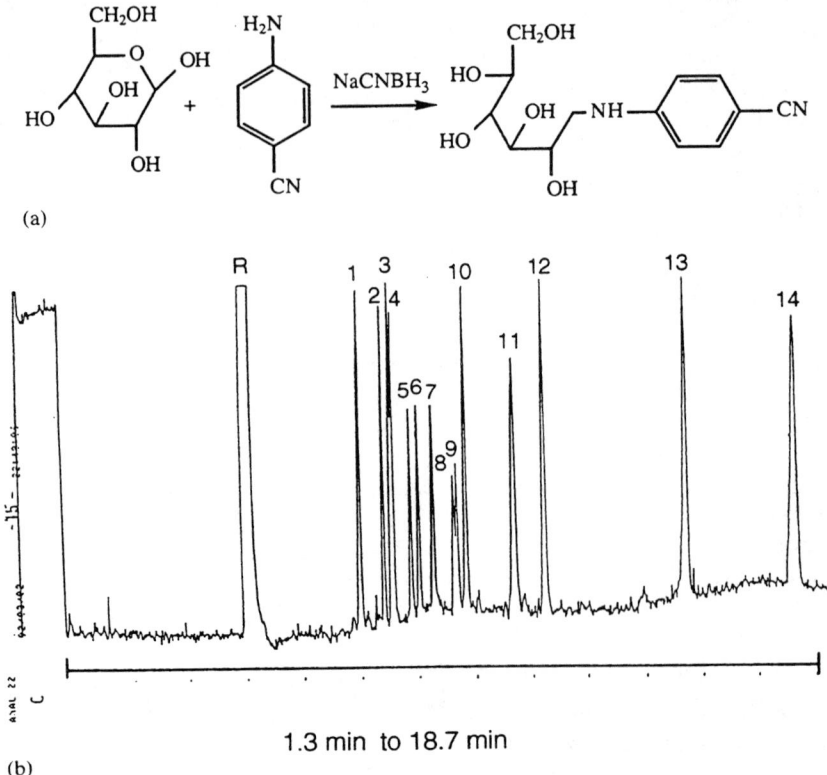

Figure 4.13 (a) Reductive amination of glucose with p-aminobenzonitrile and sodium cyanoborohydride. (b) Separation of 4-aminobenzonitrile derivatives of carbohydrates. Fused silica capillary: 72 × 50 cm, 75 μm I.D. Carrier eletrolyte: 175 mM borate, pH = 10.5. Voltage: 24 kV. Vacuum sampling for 1 sec. UV detection wavelength: 285 nm. Peak identities: (1) maltotriose, (2) rhamnose, (3) lyxose, (4) cellibiose, (5) melibiose, (6) sorbose, (7) ribose, (8) glucose, (9) fructose, (10) arabinose, (11) fucose, (12) galactose, (13) glucoronic acid, (14) galacturonic acid).

4.1.1.7 Micellar Electrokinetic Chromatography of 4-Aminobenzonitrile Derivatives of Carbohydrates

The new precolumn derivatization of carbohydrates by 4-aminobenzonitrile, described in Section 4.1.1.6, makes possible not only the conventional counterelectroosmotic separations discussed so far for those species, it can also be used to prepare the carbohydrate analytes for analysis by micellar electrokinetic chromatography (MECC). The general principles of MECC are explained in Section 2.4. Without suitable pretreatment, carbohydrates are too hydrophilic to be solubilized in ionic surfactants. The precolumn derivatization by 4-aminobenzonitrile introduces a hydrophobic moiety into the

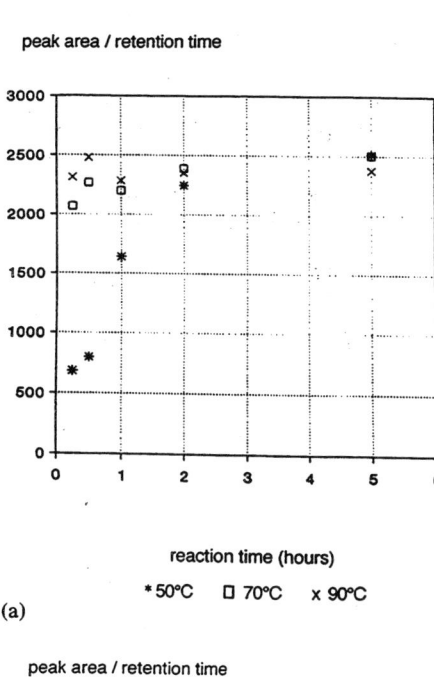

(a)

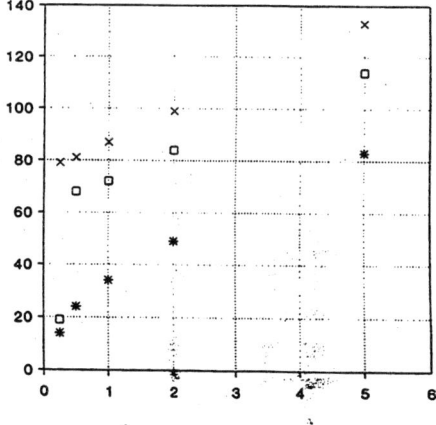

(b)

Figure 4.14 Influence of reaction temperatures on the yields of derivatization reaction between 4-aminobenzonitrile with adoses and ketoses. (a) Reaction yields versus reaction times for derivatization of glucose. (b) Reaction yields versus reaction times for derivatization of fructose.

sugar molecules (see Figure 4.13), which permits a separation to take place as a result of differences in the partitioning of various derivatized carbohydrates into the micelle.[24]

In a pressure-stable reaction vessel, which prevents evaporation of the solvent, derivatization by 4-aminobenzonitrile can be carried out at 90°C instead of 50°C, which reduces the reaction time from 2 h to 15 min.

The reaction mixture can then be introduced directly into fused silica capillaries, filled with a carrier electrolyte consisting of 100 mM sodium dodecyl sulfate (SDS) as the anionic surfactant in 15 mM phosphate buffer, pH 7.6. The MECC method for the 4-aminobenzonitrile derivatives reduces the time of analysis from 15 min required by the standard counterelectroosmotic CE in borate electrolytes (Section 4.1.1.6) to less than 10 min. Moreover, the MECC method separates the diastereomers formed during derivatization of ketoses, the existence of which has been confirmed by nuclear magnetic resonance spectroscopy.[24] An MECC separation of a mixture of 18 derivatized carbohydrates is shown in Figure 4.15.

As explained in more detail in Section 2.4, the separation principle of

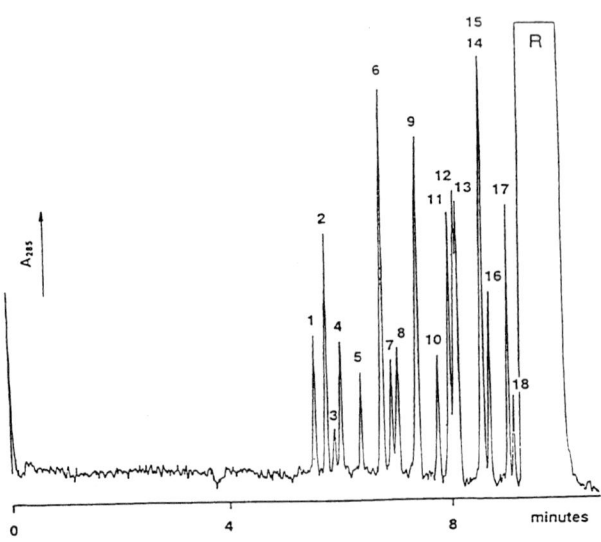

Figure 4.15 Micellar electrokinetic chromatography of 4-aminobenzonitrile derivatives of carbohydrates. Fused silica capillary: 72 × 50 cm, 75 μm I.D. Carrier electrolyte: 100 mM SDS in 15 mM phosphate buffer, pH 7.6; voltage: +30 kV; current: 61 μA. Direct UV detection at 285 nm. Vacuum sample introduction: 1 sec. Peak identification: 1, fructose; 2, sorbose; 3, fructose; 4, lactose; 5, sorbose; 6, melibiose; 7, maltotriose; 8, cellibiose; 9, maltose; 10, mannose; 11, glucose; 12, galactose; 13, ribose; 14, lyxose; 15, rhamnose; 16, arabinose; 17, fucose; 18, xylose; R, p-aminobenzonitrile.

MECC relies on the use of anionic surfactants at a concentration that is higher than the critical micellar concentration (CMC). The charged micelles that are formed above the CMC have their own mobility and will show a tendency to move toward the anode. The electroosmotic flow, however, with its greater velocity in the opposite direction, will cause the micelles to be swept toward the cathode. The derivatization reagent (4-aminobenzonitrile), which is the most hydrophobic molecule in the reaction mixture, reaches the detector in the last of the detected peaks because of its high solubility in the micelles.

Resolution of carbohydrates shows a steplike dependence on the SDS concentration. It is very poor at concentrations of SDS below 40 mM ($R = 1.005$ for melibiose/maltotriose, $R = 0.087$ for glucose/galactose at 40 mM SDS), despite the fact that the CMC value for SDS is 8.2 mM.[25] The resolution improves rapidly between 40 and 80 mM of SDS, but remains nearly constant at concentrations above 80 mM SDS ($R = 2.895$ for melibiose/maltotriose, $R = 1.981$ for glucose/galactose at 100 mM SDS). The MECC technique discussed, which includes a precolumn derivatization by 4-aminobenzonitrile and an on-column UV monitoring at 285 nm, yields 1 femtomole as a mass detection limit for glucose. The concentration detection limit for glucose was found to be 1 μM. We can conclude that the micellar electrokinetic chromatography of derivatized carbohydrates represents another useful analytical tool for sugars.

4.1.2 Indirect UV Detection of Carbohydrates

As explained in Sections 3.3.3.2 and 3.3.2, indirect detection techniques appear to be of greater practical utility in photometric detection than with fluorescence. That notwithstanding, we should note that indirect fluorescence has been applied successfully to CE detection of carbohydrates. The reported detection limits are in the femtomolar range.[26]

Indirect UV detection can routinely achieve detection limits in the picomolar range, provided that a highly UV-absorbing carrier electrolyte coion can be found with an electrophoretic mobility close to those of the analytes of interest.

The selection of sorbic acid as a carrier electrolyte anion and chromophore for the indirect UV photometric detection of carbohydrates in capillary electrophoresis is based on several reasons. First, sorbic acid has a high molar absorptivity coefficient ($\varepsilon = 27800$ at 256 nm). Second, it is compatible with the solvent system used. Third, it carries a single charge. This ensures a good transfer ratio [TR, see Eq. (3.15)], which is defined as the number of chromophore molecules displaced by one analyte molecule. Fourth, sorbic acid interacts neither with the analytes nor with the capillary surface. And, fifth, its effective mobility matches the ionic mobilities of carbohydrates suppressing any possibility of peak spreading.

Carbohydrates are weak acids. As an example, the pK for glucose is 12.35.[27] Therefore, a carrier electrolyte has to be made very basic for the ionization of carbohydrates to occur. A dissociation of sugars is a prerequisite both for their CE separation based on differences in migration velocity, and for their indirect detection based on charge displacement. This means that the pH of carrier electrolytes must approach 12 in order to convert a substantial fraction of carbohydrates into ionized form. However, when the pH of the carrier electrolyte is adjusted that high, the concentration of hydroxide ions is no longer negligible relative to the concentration of sorbic acid or any other chromophore used for indirect detection. This results in a decrease of the transfer ratio described by:

$$TR_{total} = \alpha[sugar]/[C] + [OH^-] \qquad (4.2)$$

where TR_{total} is the total transfer ratio, $\alpha[sugar]$ the concentration of sugar molecules ionized, $[C]$ the concentration of chromophore, and $[OH^-]$ the concentration of hydroxide ions. It can be seen that at constant sugar and chromophore concentrations, α in the numerator and $[OH^-]$ in the denominator are competing functions of pH. As a consequence, TR_{total} goes through a maximum if plotted as function of pH (Figure 4.16).

The pH at this maximum is the optimal pH for detection. Similar considerations preclude the use of borate due to the high concentration required

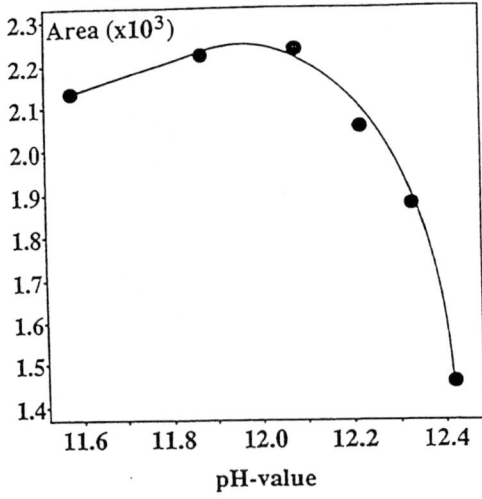

Figure 4.16 Impact of pH value on sensitivity. Electrolyte: 6 mM sorbate. Fused silica capillary, 122 × 100 cm, 50 μm I.D. Current: 6, 8, 13, 17, 20, and 24 μA at pH 11.58, 11.87, 12.08, 12.22, 12.33, and 12.42, respectively. Voltage: +28 kV. Separation temperature: 30°C. UV detection at 256 nm. Sampling by vacuum, 1.0 sec. Sample: 12.5 mM mannose.

(100–200 mM) for an efficient complexation of sugars. While the pH of the carrier electrolyte does not exert any significant impact on the resolution of sugar acids, a slight increase in the mobilities of sugars was observed at more alkaline values (Figure 4.17).

The increasing concentration of sodium ions in the carrier electrolyte with rising pH increases the thickness of the diffusion double layer at the inner capillary wall,[28] causing the electroosmotic flow to decrease. For that reason, the observed electrophoretic mobilities decreased considerably on going from pH 11.9 to 12.3, leading to an increase of migration times for mannuronic acid from 19 to 40 min. Generally, however, the resolution of sugars improved with rising pH, as depicted in Figure 4.18.

Due to the excessively long migration times, the enhancement of resolution observed above pH 12.1 is without a practical value. The pH of 12.1 was chosen for all subsequent analyses, since it offered maximal sensitivity and sufficient resolution within an acceptably short time of analysis. Moreover, despite the high pH of the carrier electrolyte, no alkaline degradation of the carbohydrates could be observed at the ambient separation temperature. This finding is in agreement with a previous study.[29]

Another major consideration in the optimization of indirect photometric detection for carbohydrates is the concentration of chromophore. As shown in Figure 4.19, the response in terms of peak areas has its maximum at approximately 6 mM.

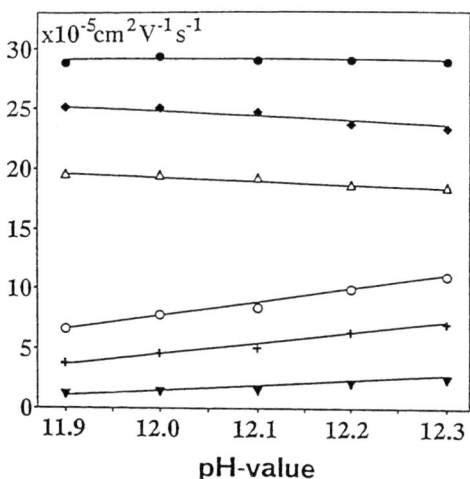

Figure 4.17 Impact of pH on mobilities and resolution of carbohydrates: (•) mannuronic acid, (♦) gluconic acid, (Δ) N-acetylneuraminic acid, (o) mannose, (+) galactose, (▼) raffinose. Carrier electrolyte: 6 mM sorbate. Fused silica capillary: 100 × 92.5 cm, 75 μm I.D. Current: 21 μA; voltage: 27.8, 23.4, 20.3, 16.4, 13.1 kV at pH 11.9, 12.0, 12.1, 12.2, 12.3, respectively. Separation temperature: ambient. UV detection at 254 nm. Hydrostatic sample introduction, 30 sec.

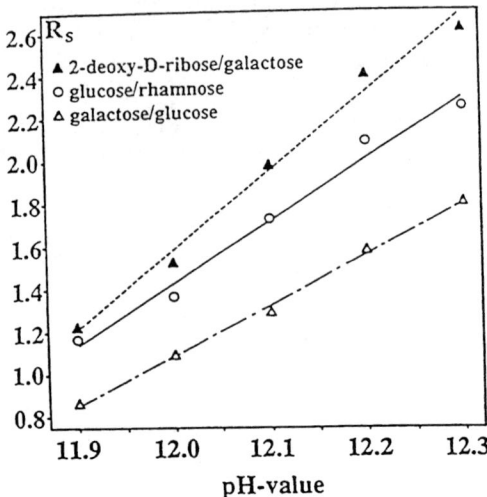

Figure 4.18 Impact of pH value on resolution of carbohydrates: (△) galactose/glucose, (o) glucose/rhamnose, (▲) 2-deoxy-D-ribose/galactose. Electrophoretic conditions were as indicated in Figure 4.17.

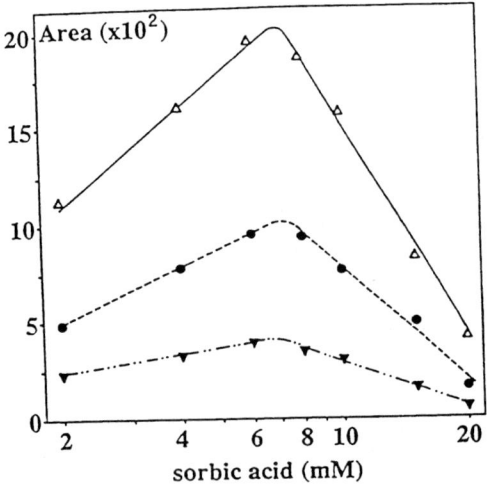

Figure 4.19 Impact of the concentration of sorbic acid on response: △, 33.3 pmol mannose; ●, 16.6 pmol mannose; ▼, 8.3 pmol mannose. Carrier electrolyte: sorbate, pH 12.1. Fused silica capillary: 122 × 100 cm, 50 μm I.D. Voltage: 28 kV. Separation temperature: 30°C. UV detection at 256 nm. Sampling by vacuum, 0.8 sec.

At its optimum concentration of 6 mM, the sorbate-containing electrolyte enables a detection limit of 2 picomoles of glucose at a signal-to-noise ratio of 3. The concentration sensitivity, however, is comparatively low (~0.5 mM) due to the small portion of dissociated carbohydrate molecules. At lower concentrations of sorbic acid, the sensitivity decreases due to the relatively high content of hydroxide ions in the background electrolyte. At concentrations above 6 mM, the response deteriorates because less light reaches the photodiode. This reduces the dynamic range available for indirect detection[30] (see also Section 3.3.2). Figure 4.20 shows the separation of a mixture of eleven sugars and sugar acids with concentrations ranging from 0.95 to 2.66 mM in the sample.

At the pH value selected, and the positive polarity of the sampling side of the capillary, the carbohydrates show the tendency to migrate away from the detector toward the anode. The separation mode can again be described as counterelectroosmotic (see Section 2.4) due to the large electroosmotic flow in which the analytes are propelled together with the bulk of the carrier electrolyte solution toward the cathode. Under counterelectroosmotic separation conditions, sugars dissociated least are detected first, since they are least able to migrate upstream (Figure 2.8B). Sugar acids, on the other hand,

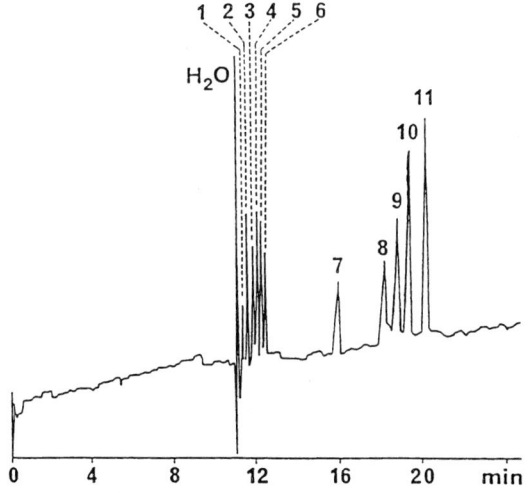

Figure 4.20 Capillary electrophoresis and indirect UV detection of carbohydrates. Carrier electrolyte: 6 mM sorbate, pH 12.1. Fused silica capillary: 122 × 100 cm, 50 μm I.D. Current: 13 μA; voltage: +28 kV. Separation temperature: 30°C. Indirect UV detection at 256 nm. Sampling by vacuum, 2.0 sec. Sample concentration: (0.95–2.66 mM). Peak identification: 1, raffinose; 2, 2-deoxy-D-ribose; 3, galactose; 4, glucose; 5, rhamnose; 6, mannose; 7, N-acetylneuraminic acid; 8, gluconic acid; 9, galacturonic acid; 10, glucuronic acid; 11, mannuronic acid.

are detected last due to their greater upstream migration rate as a result of the complete dissociation of the carboxyl group.

In Section 3.3.2, we discussed how the range of linearity in indirect photometric detection is limited by the concentration of the chromophore. For weak acids such as carbohydrates, the linear range is limited also by the degree of dissociation of the analyte. The calibration plots of all carbohydrates separated in the 8 mM sorbate electrolyte, under the conditions given in Figure 4.20, showed good correlation coefficients over two orders of magnitude and negligible Y intercepts. Interestingly, both separation efficiency and resolution reached their optimum at relatively high carbohydrate concentrations, roughly equal to that of sorbic acid (Table 4.3). Below and above that concentration, both parameters were found to decrease.

A comparison with ion partition chromatography[31] shows that sugars such as galactose, mannose, and xylose, which cannot be separated by chromatography, are resolved by means of capillary electrophoresis. Glucose and arabinose, on the other hand, can be well resolved in ion partition chromatography, but not by capillary electrophoresis. As in Section 4.1.1.5, we can again conclude that HPLC and CE are not necessarily two competing, but rather two complementary, techniques in carbohydrate analysis.

The practical utility of CE with indirect photometric detection for the determination of sugars is demonstrated by the analysis of an orange juice sample shown in Figure 4.21.

Table 4.3. EFFECT OF CARBOHYDRATE CONCENTRATION ON SEPARATION EFFICIENCY (THEORETICAL PLATES PER METER) AND RESOLUTION

Sampling Time [sec]	Sample Amount [pmol]	Carbohydrate Concentration [mmol/l]	Number of Theoretical Plates [10^5/m]		Resolution
0.4	12.5	12.5	galactose	1.069	2.38
	12.5	12.5	lactulose	0.767	
0.6	12.5	8.3	galactose	1.405	2.87
	12.5	8.3	lactulose	0.908	
1.0	12.5	5.0	galactose	1.732	2.96
	12.5	5.0	lactulose	1.224	
1.6	12.5	3.1	galactose	1.599	2.84
	12.5	3.1	lactulose	1.036	
2.2	12.5	2.3	galactose	1.274	2.39
	12.5	2.3	lactulose	0.791	
2.8	12.5	1.8	galactose	1.144	2.25
	12.5	1.8	lactulose	0.728	
3.2	12.5	1.5	galactose	0.926	2.15
	12.5	1.5	lactulose	0.620	

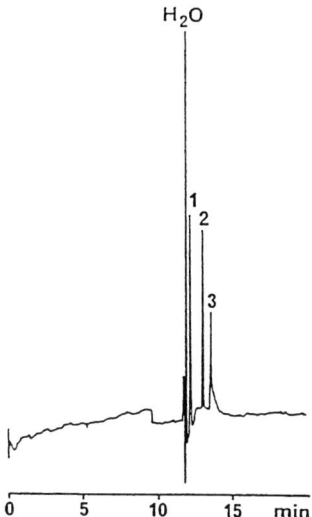

Figure 4.21 CE analysis of an orange juice sample. Electrolyte: 6 mM sorbate, pH 12.1. Fused silica capillary: 112 × 90 cm, 50 μm I.D. Current: 11 μA; voltage: +24 kV; separation temperature: 30°C. Indirect UV detection at 256 nm. Sampling by vacuum, 2.0 sec. Sample: orange juice, diluted 1:25 with bidistilled water. Peak identification: 1, saccharose; 2, glucose; 3, fructose.

4.1.3 Relative Value of CE Methods for Carbohydrates

CE methods for carbohydrates have various degrees of usefulness, not only as compared among themselves, but also relative to other techniques currently in use for sugars, such as, for example, HPLC with amperometric or refractive index detection. In the following text, we shall discuss their applicability to various classes of carbohydrates, freedom from interference, selectivity of separation, and sensitivity of detection as the most important criteria for choosing among the available instrumental techniques for the analysis of carbohydrates.

4.1.3.1 Applicability to Different Types of Carbohydrates

Unless restricted by a separation mode, HPLC methodologies, with refractive index or amperometric detection, are useful for all classes of sugars. Among the CE methods, limited applicability decreases the practical usefulness of fluorescence tagging (see Figure 4.2, aldoses only) and of the precolumn derivatization by 2-aminopyridine (Figure 4.3, aldoses and uric acids, no ketoses). Three of the discussed CE precolumn techniques (4-aminobenzoate, Figure 4.6; ethyl-4-aminobenzoate, Figure 4.12; and 4-aminobenzonitrile, Figure 4.14) are capable of analyzing all classes of carbohydrates.

4.1.3.2 Freedom from Interference

The excess of reagent, required to drive the derivatization reaction to a high yield and to make the precolumn derivatization method applicable over a wide range of concentrations, represents the main possibility of interference in UV-detected electropherograms. Fluorescence tagging does not usually generate any major peaks due to the tagging reagent; therefore, direct fluorescence detection after a precolumn derivatization appears to be the most interference-free capillary electrophoretic method for carbohydrates. In three of the discussed CE methods (2-aminopyridine, Figure 4.6; ethyl-4-aminobenzoate, Figure 4.12; and 4-aminobenzonitrile, Figure 4.14) combining precolumn derivatization with direct UV detection, the reagent peak appears in the electropherograms, but is well separated from the range of migration times reserved for the analyte peaks. These three approaches are, for practical purposes, as interference free as the fluorescence tagging methods.

4.1.3.3 Selectivity of Separation

From Table 4.4 it is apparent that the discussed CE methods for carbohydrates differ in selectivity.

Glucose and mannose, for instance, are well resolved when derivatized with 2-aminopyridine or ethyl-4-aminobenzoate. Their 4-aminobenzoic acid derivatives, however, cannot be resolved under the electrophoretic conditions used. On the other hand, it has not been possible to separate the ethyl-4-amino-benzoate derivatives of fructose, mannose, and arabinose, while their 4-aminobenzoic acid derivatives are resolved successfully. This indicates that the additional alkyl function of ethyl-4-aminobenzoate affects borate complexation. Therefore, selectivity and resolution may not be optimized only by varying pH and carrier electrolyte concentration, but also by choosing the derivatizing agent most appropriate for the carbohydrates to be separated.

Additionally, it should also be noted that the selectivity of all CE separations of carbohydrates is different from the selectivity observed in chromatographic separations on anion-exchange columns.

4.1.3.4 Sensitivity of Detection

The concentration sensitivity of indirect photometric UV detection compares favorably to that obtained with refractive index detection in high-performance liquid chromatography,[18] but it does not match the sensitivity that can be achieved with precolumn derivatization and direct UV detection in capillary electrophoresis (Table 4.5). The mass detection limits of 4-amino-

Table 4.4. ELECTROPHORETIC MOBILITIES [μ_{ion}, EQ. (4.1)] OF UNDERIVATIZED AND DERIVATIZED CARBOHYDRATES

No.	Carbohydrates	Mobility[a] [10^{-5} cm^2 V^{-1} sec^{-1}]	Mobility[a] [10^{-4} cm^2 V^{-1} sec^{-1}]	Mobility[c] [10^{-4} cm^2 V^{-1} sec^{-1}]	Mobility[d] [10^{-4} cm^2 V^{-1} sec^{-1}]
1	Raffinose	1.312	n.d.[e]	n.d.	n.d.
2	Saccharose	1.356	n.d.	n.d.	n.d.
3	2-Deoxy-D-galactose	2.419	1.540	1.486	2.912
4	2-Deoxy-D-ribose	2.798	1.230	1.240	2.704
5	D-Fucose	3.096	2.210	2.138	3.393
6	Lactose	4.175	1.830	1.855	2.849
7	Maltotriose	4.313	1.310	1.363	2.178
8	D-Galactose	4.358	2.350	2.257	3.509
9	Melibiose	4.623	1.710	1.705	2.746
10	Cellobiose	4.794	1.560	1.602	2.662
11	Maltose	4.813	1.490	1.540	2.619
12	L-Arabinose	5.121	2.000	1.913	3.274
13	D-Glucose	5.135	1.920	1.871	3.204
14	Lactulose	5.403	n.d.	n.d.	n.d.
16	D-Xylose	6.268	1.740	1.664	3.159
17	D-Lyxose	6.440	1.610	1.522	3.041
18	L-Sorbose	6.468	n.d.	1.731	3.121
19	L-Rhamnose	6.599	1.570	1.490	n.d.
21	D-Fructose	7.140	n.d.	1.898	3.218
22	D-Ribose	7.419	1.770	1.753	2.949
23	D-Mannose	7.462	1.970	1.908	3.204
25	D-Galactonic acid	24.950	n.d.	n.d.	n.d.
26	D-Gluconic acid	25.518	n.d.	n.d.	n.d.
27	D-Mannonic acid	25.677	n.d.	n.d.	n.d.
28	D-Galacturonic acid	26.789	2.990	2.841	4.034
29	D-Arabonic acid	27.593	n.d.	n.d.	n.d.
30	D-Glucuronic acid	27.796	2.820	2.626	3.989
31	D-Ribonic acid	28.139	n.d.	n.d.	n.d.
32	D-Mannuronic acid	29.315	2.800	2.585	3.835

[a] Underivatized carbohydrates, 6 mM sorbate, pH 12.1, 28 kV, L_t = 122 cm, L_D = 100 cm.
[b] N-2-pyridylglycamines, 150 mM borate, pH 10.0, 25 kV, L_t = 72 cm, L_D = 50 cm.
[c] Ethyl p-aminobenzoate derivatives, 175 mM borate, pH 10.5, 25 kV, L_t = 72 cm, L_D = 50 cm.
[d] p-aminobenzoate derivates, 150 mM borate, pH 10.0, 28 kV, L_t = 72 cm, L_D = 50 cm.
[e] n.d. = not detectable.

Table 4.5. MASS AND CONCENTRATION DETECTION LIMITS (3 TIMES S/N) FOR GLUCOSE

Derivatization or Detection Method	λ_{max} [nm]	Mass Detection Limit [fmole]	Concentration Detection Limit [μM]
4-Aminobenzonitrile	285	1	1
Ethyl-4-aminobenzoic acid	305	7	2
4-Aminobenzoic acid	285	15	4
2-Aminopyridine	237	29	8
Indirect UV, sorbic acid	256	2000	500
Direct UV detection of borate complexes	195	ca. 10^6	—
HPLC with RI detection	—	2.2×10^6	110
HPLC with PAD detection	—	0.44×10^6	2

benzoate, ethyl-4-aminobenzoate, and 4-aminobenzonitrile derivatives are not only 2 to 27 times lower than those obtained for N-2-pyridylglycamines, but they also approach the upper attomole levels achieved for electrophoretically separated mono- and oligosaccharides derivatized with various fluorophores.[32] Moreover, they are four to five orders of magnitudes better than the detection limits for carbohydrates separated by high-performance anion-exchange chromatography and detected by amperometric detection using a gold-working electrode.[18] With regard to the concentration sensitivity, amperometric detection for a 200-μl injection matches approximately that of direct UV detection with a 4-nl sample of glucose derivatized by ethyl-p-aminobenzoate. In the opinion of the authors, precolumn derivatization with 4-aminobenzonitrile in combination with capillary electrophoresis and direct UV detection becomes the most sensitive method currently available for the analysis of carbohydrates.

4.2 Counterelectroosmotic CE of Flavonoids

Flavonoids are ubiquitous secondary plant metabolites, which are widely used as remedies because of their spasmolytic, antiphlogistic, antiallergic, and diuretic properties. Their structure being based on 2-phenylbenzopyrone (Figure 4.22), flavonoids differ in the degree of saturation and position of hydroxyl, methoxy, and sugar residues.[33]

As analogues of cinnamic acid often accompany flavonoids,[34] their simultaneous determination is of relevance to phytopharmaceutical chemistry. So far, their analysis has been accomplished by high-performance liquid chromatography,[35–39] thin-layer chromatography,[40–43] gas chromatography,[44] and isotachophoresis.[45] The following text discusses the use of capillary electrophoresis for the rapid separations of flavonoids. The electropherogram in

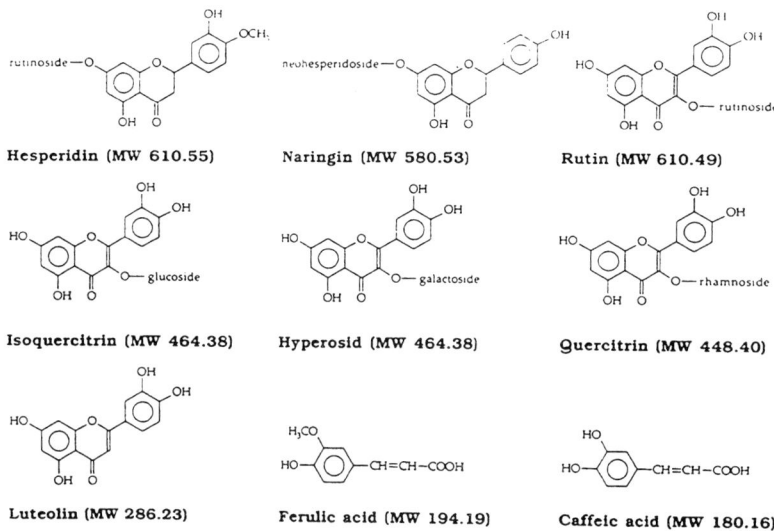

Figure 4.22 Structures of the investigated flavonoids and phenolcarboxylic acids.

Figure 4.23 illustrates the separation of a standard mixture of flavonoids and phenolcarboxylic acids.

At the pH of the carrier electrolyte selected, the analytes are negatively charged and migrate away from the detector toward the anode. However, due to the large electroosmotic flow in the system, which is faster than their electrophoretic migration, flavonoids are propelled together with the bulk of electrolyte solution toward the cathode. Under these counterelectroosmotic conditions (see Section 2.4), the flavonoids dissociated least and highest in molecular weight are detected first, since they are less able to migrate upstream. The analogues of cinnamic acid, however, are detected last due to their greater upstream mobility as a result of the dissociation of the carboxyl and phenolic hydroxyl groups. This is in accordance with the observation by Fujiwara and Honda,[46] who demonstrated counterelectroosmotic capillary electrophoretic separations of various derivatives of cinnamic acid using a phosphate carrier electrolyte. Figure 4.24 compares the electrophoretic mobilities of analytes at different borate concentrations.

With increasing borate concentration, electroosmotic flow decreases gradually, due to lower zeta potentials. The decreased values of zeta potentials are a consequence of increasing thickness of double layer at the inner capillary wall.[28] However, the decrease of electroosmotic flow rate is not linear as expected. This is probably due to the fact that the increased borate concentration leads to a current rise from 52 to 98 µA. Increasing current is known to cause a proportional increase in electroosmotic velocity.[47] Also the improved selectivity at higher borate concentrations achieved by increased

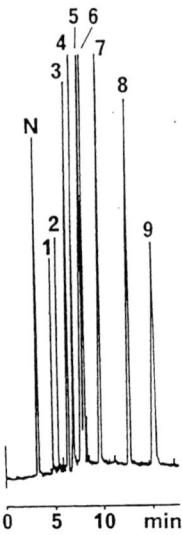

Figure 4.23 Separation of a standard solution of flavonoids and phenolcarboxylic acids. Fused silica capillary, 52 × 30 cm, 50 μm I.D., carrier electrolyte: 0.15 M borate buffer, pH 10. Separation voltage: +18 kV; current: 78 μA. Direct UV detection at 254 nm. Separation temperature: 32°C; vacuum sampling: 0.7 sec. Peak identification: (N) neutral marker, (1) hesperidin, (2) naringin, (3) rutin, (4) isoquercitrin, (5) hyperosid, (6) quercitrin, (7) luteolin, (8) ferulic acid, (9) caffeic acid.

differences in electrophoretic mobilities of the investigated flavonoids and phenolcarboxylic acids should be noted. As in the case of electrophoretic separations of carbohydrates,[48-51] this may be derived from an increase in concentration and stability of the complexes formed by the vicinal hydroxyl groups of flavonoids and phenolcarboxylic acids with borate.[52] However, the improvement in resolution (R_s) observed at higher borate concentrations was not only due to enhanced selectivity but also to the increase in the number of theoretical plates with increasing current.[53]

The reproducibility of solute mobilities was tested by six repeated separations of an identical standard solution. The coefficient of variation was found to be 1.76%.

4.2.1 Simultaneous Determination of Flavonoids and Carbohydrates

Capillary electrophoresis allows the simultaneous separation and detection of phytochemically relevant flavonoids, carbohydrates, and phenylcarboxylic acids. It was found that, even in a relatively simple borate-containing

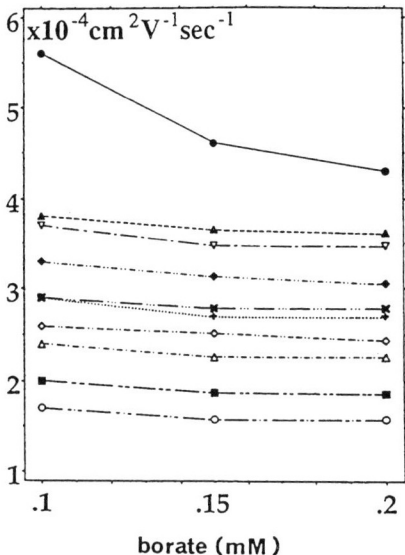

Figure 4.24 Effect of borate concentration on electroosmotic flow (•) and electrophoretic mobilities of various flavonoids and phenolcarboxylic acids: (o) hesperidin, (■) naringin, (Δ) rutin, (◊) isoquercitrin, (+) hyperosid, (x) quercitrin, (♦) luteolin, (∇) ferulic acid, (▲) caffeic acid. Fused silica capillary, 52 × 30 cm, 50 μm I.D. Carrier electrolyte: 0.10, 0.15, and 0.2 M borate, pH 10. Voltage: +18 kV; current: 52, 75, and 98 μA. Direct UV detection at 254 nm. Separation temperature: 32°C. Sampling by vacuum, 0.2 sec.

carrier electrolyte, the electrophoretic mobilities were different enough to allow good separations. A recent report describes a large number of applications for phytochemistry and food chemistry, using the separation method illustrated in Figure 4.25.[54]

4.3 Counterelectroosmotic CE of Sulfonamides

New methods for the analysis of pharmaceutically relevant compounds are of great interest, especially as a possible alternative to liquid chromatography. Let us consider, for illustration, a counterelectroosmotic separation of sulfonamides. This group of drugs consists of a great variety of chemically related compounds, and their fast analyses in different matrices is of importance. The electropherogram in Figure 4.26 shows a separation of a mixture of sulfonamides by a method allowing determinations of sulfonamides from meat extracts.[55]

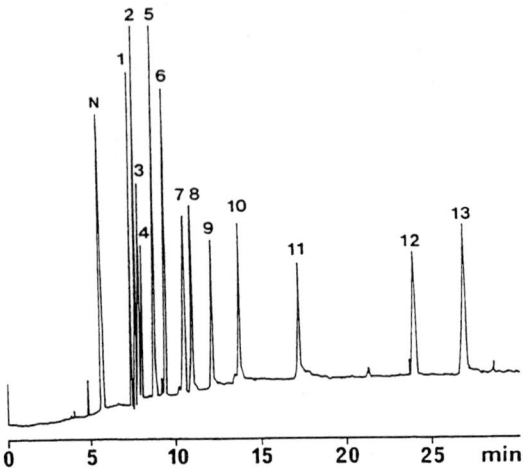

Figure 4.25 Simultaneous analysis of flavonoids and carbohydrates. Carrier electrolyte: 0.15 M borate, pH 10. Voltage: +25 kV. Vacuum sampling, 1 sec. UV detection at 254 nm. Peak identity: N, mesityl oxide; 1, 2-deoxyribose; 2, maltotriose; 3, gentibiose; 4, maltose; 5, ribose; 6, glucose; 7, fucose; 8, rutin; 9, isoquercitrin; 10, hyperosid; 11, luteolin; 12, ferulic acid; 13, caffeic acid.

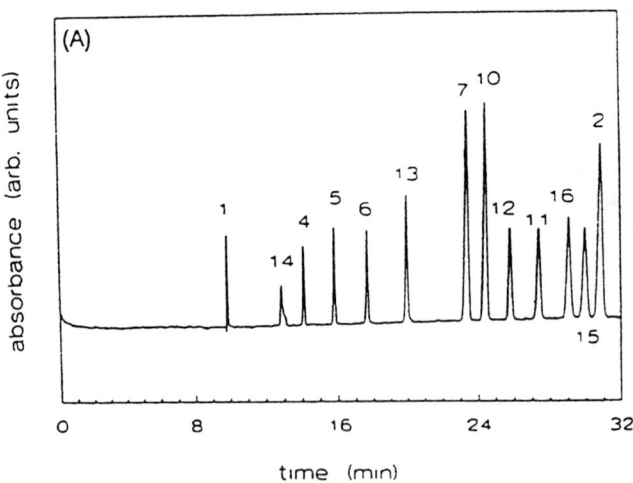

Figure 4.26 CE separation of eleven sulfonamides (each at 0.01 mg/ml). Fused silica capillary, $L_d = 50$ cm. Sample 39 nl, sampling by pressure. Voltage +10 kV. Carrier electrolyte, 0.02 M imidazole–acetate at pH 7. Peak identity: (1) trimethoprim, (2) sulfamethoxazole, (4) sulfadimidine, (5) sulfathiazole, (6) sulfamerazine, (7) sulfadiazine, (10) sulfadimethoxine, (11) sulfatroxazole, (12) sulfaquinoxaline, (14) sulfaguanidine, (16) sulfachloropyridazine. Reproduced with permission.[55]

4.4 Counterelectroosmotic CE of Biomass Degradation Products

Biomass is of great interest as a raw material for the pulp and paper industry, as well as a key source for chemicals. Due to the increasing need for environmental protection, the analysis of compounds, derived both from pulping and bleaching liquors as well as from different biomass hydrolysis products, is of importance.

Of special interest are phenols and phenolic components with different functional groups.

Rapid and reproducible methods for CE separations of phenolic compounds have been reported in the literature (Figure 4.27).[56,57]

The CE method illustrated in Figure 4.27 yields informative results for many types of biomass hydrolysis products, for example, for a degradation solution after hydrothermolysis of rice straw (see Figure 4.28). The electropherogram clearly identifies compounds with three basic phenylpropane units of the 4-hydroxyphenyl ("*p*-hydroxyphenyl propane"), the 3-methoxy-4-hydroxyphenyl ("guajacyl propane"), and the 3,5-dimethoxy-4-hydroxyphenyl ("syringyl propane") type. Special attention is paid to the presence of phenolic acids, especially to ferulic acid, syringic acid, vanillic acid, 4-

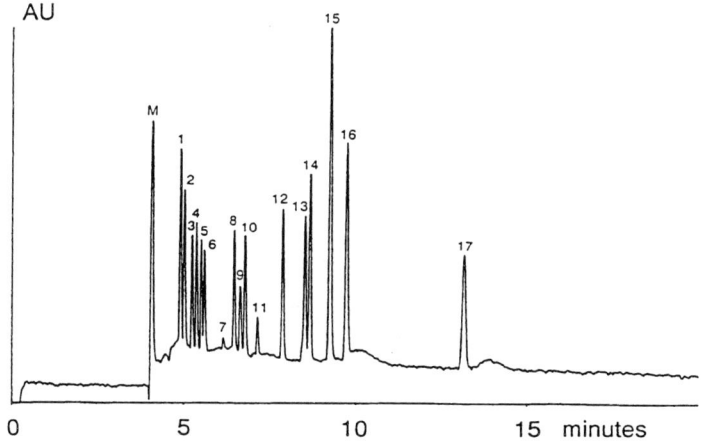

Figure 4.27 CE separation of a phenolic standard mixture containing: 1, salicyclic aldehyde; 2, 4-hydroxypropiophenone; 3, 4-hydroxyacetophenone; 4, α-hydropropiovanillon; 5, α-methoxypropioguajacon; 6, α-acetoxypropioguajacon; 7, 4-hydroxybenzylalcohol; 8, 4-hydroxybenzaldehyde; 9, vanillin; 10, coniferylalcohol; 11, sinapinic acid; 12, syringa acid; 13, benzylalcohol; 14, vanillic acid; 15, 4-hydroxybenzoic acid; 16, cinnamic acid; 17, pyromucic acid. Conditions: Fused silica capillary, 60 × 52.5 cm, 75 μm I.D. Voltage: +22 kV. Carrier electrolyte: 45 mM phosphate/15 mM borate, pH 7.7; M, neutral marker (mesityl oxide).

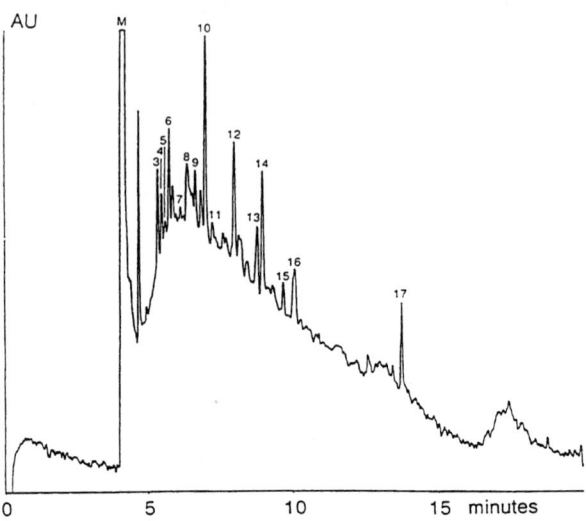

Figure 4.28 CE separation of a liquid eluate of rice straw after hydrothermolysis. Conditions and peak identities are the same as in Figure 4.27.

hydroxybenzoic acid, *p*-coumaric acid, and sinapic acid, as these acids are known to be main components of grass lignins, lignins of annual plants, and of other phenolic compounds present in the raw material for hydrothermal pretreatment processes.[58–60]

4.5 Counterelectroosmotic CE of Tannins

The pharmaceutical and cosmetic industries are utilizing tannins, mainly because of their astringent and hemostatic astringent properties. Tannins belong to the class of polyhydroxycompounds and are derivatives of gallic acid and catechin. The catechin molecule shows a positive effect in therapy of infections caused by the hepatitis virus.

TLC and HPLC have acquired great importance in phytochemistry for the qualitative and quantitative analysis and enrichment of a number of compounds, such as tannins, from plant material. Despite the fact that these methods are quite popular, they require a lot of time and work. Capillary electrophoresis is capable of separating a wide range of different molecules with a minimum of sample pretreatment and within very short run times. When an application of CE to the analysis of tannins was developed and applied for the first time, five tannin-related substances could be separated in less than 12 min (see Figure 4.29).[61]

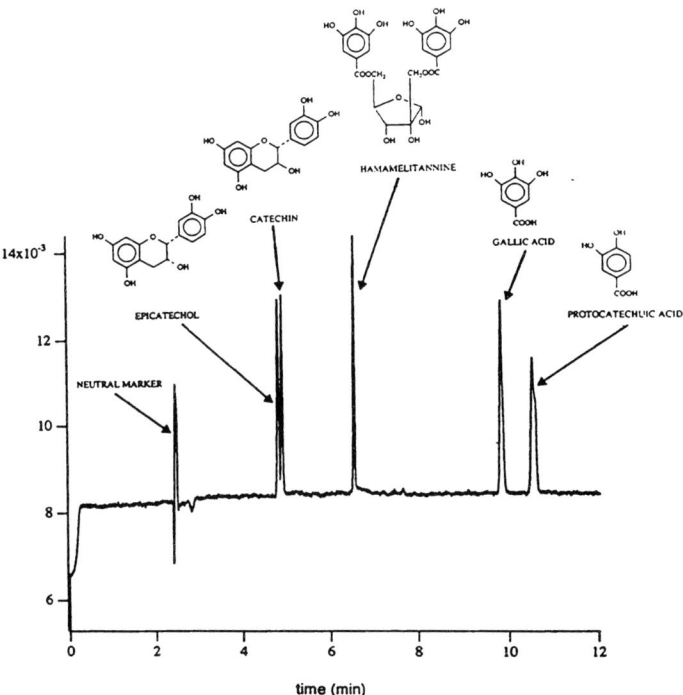

Figure 4.29 Separation of a standard mixture of tannins. Fused silica capillary, 52 × 44.5 cm, 50 μm I.D.; carrier electrolyte: 0.15 M borate buffer, pH = 9.8. Voltage: +25 kV; current: 92 μA. Direct UV detection at 214 nm. Temperature: ambient. Hydrostatic sampling, 9.8 cm, 10 sec.

4.6 Counterelectroosmotic CE of Chiral Compounds

A large number of chiral separations has been reported using capillary electrophoresis or micellar electrokinetic chromatography (MECC). One of the first chiral separations has been described by Gassmann et al.,[62] based on the complexation of D,L-amino acids with the Cu(II) complex of L-histidine.

In the meantime, a lot of other chiral reagents have been used for derivatization of amino acid enantiomers and peptides.

Recently, efficient separations of amino acids have been made possible through derivatization with L-Marfey's or D-Marfey's reagent (1-fluoro-2,4-dinitrophenyl-5-alanine amide).[63] Figure 4.30 shows a typical separation of the amino acids Ala, Asp, Glu, and Leu obtained under counterelectroosmotic CE conditions, whereas Figure 4.31 shows the separation under MECC conditions.

252 CAPILLARY ELECTROPHORESIS OF SMALL MOLECULES AND IONS

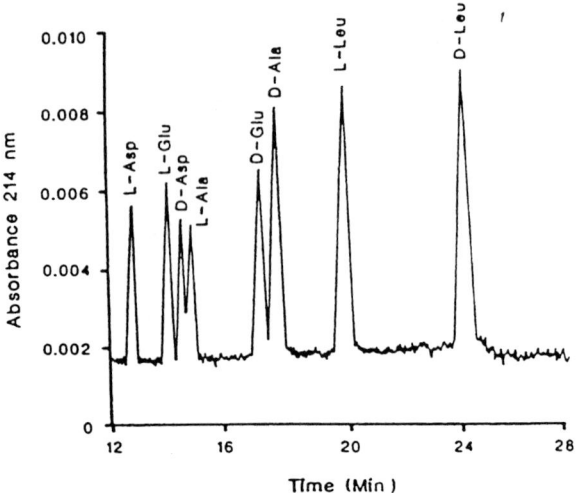

Figure 4.30 Electropherogram of racemic mixture of Ala, Asp, Glu, and Leu derivatized with L-Marfey's reagent. Fused silica capillary: 75 μm I.D. 57 × 50 cm. Carrier electrolyte: 50 mM ammonium phosphate, pH 3.3; voltage +20 kV, current; 120 μA. Separation temperature: 25 ± 0.1°C; direct UV detection at 214 nm. Reproduced with permission.[63]

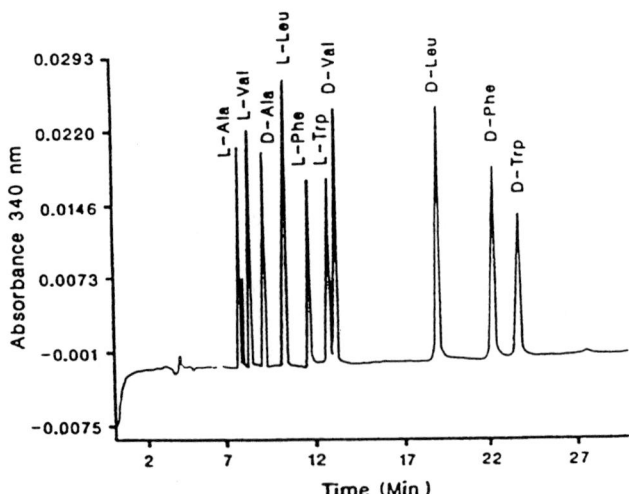

Figure 4.31 Electropherogram of racemic mixture of Ala, Val, Leu, Phe, and Trp, derivatized with L-Marfey's reagent. The electropherogram was obtained by MECC. Fused silica capillary: 50 × 75 cm, 75 μm I.D. Carrier electrolyte: 100 mM sodium borate, 200 mM SDS, pH 8.5. Voltage: +20 kV; current: 20 μA. Separation temperature: 25 ± 0.1°C. Direct UV detection at 340 nm. Reproduced with permission.[63]

Further chiral separations are possible by including cyclodextrins in the carrier electrolyte. This allows separations via inclusion complexation of, for example, amino acids[64] or catecholamines.[65]

References

1. B. L. Ling, W. R. G. Baeyens, and C. Dewaele, *HRC* **14**, 169 (1991).
2. H. J. Issug, G. M. Janini, I. Z. Atamna, and G. M. Muschik, *J. Liq. Chromatogr.* **14**, 817 (1991).
3. E. Stahl and U. Kaltenbach, *J. Chromatogr.* **5**, 351 (1961).
4. R. A. Laine, W. J. Esselman, and C. C. Sweeley, *Methods Enzymol.* **28**, 159 (1972).
5. R. Pecina, G. Bonn, E. Burtscher, and O. Bobleter, *J. Chromatogr.* **287**, 245 (1984).
6. G. Bonn, *J. Chromatogr.* **322**, 411 (1985).
7. S. C. Churms, *J. Chromatogr.* **500**, 555 (1990).
8. S. Hoffstetter-Kuhn, A. Paulus, E. Gassmann, and H. M. Widmer, *Anal. Chem.* **63**, 1541 (1991).
9. J. Böseken, *Adv. Carbohydr. Chem.* **4**, 189 (1949).
10. H. Weigel, *Adv. Carbohydr. Chem.* **18**, 61 (1963).
11. H. Scherz, *Electrophoresis* **11**, 18 (1990).
12. J. Liu, O. Shirota, and M. Novotny, *J. Chromatogr.* **223** (1991).
13. P. J. Oefner, A. E. Vorndran, E. Grill, C. Huber and G. Bonn, *Chromatographia* **34**, 308 (1992).
14. C. F. Bell, R. D. Beauchamp, and E. L. Short, *Carbohydr. Res.* **185**, 39–50 (1989).
15. B. Pettersson and O. Theander, *Acta Chem. Scand.* **27**, 1900 (1973).
16. A. E. Vorndran, P. Oefner, H. Scherz, and G. K. Bonn, *Chromatographia* **33**, 163 (1992).
17. J. P. Liu, O. Shirota, D. Wiesler, and M. V. Novotny, *Proc. Natl. Acad. Sci. U.S.A.* **88**, 2302 (1991).
18. D. A. Martens and W. T. Frankenberger, *Chromatographia* **29**, 7 (1990).
19. W. T. Wang, N. C. LeDonne, B. Ackerman, and C. C. Sweeley, *Anal. Biochem.* **141**, 366 (1984).
20. T. Akiyama, *J. Chromatogr.* **588**, 53 (1991).
21. A. E. Vorndran, E. Grill, C. Huber, P. J. Oefner, and G. K. Bonn, *Chromatographia*, **34**, 109 (1992).
22. S. Honda, S. Iwase, A. Makino, and S. Fujiwara, *Anal. Biochem.* **176**, 72 (1989).
23. H. Schweiger, P. Oefner, C. Huber, and G. K. Bonn, Submitted to *J. Chromatogr.*
24. H. Schweiger, Optimierung kapillarelektrophoretischer Trennmethoden für Kohlenhydrate, Diplomarbeit, Innsbruck (1992).

25. J. Vindevogel and P. Sandra, *Introduction to Micellar Electrokinetic Chromatography*, Hüthig Verlag, Heidelberg (1992).
26. T. W. Garner and E. S. Yeung, *J. Chromatogr.* **515**, 639 (1990).
27. J. A. Rendle, Jr., in *Carbohydrates in Solution*, H. S. Isbell, ed., American Chemical Society, Washington, DC (1973), p. 54.
28. T. Tsuda, K. Nomura, and G. Nakagawa, *J. Chromatogr.* **248**, 241 (1982).
29. G. K. Bonn, H. Binder, H. Leonhard, and O. Bobleter, *Monatsh. Chem.* **116**, 961 (1985).
30. E. S. Yeung, *Acc. Chem. Res.* **22**, 125 (1989).
31. R. Pecina, G. Bonn, E. Burtscher, and O. Bobleter, *J. Chromatogr.* **287**, 245 (1984).
32. S. Nathakarnkitkool, P. Oefner, G. Bartsch, M. A. Chin, and G. K. Bonn, *Electrophoresis* **13**, 18 (1992).
33. E. Steinegger and R. Hänsel, *Lehrbuch der Pharmakognosie und Phytopharmazie*, Springer, Berlin, p. 394 (1988).
34. H. Wagner, S. Bladt, and E. M. Zgainski, *Drogenanalyse*, Springer, Berlin, p. 164 (1983).
35. P. Pietta, E. Manera, and P. Ceva, *J. Chromatogr.* **357**, 233–237 (1986).
36. F. A. Blouin and Z. M. Zarins, *J. Chromatogr.* **441**, 443–447 (1988).
37. B. Heimhuber, R. Galensa, and K. Herrmann, *J. Chromatogr.* **439**, 481–483 (1988).
38. P. G. Pietta, P. L. Mauri, E. Manera, and P. L. Ceva, *Chromatographia* **28**, 311–312 (1989).
39. M. C. Pietrogrande, C. Bighi, G. Blo, Y. D. Kahie, P. Reschiglian, and F. Dondi, *Chromatographia* **27**, 625–627 (1989).
40. P. Spiegel, Ch. Dittrich, and K. Jentzsch, *Sci. Pharm.* **44**, 129–140 (1976).
41. A. Hiermann and Th. Kartnig, *J. Chromatogr.* **140**, 322–326 (1977).
42. P. P. Schmid, *J. Chromatogr.* **157**, 217–225 (1978).
43. A. Hiermann, *J. Chromatogr.* **174**, 478–482 (1979).
44. W. Greenaway, S. English, E. Wollenweber, and F. R. Whatley, *J. Chromatogr.* **481**, 352–357 (1989).
45. U. Seitz, G. K. Bonn, P. Oefner, and M. Popp, *J. Chromatogr.* **559**, 499–504 (1991).
46. S. Fujiwara and S. Honda, *Anal. Chem.* **58**, 1811–1814 (1986).
47. T. Tsuda, K. Nomura, and G. Nakagawa, *J. Chromatogr.* **264**, 385–392 (1983).
48. R. Consden and W. M. Stanier, *Nature* **169**, 783–785 (1952).
49. A. B. Forster, *Adv. Carbohydr. Chem.* **12**, 81–115 (1957).
50. B. Pettersson and O. Theander, *Acta Chem. Scand.* **27**, 1900–1906 (1973).
51. S. Hoffstetter-Kuhn, A. Paulus, E. Gassmann, and H. M. Widmer, *Anal. Chem.* **63**, 1541–1547 (1991).
52. J. G. Dawber and G. E. Hardy, *J. Chem. Soc., Faraday Trans.* **180**, 2467–2478 (1984).
53. U. Seitz, P. Oefner, S. Nathakarnkitkool, M. Popp, and G. K. Bonn, *Electrophoresis* **13**, 35–38 (1992).

54. M. Popp, A. Vorndran, U. Seitz, P. Oefner, and G. K. Bonn, Presentation at the "Electrophorese-Forum," Munich (1991).
55. M. T. Ackermanns, J. L. Beckers, F. M. Everaerts, H. Hoogland, and M. J. H. Tomassen, *J. Chromatogr.* **596,** 101–109 (1992).
56. Ch. D. Gaitonde and P. V. Pathak, *J. Chromatogr.* **514,** 389 (1990).
57. A. Zemann, Thesis, University of Innsbruck, Austria, 1992.
58. R. Concin, P. Burtscher, E. Burtscher, and O. Bobleter, *Int. J. Mass Spectr. Ion Phys.* **48,** 63 (1983).
59. R. Pecina, E. Burtscher, G. K. Bonn, and O. Bobleter, *Fresenius Z. Anal. Chem.* **325,** 461 (1986).
60. P. Pfeifer, G. K. Bonn, and O. Bobleter, *Analytical and Preparative Isotachophoresis,* C. J. Holoway, ed. (1984), Walter de Gruyter & Co., Berlin–New York, p. 89.
61. B. Thalhamer, Thesis University of Linz, Austria (1992).
62. E. Gassmann, J. Kuo, and R. Zare, *Science* **230,** 813 (1985).
63. A. D. Tran, T. Blanc, and E. J. Leopold, *J. Chromatogr.* **516,** 241 (1990).
64. A. Guttmann, A. Paulus, A. Cohen, N. Grinberg, and B. Karger, *J. Chromatogr.* **448,** 41 (1988).
65. S. Fanali, *J. Chromatogr.* **474,** 441 (1989).

CHAPTER

5

Application Examples of Coelectroosmotic Capillary Electrophoresis

Coelectroosmotic separations had only been scarcely used during the first decade of the development of capillary electrophoresis. As explained in Sections 2.3 and 2.4, the reason for this was in the biological orientation of leading researchers in the field and also in the fact that the coelectroosmotic mode usually gives good results only with small molecules and ions. The coelectroosmotic mode of capillary electrophoresis relies on mobility differences of low-molecular-weight ions, which are usually more pronounced, or can be made larger, than in the case of biologically relevant molecules. In contrast to the rules postulated for macromolecules and other low-mobility analytes (see Section 2.2), coelectroosmotic separations make use of electroosmotic flow to enhance electrophoretic mobility.

The impetus for the development of coelectroosmotic separations was provided by the limitations of ion chromatography in the analysis of some complex ionic mixtures. One such limitation is encountered in anion-exchange separations of low-molecular-weight carboxylic acids, usually eluting in a narrow range of retention volumes closely coinciding with the retention time of fluoride. (See Figure 5.1.)

Practical solutions to this problem were found in the form of coupling of ion exclusion/anion exchange[1,2] or with gradient elution in anion-exchange separations.[3,4] However, these methods are either time consuming (gradients) or require a complex chromatographic apparatus (coupled separation modes). Furthermore, neither of the two ion chromatographic methodologies provides a complete separation. Coupled systems are incapable of ana-

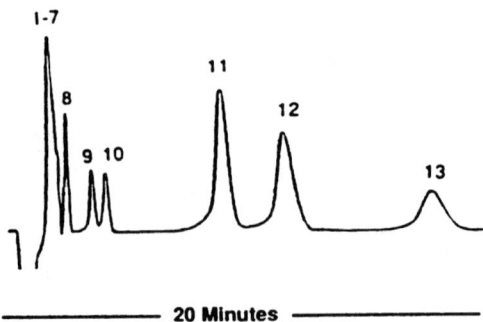

Figure 5.1 Coelution of early eluting peaks (peaks (1–7) in ion chromatography. Mobile phase: 3 mM sodium octanesulfonate. Flow rate: 1 ml/min. Column: polyacrylate based anion exchanger, 0.46 × 5 cm. Detection: conductivity. (1) Fluoride, 0.5 ppm; (2) glycolate, 2.0 ppm; (3) formate, 1.0 ppm; (4) acetate, 10.0 ppm; (5) propionate, 20.0 ppm; (6) butyrate, 20.0 ppm; (7) iodate, 50 ppm; (8) chloride, 2.0 ppm; (9) bromide, 2.0 ppm; (10) nitrate, 2.0 ppm; (11) iodide, 20 ppm; (12) sulfate, 4.0 ppm; (13) thiocyanate, 5.0 ppm. Reproduced with permission.[1]

lyzing phosphate together with carboxylic acids and common inorganic anions.

Gradient elution can resolve only formate and acetate from the fluoride peak in a separation also including common inorganic anions. If propionate, glycolate, or other weakly retained ions are present, they still coelute with fluoride. The true identity of the fluoride peak thus remains doubtful, even with the gradient separations. A definitive solution was found only with the help of coelectroosmotic capillary electrophoresis. Typical coelectroosmotic separations are capable of resolving all relevant carboxylic acids from inorganic anions.

The electropherogram in Figure 5.2 was generated for a direct comparison with the coupled mode chromatogram, and, for that reason, the standard mixture did not contain phosphate, which, if present, would have been separated as a well-resolved peak following formate. Electropherograms of mixtures of carboxylic acids including phosphate along with other inorganic anions are discussed in Section 5.2. The coelectroosmotic approach to electrophoretic separations of ions was recently developed into a comprehensive methodology, which also includes indirect UV detection. A new name was proposed for the technique—electrophoretic capillary ion analysis (CIA).[5] A CIA method is defined by optimal realization of several independent conditions:

1. *The Electroosmotic Flow Condition.* To satisfy this condition, the analyte must be forced to migrate in the direction of the electroosmotic flow. For the analysis of cations (Sections 5.4 and 5.5), this is realized by assigning positive polarity to the sample introduction side and utilizing the natural

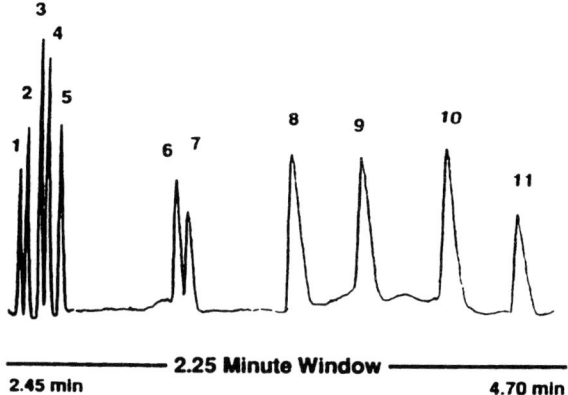

Figure 5.2 Pure water sample spiked with seven inorganic and four organic anions. Carrier electrolyte: 5 mM chromate, 0.5 mM CIA OFM BT Anion (Waters), pH 8.0. Capillary: 75 μm × 52 × 60 cm, fused silica. Separation voltage: −20 kV. Detection: indirect UV at 254 nm. (1) Bromide, 4 ppm; (2) chloride, 2 ppm; (3) sulfate, 4 ppm; (4) nitrite, 4 ppm; (5) nitrate, 4 ppm; (6) fluoride, 1 ppm; (7) formate, 2 ppm; (8) carbonate, 4 ppm; (9) acetate, 4 ppm; (10) propionate, 4 ppm; (11) butyrate, 4 ppm. Reproduced with permission.[1]

direction of electroosmosis in fused silica capillaries, which is toward the negative side of the system. In order to analyze for anions, the injection side is made negative, and the direction of electroosmosis in a fused silica capillary is changed by a suitable additive or by a chemical modification of the capillary wall. The direction of the electroosmotic flow has to be toward the positive side of the system for the analysis of anions.

2. *The First Coion Condition.* The mobility of analyte ions must be closely matched by that of the carrier electrolyte coions (electrolyte ions having the same charge polarity as analytes). The mismatch of mobilities results in deformed peak shapes and deteriorating resolution of closely comigrating peaks (Section 2.7.3).

3. *The Second Coion Condition.* The wavelength of the electrolyte coion's UV-absorption maxima must be sufficiently greater or less than those of the analyte ions. Unless the carrier electrolyte coion generates a UV absorption background at a wavelength at which the analyte absorption is nonexistent or minimal, sensitivity is compromised (Section 3.3.2).

4. *The Electrostacking Condition.* Sensitivity of detection is optimized and system peaks are eliminated by keeping the total ionic strength of the sample below that of the carrier electrolyte (Section 3.1.1).

5. *The Kohlrausch Condition.* Electromigrative sample introduction can be optimized beyond the improvements achievable by electrostacking, if analytical conditions approximate an isotachophoretic steady state (Section 3.1.5.2).

Because of the prevalence of UV detectors in commercial CE instruments, electrophoretic capillary ion analysis has become a frequently utilized coelectroosmotic methodology. In the following sections, we discuss additional rules for optimization and also some application examples of that method.

The general capabilities of electrophoretic capillary ion analysis are illustrated in Figure 5.3, and the migration order of anions is listed in Table 5.1.

5.1 Adjusting Selectivity of Anion Separations

One of the frequently voiced concerns in the beginnings of capillary electrophoresis was the perceived inflexibility of the method. The minimal differences in electrophoretic mobilities of ions, it was argued, did not allow separations of more complex mixtures. Only gradually had this perception been replaced by the realization that numerous auxiliary separation mechanisms (Section 2.4) could be utilized to enhance selectivity beyond the degree achievable by a mere addition of electrophoretic and electroosmotic vectors.

The influence of carrier electrolyte pH on the migration behavior of weak acids was one of the first mechanisms utilized for adjustments of selectivity.

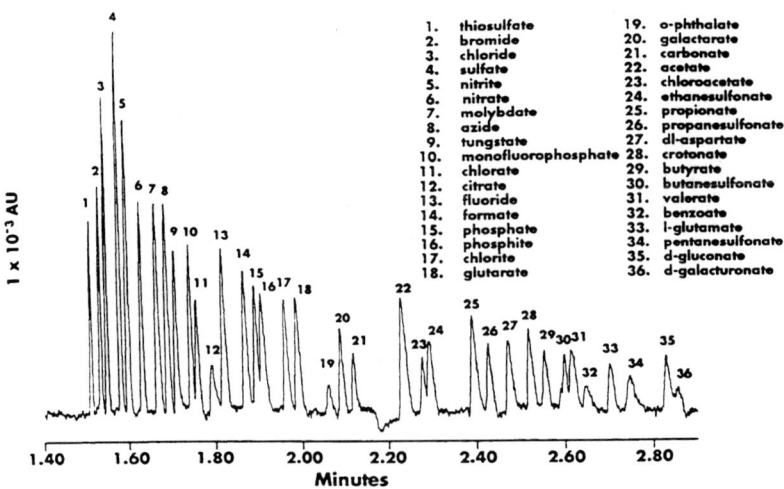

Figure 5.3 Separation of 36 anions by capillary ion analysis. The concentrations of all components were in the range of 0.3 to 3.3 ppm. The sample was introduced by electromigration for 15 sec at 1 kV. Carrier electrolyte: 5 mM chromate, 0.4 mM OFM BT Anion (Waters), pH 8.0. Capillary: 50 μm × 60 × 52 cm, fused silica. Separation voltage: −30 kV. Indirect UV detection at 254 nm. Reproduced with permission.[6]

Table 5.1. MIGRATION ORDER OF ANIONS IN CAPILLARY ION ANALYSIS[a]

1	thisosulfate	19	*o*-phthalate
2	bromide, *chromate*	20	galactarate, **perchlorate**
3	chloride	21	carbonate, *methanesulfonate, glycolate, selenite*
4	sulfate, <u>*sulfide*</u>	22	acetate, *oxalacetate, pyruvate*
5	nitrite, **iodide**	23	chloroacetate
6	nitrate, *selenate,* <u>*oxalate*</u>	24	ethanesulfonate, *trifluoroacetate*
7	molybdate, *orthovanadate*	25	propionate, *dichloroacetate, lactate*
8	azide	26	propanesulfonate, *glycerate*
9	tungstate	27	DL-aspartate
10	monofluorophosphate	28	crotonate
11	chlorate	29	butyrate, α-*hydroxybutyrate*
12	citrate, *isocitrate, transacotinate, metavanadate,* <u>*sulfite*</u>	30	butanesulfonate
13	fluoride, *maleate, malonate, fumarate*	31	valerate, *2-hydroxyvalerate*
14	formate, <u>*cyanide*</u>, *bromate*	32	benzoate
15	phosphate, α-*ketoglutarate, succinate, tartrate, arsenate, trimesate*	33	L-glutamate, *sorbate*
16	phosphite	34	pentanesulfonate
17	chlorite	35	D-gluconate, *shikimate*
18	glutarate, **thiocyanate**	36	D-galacturonate, *glucuronate, hexane, heptanesulfonate*

[a] Anions printed in plain letters correspond to peaks 1 through 36 in Figure 5.3. Bold print indicates anions with migration behavior sensitive to changes of EOF modifier. Anion names in italics: approximate migration of anions not included in the electropherogram in Figure 5.3. Underlined anions are better separated at a pH different from 8.0. Adapted from Reference 6.
The migration data for italicized and bold-lettered anions are only approximate and strongly influenced by pH and/or other changes in analytical conditions. Many of the neighboring peaks are resolved only in 50 μm and not in 75 μm capillaries (e.g., glycolate and chloroacetate). Additional ways of influencing migration sequence or spacing of the peaks are discussed in section 5.1.

The plot in Figure 5.4 illustrates the dependence of migration times of weakly and strongly acidic anions on pH.

As expected, only the weakly acidic anions, borate and carbonate, exhibit large shifts of migration times upon changes in pH. A more subtle effect is discernible for phosphate and fluoride. These two anions are fully resolved only at pH 8, showing a complete lack of resolution between pH 9 and 11.5. Above pH 11.5, the phosphate approaches its third dissociation equilibrium, and the increasing contribution of the trivalent ion to the observed mobility shifts this anion to migration times that are shorter than those for fluoride. The migration sequence of the fluoride/phosphate pair can thus be completely reversed in going from pH 8 to 11.5. The uniform decrease in migration times beyond pH 11.5 is caused by an increasing contribution of electroosmotic flow to the observed mobilities for the respective analyte anions [Eq. (2.8)]. The electroosmotic flow rate increases with pH because of a

262 CAPILLARY ELECTROPHORESIS OF SMALL MOLECULES AND IONS

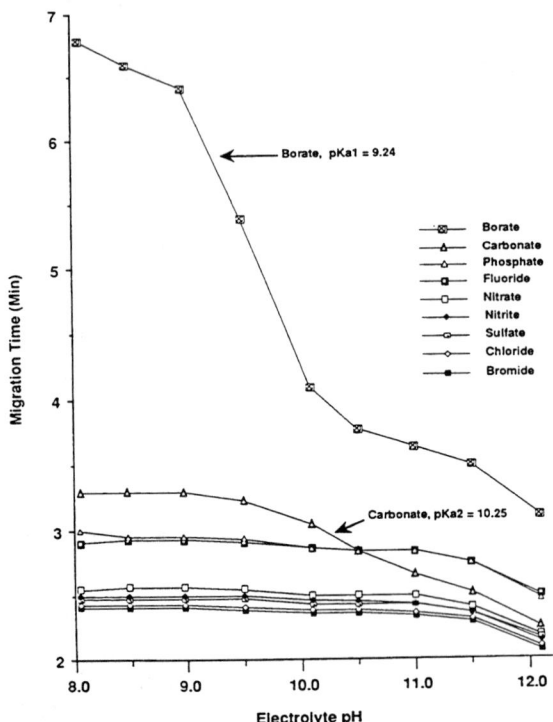

Figure 5.4 Migration times of nine anions are plotted against the pH of carrier electrolyte. Except for pH variations, the conditions are as described in Figure 5.2. Reproduced with permission.[9]

larger degree of dissociation of silanol groups on the capillary wall. This increase in the number of dissociated silanol groups leads in turn to a corresponding increase of hemimicellar density (Section 2.3) and to higher positive values of the zeta potential.

Somewhat unexpected, on the other hand, are the occasional changes in selectivity due to changed concentration of carrier electrolytes. One example of such a selectivity change is illustrated in Figure 5.5.

At low electrolyte concentrations, the sulfate anion is migrating close to chloride. With increasing chromate concentrations, it gradually moves away from chloride until it merges with the nitrite peak. The shift in the mobility of sulfate is likely to be caused by the change in effective charge with increasing ionic strength. The general trend to longer migration times with increasing concentrations of chromate can be explained by increases in separation currents and the corresponding weakening of the electric field in the capillary filled with carrier electrolyte.

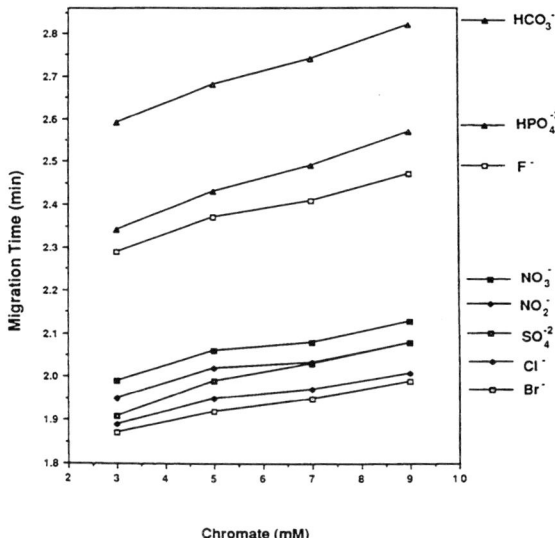

Figure 5.5 Plot of migration times for eight common anions against the concentration of a carrier electrolyte. Except for the changes in the concentration of chromate, the analytical conditions are as described in Figure 5.2. Reproduced with permission.[9]

Pronounced selectivity changes can be induced by different concentrations of electroosmotic flow modifier, which is usually an alkylammonium salt.

Before considering the plot in Figure 5.6, we should note that it uses normalized values of migration times; the respective migration times were divided by a migration time of bromide obtained at the same concentration of modifier. This was done to separate the gradual lengthening of all migration times with an increasing conductivity of electrolyte from the specific effects observed for bromide, sulfate, and nitrate. These three seemingly unrelated anions behave as a group and show a distinct shift toward longer migration times, if the modifier concentration is increased. The difference in migration data between the three anions on the one hand and all the remaining anions in Figure 5.6 on the other hand is most likely due to different tendencies to undergo ion pairing with the EOF modifier.[6] Following the rules explained in Section 2.4, we could thus describe the conditions above ~ 1.0 mM of EOF modifier as coelectroosmotic capillary electrophoresis with an auxiliary ion pairing separation mode.

Addition of organic solvents to carrier electrolytes can also lead to pronounced changes of migration order. Buchberger and Haddad[7] investigated the influence of a series of organic solvents on the migration behavior of

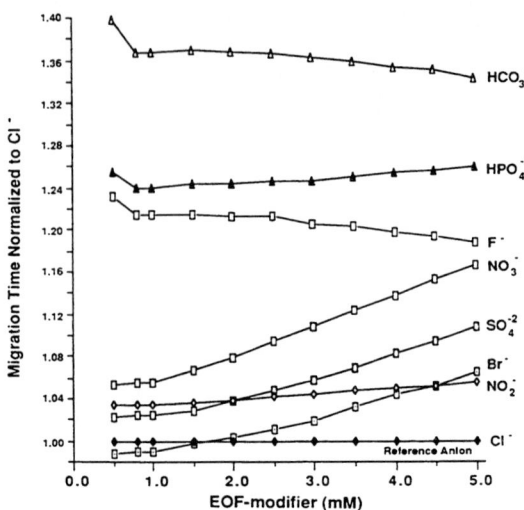

Figure 5.6 Plot of migration times for eight common anions against the concentration of an electroosmotic flow (EOF) modifier. Except for changes in the concentration of EOF modifier, analytical conditions are as described in Figure 5.2. Reproduced with permission.[9]

anions in a carrier electrolyte of similar composition as the one in Figure 5.2. Only the water–to–solvent ratio was varied; the concentrations of chromate and EOF modifier were held constant. (See Figure 5.7.)

In all four solvents, sulfate and fluoride are the most affected anions. Addition of methanol causes a large change in the migration of chlorate,

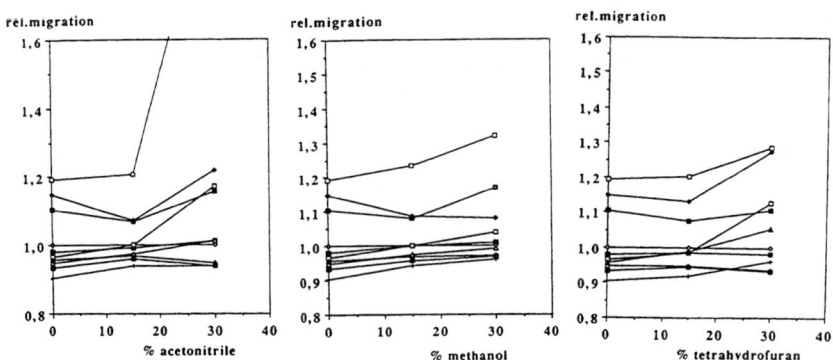

Figure 5.7 Effect of three organic solvents on selectivity. Relative migration times were calculated as ratios with the migration time of nitrate. Reproduced with permission.[7] -□- = fluoride, -◆- = thiocyanate, -■- = chlorate, -◇- = nitrate, -■- = nitrite, -□- = sulphate, -▲- = iodide, -△- = chloride, -■- = bromide, + = thiosulphate

whereas iodide is shifted strongly by an addition of tetrahydrofuran. The general trend toward longer migration times with increasing concentration of a solvent[7] seems to justify an explanation based on the influence of ion pairing, also assuming that organic solvents tend to shift ion pairing equilibria to the right. Overall, the variations achievable by the addition of solvents represent a very useful tool for adapting analytical conditions to the requirements of different types of samples.

A change of the main carrier electrolyte component rarely changes the migration order. However, separations can be influenced by improving or deteriorating resolution between the neighboring peaks.

The migration sequence remains essentially the same in all nine electrolytes (Figure 5.8). The first three electrolytes even exhibit an identical spacing of peaks. However, a considerable change in resolution can be achieved, for example, by a change from benzoate to mellitate.

Shifts from one electrolyte to another are more commonly dictated by one of the conditions of electrophoretic capillary electrophoresis, prescribing the match of mobilities between analyte ions and electrolyte coions (the first coion condition, see previous discussion). Distinct tailing, usually observed for most carboxylates in chromate, can be avoided by choosing a phthalate-containing electrolyte. Alkylsulfonates with more than three carbon atoms

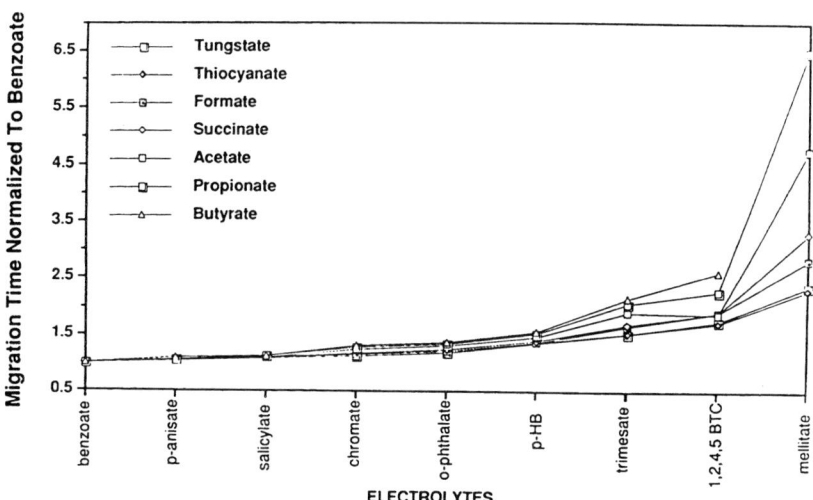

Figure 5.8 Selectivity changes in different electrolytes. All electrolytes were used at 5 mM concentrations and also contained 0.5 mM CIA OFM BT Anion (EOF modifier from Waters). The respective pH values were 8.0 for chromate and 6.0 for the aromate containing electrolyte. Migration times are relative to those in benzoate. Abbreviations: *p*-HB: *p*-hydroxybenzoate; 1,2,4,5 BTC: benzenetetracarboxylic acid. Reproduced with permission.[6]

show strong tailing even in the phthalate electrolyte, and benzoate or p-hydroxybenzoate carrier electrolytes have to be used instead.

Even within a single class of compounds, such as alkylsulfonates, a practitioner may be well advised to consider an application of more than one carrier electrolyte for a sample—see the strong fronting for the methanesulfonate in the lowest-mobility electrolyte in Figure 5.9. A detailed explanation of mechanisms leading to asymmetry of peaks in capillary electrophoresis is given in Section 2.7.3.

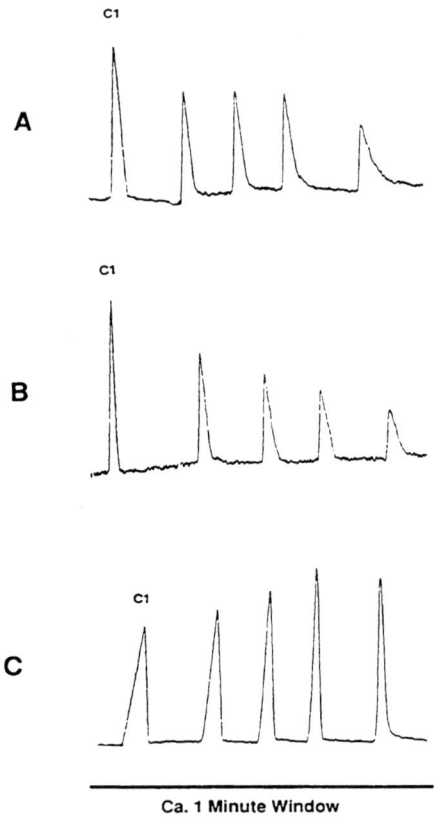

Figure 5.9 Coelectroosmotic separations of a mixture of linear alkylsulfonates. Three electrolytes of different mobility are used. A: 5 mM Chromate, B: 5 mM phthalate, C: 5 mM p-hydroxybenzoate. All three electrolytes contain 0.5 mM EOF modifier (OFM BT Anion from Waters). Peak identities from left to right: methane-, ethane-, propane-, butane-, and pentanesulfonate. Hydrostatic injection for 30 sec at 9.8 cm, all five components are at 10 ppm. Capillary: 75 μm × 60 × 52 cm, fused silica. Separation voltage: −20 kV. Indirect UV detection at 254 nm.

Of great practical value is the possibility to adjust migration order with the help of metal–ligand interactions. Glycolate and monochloracetate exhibit close comigration under standard coelectroosmotic analytical conditions. This separation problem is solved by adding a small amount of calcium to the carrier electrolyte. Use is made of the better ability of glycolate relative to monochloroacetate to function as a complexing ligand in a complexation equilibrium with an added cation. Since at any given time a portion of glycolate is contained in a slow-moving complex, the resulting apparent mobility for that anion is slowed down. The separation between glycolate and monochloroacetate is thus achieved. (See Figure 5.10.)

The selectivity of separations of low-mobility anions (carbonate, fluoride, carboxylates, sulfonates, etc.) can be modified even to the complete reversal of order of migration by switching from the coelectroosmotic to the counterelectroosmotic mode.

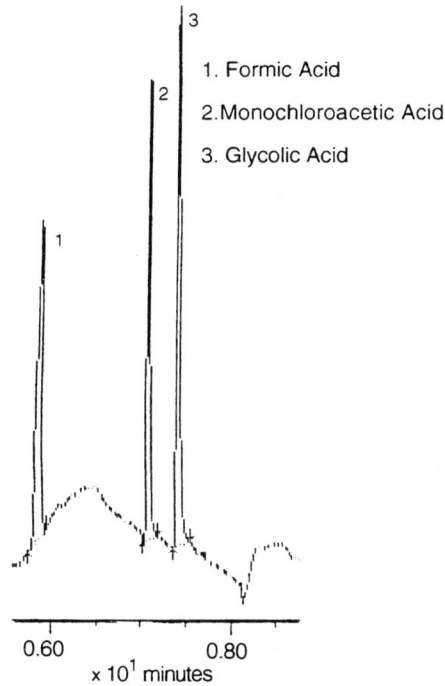

Figure 5.10 Example of coelectroosmotic conditions utilizing a complexing reaction as an auxiliary separation mode. Carrier electrolyte: 7 mM phthalate, 0.25 mM OFM BT Anion (Waters) with an addition of 50 mg of $Ca(OH)_2$ to 100 ml. pH was adjusted to 9.5. Capillary: 75 μm × 60 × 52 cm, fused silica. Separation voltage: −12 kV. Indirect UV detection at 214 nm. All sample components at 5 ppm. Sample was introduced by electromigration for 5s at 10 kV. Peak identities: 1) formate, 2) monochloroacetate, 3) glycolate. Courtesy: J. Burg, Millipore, Erkrath, Germany.

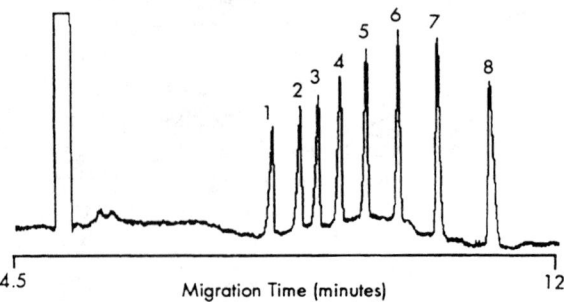

Figure 5.11 Counterelectroosmotic separation of a mixture of alkylsulfonates. Carrier electrolyte: 10 mM naphthalenesulfonic acid in 30% acetonitrile. No EOF modifier is used. Capillary: 75 μm × 60 × 52 cm, fused silica. Separation voltage: +20 kV. Hydrostatic injection for 30 sec at 9.8 cm. All components at 25 ppm. Peak identities: (1) dodecanesulfonate, (2) undecanesulfonate, (3) nonanesulfonate, (4) octanesulfonate, (5) heptanesulfonate, (6) hexanesulfonate, (7) pentanesulfonate, (8) butanesulfonate.

Note that the electropherogram in Figure 5.11 includes two of the alkylsulfonates from Figure 5.9 (butane-, pentane-), only this time, their order of migration is reversed. The change to counterelectroosmotic conditions makes it possible also to separate the alkyl homologues higher than C7. A peak for octanesulfonate is difficult to obtain under coelectroosmotic conditions. A coelectroosmotic separation of higher alkylsulfonates is impossible due to close comigration with the solvent peak. Addition of acetonitrile to the carrier electrolyte was made to improve the solubility of some of the higher sulfonates, not for adjustment of selectivity.

In summary: A practitioner of coelectroosmotic capillary electrophoresis has at his or her disposal a variety of tools for adjustments of selectivity. Resolution of neighboring peaks is optimized by a choice of different electrolytes or by employment of complexation as an auxiliary separation mode. A reversal of migration order of peak pairs is readily accomplished either by changed concentrations of EOF modifier or by adding organic solvents to the carrier electrolyte. An encompassing change of order of migration can be accomplished for low-mobility analytes by a change from the coelectroosmotic to the counterelectroosmotic mode.

5.2 Electromigrative Preconcentration of Anions

In terms of concentration detection limits, capillary electrophoresis had been lagging behind the other more established separation techniques for quite a number of years. Until recently, many authors felt that an excellent mass sensitivity notwithstanding, the poorer concentration sensitivity was inherent to capillary electrophoresis because of the low "injection volumes." During a time period between 1979 and 1990, the achieved levels of sensitivity for inorganic anions seemed to be confirming this view.

Based on experience with liquid chromatography, many researchers focused their efforts at first on the search of a sensitive detection method for capillary electrophoresis. This is mirrored clearly in Table 5.2. Virtually all improvements of the sensitivity in liquid chromatography or gas chromatography had been achieved by enhancing detectors, either by redesigning existing models or by introducing new detection techniques. The first breakthrough in improving the sensitivity of capillary electrophoresis was achieved, perhaps surprisingly, not by a new detection method, but by an improved method of sample introduction. As described in detail in Section 3.1.5.2, a method of electromigrative sampling could be found that makes possible detection limits for anions in the nanomolar range.[19,20] Unlike in some previous attempts, the new method does not require a separate set of isotachophoretic electrolytes. It is performed on line, with a conventional CE instrument utilizing the same indirect detection technique as the 1990 reference in Table 5.2. A trace enrichment of anions in ion chromatography usually requires 10 to 20 min; the preconcentration carried out for the electropherogram in Figure 3.15 takes only 45 sec.

Since the first report in 1991, the method was further optimized and applied to the types of water samples similar to those common in the nuclear power industry.

The main differences between the initial work represented by the electropherogram in Figure 3.15 and the optimized version of separation shown in Figure 5.12 are first, the increased concentration of chromate (10 vs. 5 mM) and, second, the lower separation voltage (-15 vs. -20 kV). The higher chromate concentration was chosen to prevent a close comigration of fluoride and phosphate peaks in the electropherogram after an electromigrative trace enrichment (see Figure 3.15). The carrier electrolyte inside the capillary becomes more alkaline in the course of trace enrichment, shifting the dissociation equilibrium of phosphate ($pK_2 = 7.2$) increasingly in the direction of divalent anion, which in turn leads to a shorter migration time for that species. Even though the original separation provides a sufficient space between phosphate and fluoride at pH 8.0, at higher pH values the quality of resolution is compromised. Higher chromate concentration increases the distance between the two peaks at pH 8.0 (see Figure 5.5), so that even a shift of the migration time of phosphate does not cause the fluoride and phosphate peaks to comigrate. Indeed, the optimized separation even provides enough resolution for an additional peak to be placed between the fluoride peak and phosphate peak (formate $t = 3.81$ in Figure 5.12). There is an additional benefit brought about by the increased chromate concentration. According to the Kohlrausch Regulation Function [Eq. (3.10)], concentrations of analytes after an isotachophoretic trace enrichment are directly proportional to the concentration of the leading electrolyte, which in this

Table 5.2. REPORTED DETECTION LIMITS FOR ANIONS BETWEEN 1979 AND 1990

Reference	Year	Detection Method	Anions	Estimated Limits of Detection
[8]	1979	conductivity	chloride, sulfate, chlorate, chromate, adipate, benzoate, malonate, glutamate, etc.	10^{-4} M
[9]	1983	conductivity	nitrate, sulfate, chloride	10^{-4} M
[10]	1986	conductivity	chloride, sulfate, perchlorate, nitrate, fluoride, phosphate, formate, maleate, glycolate, etc.	
[11]	1988	conductivity	chloride, formate, acetate, propionate	10^{-5} M 10^{-5} M
[12]	1989	conductivity	seven alkylsulfonates (C1–C7)	10^{-5} M
[13]	1989	direct UV	metal–cyanide complexes	10^{-5} M
[14]	1987	indirect UV	bromide, cacodylate, four alkylcarboxylates (C1–C4)	10^{-4} M
[15]	1989	indirect UV	chloride, chlorate, phosphate, diverse carboxylic acids	10^{-4}–10^{-5} M
[16]	1988	laser-induced fluorescence	bicarbonate, iodate	10^{-4}–10^{-5} M
[17]	1989	laser-induced fluorescence	chloride, nitrate, perchlorate, permanganate, chromate, iodate, phosphate	10^{-4}–10^{-5} M
[1]	1990	indirect UV	53 inorganic and organic anions	10^{-6} M

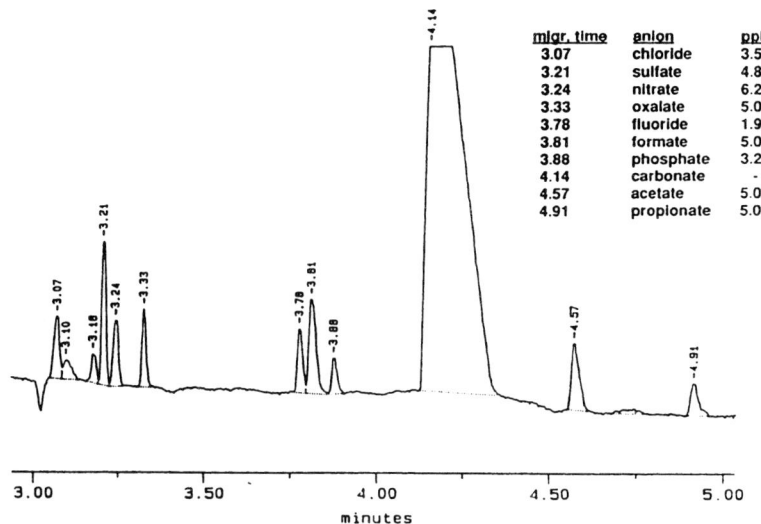

Figure 5.12 Trace analysis of inorganic and organic anions by capillary ion analysis. The electropherogram was recorded after an electromigrative trace enrichment for 45 sec at 5 kV and with 75 μM octanesulfonate additive in the sample. Peak identities: 3.07 min, 3.5 ppb chloride; 3.21 min, 4.8 ppb sulfate; 3.24 min, 6.2 ppb nitrate; 3.33 min, 5 ppb oxalate; 3.78 min, 1.9 ppb fluoride; 3.81 min, 5 ppb formate; 3.88 min, 3.2 ppb phosphate; 4.57 min, 5 ppb acetate; 4.91 min, 5 ppb propionate. The large peak at 4.14 min corresponds to carbonate. The concentration of carbonate was not controlled. Carrier electrolyte: 10 mM sodium chromate and 0.5 mM OFM BT Anion (Waters), pH 8.0. Capillary: 75 μm × 60 × 52 cm, fused silica. The separation voltage: −15 kV. Indirect UV detection at 254 nm. Reproduced with permission.[19]

particular case is the chromate carrier electrolyte fulfilling the role of the leading electrolyte for the purposes of trace enrichment. This role of the chromate carrier electrolyte was confirmed in an experiment.[19] For a constant analyte concentration, peak areas nearly doubled in going from 5 to 10 mM chromate in the carrier electrolyte. A minor disadvantage of higher chromate concentrations is the loss of resolution between sulfate and nitrite peaks (see Figure 5.5). However, since nitrite was not expected to be present in the targeted water samples, this was deemed acceptable. The use of −15 instead of −20 kV as a separation voltage was found to lead to lower levels of baseline noise.[19] The combined effect of both optimization steps, the increased chromate concentration and lowered separation voltage, is an improvement in detection limits in comparison with the initial experiments.[20]

The detection limits listed in Table 5.3 make the new method suitable for the monitoring of anion concentrations in deionized feed water in a nuclear

Table 5.3. DETECTION LIMITS (2 × NOISE IN PPB) WITH OPTIMIZED PRECONCENTRATION METHOD[a]

Anion	Detection Limit [ppb]
chloride	0.5
sulfate	0.3
nitrate	0.8
oxalate	0.6
fluoride	0.3
formate	0.5
phosphate	0.3
acetate	0.8
propionate	0.6

[a]Adapted from Reference 19.

power plant. Typical admissible levels of 2 ppb sulfate and 2 ppb chloride[20] are sufficiently high above the detection limits listed in Table 5.3.

The precision of the optimized method was tested for six analytical runs at a constant concentration of analyte anions. RSD values for nitrate, oxalate, fluoride, formate, and phosphate were lower than 3%. RSD values for sulfate, chloride, acetate, and propionate were 3.3, 4.3, 7.5, and 7.2%, respectively. The higher levels of RSD for sulfate and chloride were explained as derived from contamination by these two anions in a regular laboratory environment. The relatively lower precision for acetate and propionate, on the other hand, reflects the relatively higher limits of detection for these two analytes.

Interesting observations were made during the calibration study for the optimized trace enrichment and separation method.

As could be expected from the good selectivity and sensitivity of the method, the calibration study yielded statistically significant values of correlation coefficients for all investigated anions. The calibration diagrams were plots of peak areas (arbitrary units) against concentrations in μM (0.1 to 1 μM, for example, 1.9 to 19.0 ppb fluoride). Further inspection of calibration data reveals a regular pattern in the value of slopes. The slope values for monovalent anions (chloride, nitrate, fluoride, formate, acetate, and propionate) are about half of those for the divalent ones (sulfate and oxalate). The slope of the calibration plot for phosphate seems to be outside that pattern. However, with the pH of the carrier electrolyte at 8.0, the second dissociation step for phosphate is not yet complete, and the low value of the slope reflects a fractional valency between -1 and -2 for that anion.

On the one hand, it is known that a signal in indirect detection is generated by a regular and predictable, equivalent per equivalent, displacement of UV background-providing ion (Section 3.3.2). On the other hand, the observed regular pattern of the slopes is at first surprising in view of the dependence of peak areas on the varying velocity of separated zones, as discussed in Sections 3.4.3 and 3.4.5. However, a theoretical possibility that with electromigrative sample introduction peak areas may become independent of zone velocity has been discussed in the literature.[18] The calibration data in Table 5.4 represent experimental evidence supporting those conclusions.

Of direct practical consequence is the possibility of using the regular pattern of the slopes for universal calibration. Such calibration is based on the use of only one standard and on a calculation of the concentration of all peaks with the help of only one calibration curve. This approach is not only time saving, but it also makes it possible to determine the concentration of unidentified compounds as long as their peak areas can be measured. In the power industry, measured values of the total conductivity of process water have to be frequently compared with the conductivity calculated from independently determined concentrations of ionic compounds. With ion chromatography, the identity of all peaks cannot always be verified or can be subject to misinterpretation—see, for example, the discussion of the identity of fluoride peaks in the beginning of Chapter 5. Peaks of unknown identity cannot be calibrated in IC (in the suppressed or nonsuppressed version of the method with direct conductivity detection), and an identification error often leads to large differences between measured and calculated total conductivities. In the discussed electrophoretic preconcentration and separation technique, however, even the concentrations of unknown peaks can be used reliably for calculations of total conductivity. Moreover, migration times of such unknown peaks also reveal a correct value of the equivalent conductance for use in the calculation of total conductivity (Section 2.2).

The method's evaluation was continued by applications to two typical types of water samples in nuclear pressurized water reactors (PWR). The first type was water from the primary circuit. In PWR plants, water in the primary circuits is brought into direct contact with nuclear fuel elements to heat up and is then pumped under high pressure through a heat exchanger. In the heat exchanger, the primary water exchanges thermal energy with the

Table 5.4. CALIBRATION DATA FOR ANIONS, 0.1 TO 1 μM IN WATER[a]

Anion	Correlation Coefficient	Slope
chloride	0.999	8,423
sulfate	0.998	21,159
nitrate	0.999	8,832
oxalate	0.999	18,520
fluoride	0.999	7,796
formate	1.000	9,700
phosphate	0.993	12,686
acetate	0.999	8,503
propionate	0.997	7,781

[a]Adapted from Reference 19.

secondary circuit water, which is converted to steam in the process. Most nuclear power plants maintain the following compositions of water in the two separate circuits:[21]

Primary Circuit Water: 500 to 2,000 ppm boron (as 2857 to 11,427 ppm boric acid), 3 ppm Li added as LiOH. Maximally admissible levels of anions are 40 ppb each, chloride, sulfate, nitrate, and fluoride. Normal levels are usually less than 5 pbb of each of the anions. The pH value is maintained between 6 and 10.

Secondary Circuit Water: ~ 3 ppm ammonium or morpholine. The products of heat decomposition of morpholine or heat-induced reactions of ammonium (amines, organic acids). Maximally admissible levels of anions are 40 ppb chloride, sulfate, nitrate, fluoride, phosphate, and carboxylic acids. As in the primary water, the normal levels in the secondary water are also at less than 5 ppb for each of the anions present. The secondary water is maintained at ~ pH 9.

Ion chromatography requires at least 10 ml of the radioactive primary water sample, and a typical analytical run lasts between 35 and 45 min. At an eluent flow rate of 1 ml/min and sample volume of 10 ml, each analysis generates 45 to 55 ml of radioactive waste.

On the other hand, the sample volume required for the electromigrative sample preconcentration is 2 ml, and ~ 4 ml carrier electrolyte can be made to last for six analytical runs. The volume of radioactive waste is thus only about 2.7 ml for one analysis with the CE method under discussion.

As can be seen from Figure 5.13, capillary electrophoresis offers not only the advantage of lower volume of contaminated waste, it also cuts short the total time required for analysis of a primary water sample. Instead of the usual 35 to 45 min in ion chromatography, the total run time in capillary electrophoresis is less than 5 min for primary circuit water samples. Ionic contamination of primary water can be in a broader range than in the case of deionized feed water discussed in the preceding paragraphs. For that reason, the calibration study for primary water was carried out for the range of concentrations of 0.1 to 2 μM. The correlation coefficients for all anions separated in Figure 5.13 were in the range of 0.996 to 1.000. The values of slopes of calibration plots from synthetic primary water samples (1000 ppm boron) reflected the same regular pattern as in the case of calibration plots for deionized water samples: chloride, 2,161; sulfate, 4,307; nitrate, 1,930; and fluoride, 1,875. The universal calibration discussed is thus possible even for primary water samples. Elevated levels of boric acid used for moderation of the fission process represent an important variable in primary circuit water samples. As discussed in Section 3.4.2, fluctuating levels of a major ionic component can cause variations of sample conductivity and in consequence fluctuating migration times. Concentrations of 1 μM of four anions were analyzed in several synthetic samples with a constant level of 1.7 ppm Li, added as LiOH, and with levels of boron varied in the range of 500 to 3000 ppm.

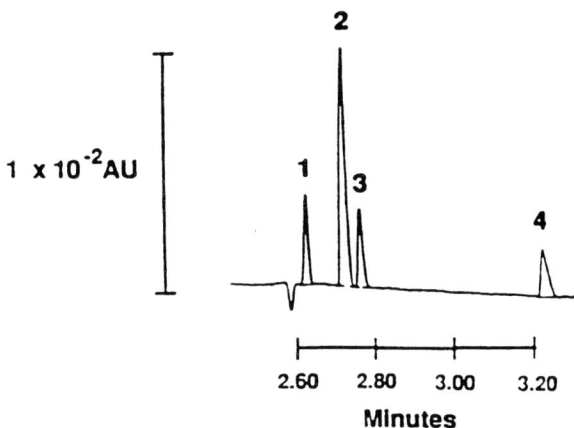

Figure 5.13 Micromolar (1 μM) levels of anions in a simulated primary water sample. The sample contained 1000 ppm boron (5714 ppm boric acid). Analytical conditions were as in Figure 5.12 with the only exception of the time interval for preconcentration, which was reduced to 30 sec for primary water samples. Peak identities and concentrations: (1) chloride, 35 ppb; (2) sulfate, 48 ppb; (3) nitrate, 62 ppb; (4) fluoride, 19 ppb.

Similarly to the experiments discussed in Section 3.4.2, the migration times differ strongly in samples with different concentrations of boric acid. Figure 5.14A also shows the migration times for carbonate, which is omnipresent in the primary water samples. The concentration of that anion cannot be controlled in samples exposed to the atmosphere. In neutral water samples, it equilibrates by absorption of carbon dioxide in water, reaching concentrations of several hundred ppb within several minutes. A signal corresponding to carbonate from atmospheric carbon dioxide is shown in Figure 5.12 but not in Figure 5.13, where the separation was interrupted after the passage of the fluoride zone through the detector.

The ubiquitous peak of carbonate can be utilized for the purpose of normalization of migration times. As illustrated in Figure 5.14B, the influence of changing boron concentrations on the migration times of anions can be eliminated by normalization with carbonate as a reference peak.

Changing concentrations of boron were also evaluated as a variable influencing peak areas of four anions of interest in the primary circuit water (Figure 5.15).

It was concluded that only the changes of nitrate and fluoride represent fluctuations according to the experimental design. The value of %RSD for nitrate and fluoride are in relatively good agreement with the variance observed in consecutive runs with spiked deionized water samples (ca. 3%). The influence of changed migration rates on peak areas is once again compensated by changes in recoveries of electromigrative sample introduction

276   CAPILLARY ELECTROPHORESIS OF SMALL MOLECULES AND IONS

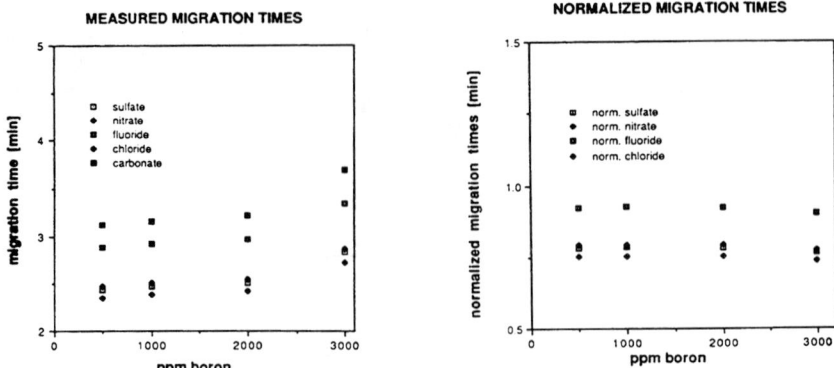

Figure 5.14 Influence of changing boron concentration on migration times of anions. A: Directly measured migration times against ppm of boron. B: Migration times divided (normalized, see Section 3.4.5) by migration time of the carbonate peak at the same concentration of boron. Analytical conditions were as described in Figures 5.13 and 5.12. Reproduced with permission.[19]

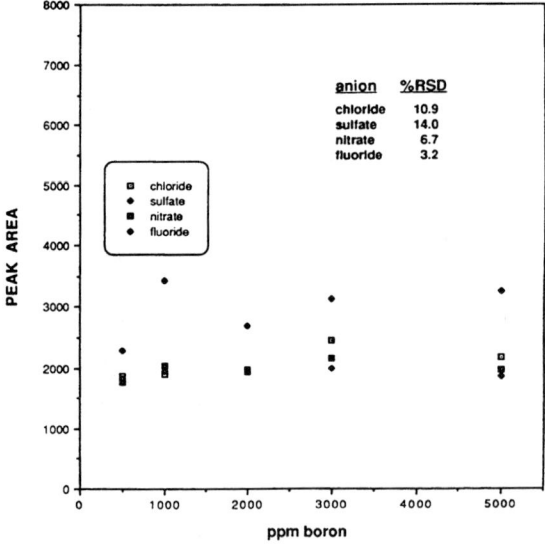

Figure 5.15 Influence of changing boron matrix on peak areas. Analytical conditions and anion concentrations were as in Figure 5.14. Reproduced with permission.[19]

due to changed sample conductivity. The relatively higher values of %RSD for chloride and sulfate, on the other hand, were interpreted as due to a contamination of boric acid by these anions. Overall, the observed imprecision for the four anions can be considered as acceptable for a low-level analysis in a laboratory environment without special equipment for preventing contamination of samples. Improved precision for sulfate and chloride, in an evaluation such as this one, can most likely be achieved with a higher-purity boric acid. It can thus be concluded that the primary water can be analyzed with less than 10% RSD, even if the exact concentration of boron is unknown. The normalization of migration times eliminates their fluctuations. As predicted by Reference 18, the influence of different migration rates on peak areas is compensated by different recoveries of electromigrative sample introduction as a function of ionic mobility of analytes.

An analysis of secondary circuit water containing morpholine, LiOH, and the anions specified invariably yields electropherograms indistinguishable in appearance from those of a low-level standard mixture in pure water. The same type of evaluation was conducted for secondary as for primary water and deionized water samples. The results of precision and calibration experiments are virtually identical with those for other types of PWR water samples and for that reason will not be discussed here.

The combination of electromigrative trace enrichment with separation and detection by electrophoretic capillary ion analysis holds great promise for the monitoring of trace anions in industrial water samples. A similar sensitivity of detection as in ion chromatography is achieved within considerably shorter analytical run times.

5.3 Samples with Disparate Levels of Anions

The ability to analyze samples with an uneven distribution of concentrations, or disparate levels of analytes, is among the important parameters defining the practical utility of a method. Samples, in which the concentration of one of the analytes does not, at least occasionally, exceed the concentrations of other components by at least a factor of 1000, are extremely rare.

The preceding section includes the first important example of samples with extreme concentration ratios of anionic species. Assuming that in most primary water samples ten to twenty percent of boric acid is present as borate anion, the typical ratios of analyte anions to borate anion in the sample matrix are in the 1 : 100 to 1 : 5000 range. The capability of ion chromatography to analyze such samples successfully is limited. Sensitive conductivity detection requires the utilization of low ion exchange capacity columns for separations. Such columns can be overloaded by samples with disparate levels of constituents. Typical ion chromatographic procedures for primary circuit water thus include a rinsing step during which an excess of boric acid is washed off the concentrator before the actual separation of preconcentrated traces of anions. This rinsing step increases not only the complexity of the ion chromatographic method, but also the volume of radioactively contaminated waste. The electrophoretic separation does not require any rinsing prior to the analysis of primary circuit samples.

A recent study[22] revealed peak distortions for fluoride, if fluoride and chloride at a ratio higher than 1 : 100 were separated on an anion-exchange column. At still higher concentration ratios, the size of the peak for fluoride decreased steadily, until this peak disappeared entirely at a ratio of 1 : 1000. The artifact was explainable by the limited ion-exchange capacity of the column. However, a large enough capacity for suppressing the artifact would have rendered detection by conductivity difficult. A comparable matrix problem could be resolved easily with help of capillary electrophoresis.

The migration sequence in CE places the signal for fluoride at a large distance from chloride. In ion exchange, the chloride is more strongly retained and, if present at a large concentration, can disturb interactions of more weakly retained anions, such as fluoride, with anion-exchange groups. The good sensitivity of CE, discussed in Section 5.2, makes possible the analysis of excessive levels of ions even in very small sample volumes. The electropherogram in Figure 5.16 can be obtained from volumes as low as 400 µl.

The purity of the sodium octanesulfonate, which is used as an additive in electromigrative preconcentration in Figure 5.16, as well for the electropherograms in the preceding section, has to be checked prior to use. A capillary electrophoretic separation (see Figure 5.17) was developed for that

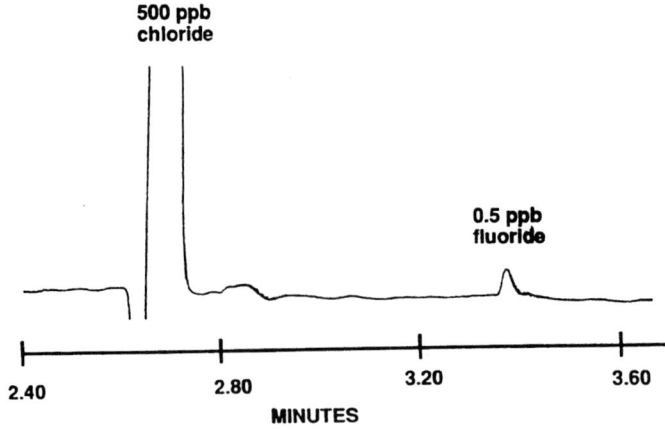

Figure 5.16 Analysis of fluoride–to–chloride ratio of 1:1000. The separation and detection were carried out under the conditions listed in Figure 5.2. The sample was preconcentrated by a procedure outlined in Figure 5.13.

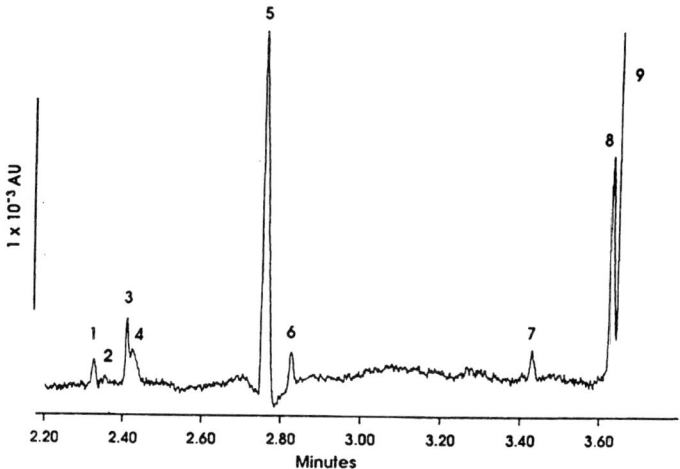

Figure 5.17 Anionic impurities in sodium octanesulfonate. Carrier electrolyte: pH 10.8, 5 mM chromate, 0.5 mM OFM BT Anion (Waters) converted to hydroxide form. Capillary: 75 μm × 60 × 52 cm, fused silica. Separation voltage: −20 kV. Indirect UV detection at 254 nm. Peak identities and concentrations: (1) bromide, 0.37 ppm; (2) chloride, below detection limit; (3) sulfate, 0.57 ppm; (4) iodide, not quantified; (5) carbonate, 1.75 ppm; (6) phosphate, not quantified; (7) and (8) unidentified organic acids; (9) octanesulfonate, 5000.0 ppm. Reproduced with permission.[6]

purpose, utilizing a 30 sec hydrostatic sample introduction of 0.5% solution of the substance.

The ratios of concentrations analyzed in Figure 5.17 are 1 : 13,513 for bromide/octanesulfonate and 1 : 8,772 for sulfate/octanesulfonate. The CE technique does not require any sample pretreatment, and the total time required for analysis is less than 5 min. The carrier electrolyte is of a standard composition, with the exception of an increased pH and the use of the electroosmotic flow modifier in its hydroxide form. For reasons of long-term stability, the EOF modifier (an alkylammonium salt) is usually obtained in bromide form. In the analysis of trace impurities, the sample zones usually show a higher conductance than the carrier electrolyte (5000 ppm and more of the main sample constituent). If bromide is introduced into the carrier electrolyte, a phenomenon termed *reversed electrostacking* occurs (see Section 3.4.2). The reverse electrostacking begins to take place not in the sample zone, but in the section of the electrolyte between the electrode and the injection end of the capillary. The result of such reverse electrostacking is the occurrence of a relatively concentrated bromide zone just inside the sample plug. At least a portion of that zone is capable of behaving as an original sample constituent, generating a distinct peak at the migration time of bromide.

Conversion of the electroosmotic flow modifier is performed by passing a small volume of the 20x concentrate through a segment of an anion-exchange resin in hydroxide form. The CE technique under discussion is also applicable to samples containing higher concentration ratios than those shown in Figure 5.17. Concentrations of up to 1% (10,000 ppm) of a major ionic component can be injected directly into the capillary without any overload effects. One example is the analysis of aromatic impurities in terephthalic acid.

The trace impurities separated from terephthalate in Figure 5.18, are important "chain terminators" and have to be monitored, as they affect the quality of polymers made from terephthalic acid. Note that the concentration ratios are now 1 : 45,454 and 1 : 8929 for p-carboxybenzaldehyde and p-toluate, respectively. Another important aspect of the terephthalate method is the utilization of direct detection at 185 nm.

The reader is referred to the corresponding discussion of the advantages of low UV detection in Section 3.3.1. Direct UV detection provides an additional reason for the removal of bromide from the EOF modifier. The sensitivity of detection at 185 nm would be strongly affected by the presence of a UV absorbing anion in the carrier electrolyte. The main electrolyte anion, hexanesulfonate, offers two main advantages. First, it enables direct photometric detection by absorbing only minimally in the UV range, and, second, it helps to generate a peak shape of the main ionic component facilitat-

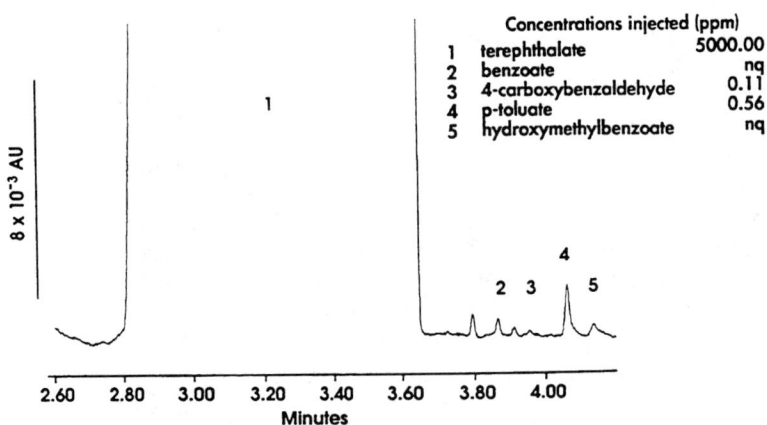

Figure 5.18 Anionic impurities in 99.95% terephthalate. Carrier electrolyte: pH 10.0, 25 mM hexanesulfonate, 0.5 mM OFM BT Anion (Waters) in hydroxide form. Capillary: 75 μm × 60 × 52 cm, fused silica. Separation voltage: −25 kV. Direct UV detection at 185 nm. Sample: 0.500 g solid terephthalic acid and 1.0 g of LiOH in 100.0 ml water. Peak identities and concentrations: (1) terephthalate, 5000 ppm; (2) benzoate, not quantified, (3) p-carboxybenzaldehyde, 0.11 ppm; (4) p-toluate, 0.56 ppm; (5) hydroxymethylbenzoate, not quantified. Reproduced with permission.[23]

ing the electrophoretic separation. The full control of the type of peak symmetry is another useful feature of capillary electrophoresis. As explained in Section 2.7.3, analyte peaks in CE can be either symmetrical or asymmetrical. The extent of asymmetry is determined mostly by analyte concentration. The type of asymmetry depends predominantly on the relative values of mobilities of the analyte ion and electrolyte coion. Since a rather excessive concentration of the main anionic constituent is always the case in purity assays, a pronounced asymmetry of the major peak has to be expected. A good resolution of trace peaks from the major one is possible only if the minor peaks are positioned alongside a sharply defined concentration boundary of the zone containing the main sample component. In capillary electrophoresis a sharply defined concentration boundary is indicated by a steep rise or fall of a peak. A discernible slope at the beginning or end of a peak suggests a diffuse concentration boundary. Fronting peaks will thus be better defined at their end, and the opposite is true for tailing peaks. The aromatic carboxylates that have to be monitored in terephthalate migrate at a slower rate than the main peak; therefore, it is desirable to create a fronting shape of the peak for the terephthalate. Hexanesulfonate exhibits a lower electrophoretic mobility than terephthalate and, if included in the electrolyte, will generate a fronting major peak (see Figure 5.19A). If impurities of lower mobility than that of terephthalate had to be analyzed, the main peak could be easily converted to a tailing shape by the use of, for example, methanesulfonate.

Figure 5.19B illustrates a case where the tailing shape of a major peak was induced by another high-mobility electrolyte, sodium chromate. The major peak belongs to citrate, which is being analyzed for chloride and sulfate impurities.

It should also be noted that the total run time for a purity assay of citrate in Figure 5.19B is only about 5 min. An IC separation achieving a comparable resolution of minor components from the major peak typically requires 1 h. Citrate is retained very strongly on anion-exchange columns. The largest portion of time in ion chromatographic analysis of mixtures, similar to one in Figure 5.19B, is spent waiting for the last remnants of the excessive citrate concentration to elute from the column.

5.4 Modulation of Selectivity for CE Separations of Cations

The first reported capillary electrophoretic separation of cations can be dated as far back as 1967. In that year, Hjerten described the separation of bismuth and copper under coelectroosmotic conditions using lactic acid as the main component of carrier electrolyte. Unlike in the case of inorganic and low-molecular-weight organic anions, the total number of papers discussing CE separations of inorganic cations[24-31] has remained relatively low. The reason for this may be in the considerable difficulty encountered in the

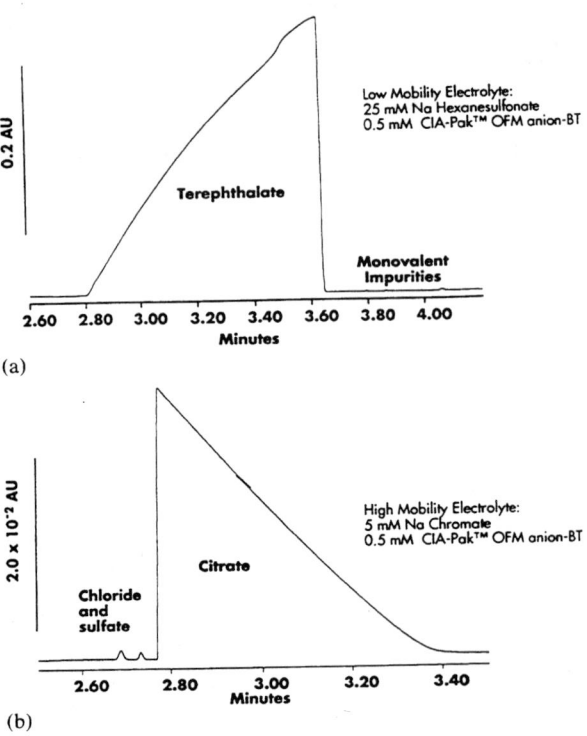

Figure 5.19 Two examples of controlled peak symmetry. A: Separation and detection method as in Figure 5.18. B: Analytical conditions as in Figure 5.17, except for pH, which was adjusted at 8.0. The chloride migrates first, followed by sulfate. Both trace anions are at 1 ppm level. The concentration of citrate was 0.15%. Sample was introduced hydrostatically for 30 sec at 9.8 cm. Reproduced with permission.[23]

initial development of cation separations. The close correlation of the migration order of small ions with values of electrophoretic mobilities (μ_i), which in turn are directly proportional to ionic equivalent conductances (λ_i) available in the literature, is discussed in Section 2.2. This correlation indicates that any two ions of identical or similar conductivities will be difficult to resolve relying only on the migration behavior of their free, uncomplexed forms. To illustrate how accurate such a prediction is, let us consider an attempted separation of ammonium ion from alkali cations in Figure 5.20a.

The equivalent ionic conductances for ammonium and potassium are identical, and the peaks for the two cations comigrate in a separation based purely on electrophoretic mobilities (Figure 5.20a). As shown in Figure 5.20b, however, such a separation problem can be improved by altering the pH to a level where the protonation of ammonium, as a weak base (pK_b = 4.75), is partially suppressed. This is accomplished by changing the pH of

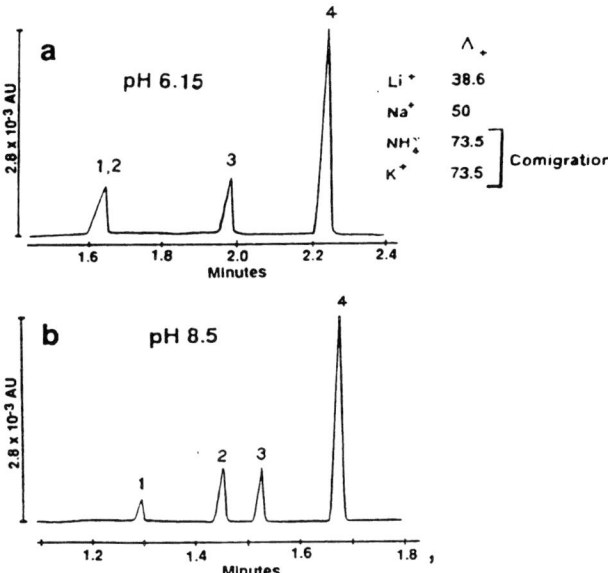

Figure 5.20 a: Separation of alkali cations in the presence of ammonium, using indirect photometric detection at 214 nm. Carrier electrolyte contains 5 mM morpholinoethanesulfonate (MES) adjusted to pH 6.15. Separation voltage: +25 kV. Capillary: 75 μm × 60 ×52 cm, fused silica. Hydrostatic sample introduction for 30 sec at 9.8 cm. Concentrations of all cations is 0.4 ppm. b: CE conditions as in a, except for pH, which is now at 8.5. Peak identities: (1) potassium; (2) ammonium; (3) sodium; (4) lithium. Reproduced with permission.[31]

the carrier electrolyte from 6.15 to 8.5. As the equilibrium between the ionized and nonionized forms of ammonium is shifted in favor of the nonionic form, the observed, apparent rate of migration of that cation is reduced, while the migration behavior of the more strongly basic cations remains unchanged. The ammonium, as a result, becomes well separated from the potassium peak. At the same time, the migration times of all four cations become shorter, due to the increased rate of the electroosmotic flow at the higher pH (see Figure 5.20b). Note that the separation is performed in the coelectroosmotic mode, with the negative electrode on the detector side attracting analyte cations and the direction of electroosmotic flow also being toward the detector. The indirect UV detection is made possible by the UV absorption background provided by MES.

The plot in Figure 5.21 illustrates a number of other possible resolution problems of mixtures of cations. The values of tabulated limiting ionic equivalent conductances are plotted against the sequential numbers for metals, assigned according to increasing values of ionic equivalent conductances within arbitrary groups.

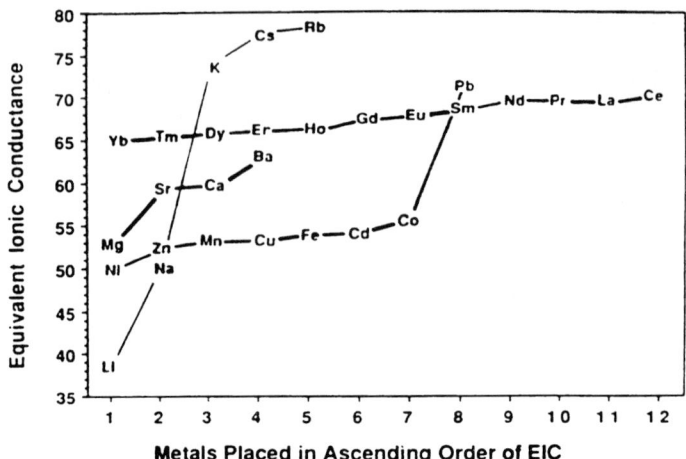

Figure 5.21 Plot of limiting equivalent ionic conductances against sequential ionic conductance numbers. Reproduced with permission.[28]

The closer any two points are in Figure 5.21, the more difficult is the separation of corresponding cations without utilization of an auxiliary separation mode (Section 2.4). As a next example, we will consider the problem of separating a mixture of alkaline earth and alkali cations. Figure 5.21 shows that the points corresponding to strontium and calcium have close equivalent conductance values. Another two points differing only slightly in their respective equivalent conductance values are those for sodium and magnesium. Predictably, from the closeness of their equivalent conductance values, strontium comigrates with calcium, and sodium cannot be resolved from magnesium in the coelectroosmotic separation in Figure 5.22A.

A change of pH has to be ruled out for improving the lack of resolution in Figure 5.22A. Increased pH would cause the analyte cations to precipitate, preventing their electrophoretic separation. Additionally, all four cations that are to be slowed down or accelerated for better resolution are relatively strongly basic, and a small change of pH is not likely to change their migration behavior. The solution of the problem is in the employment of an auxiliary separation mechanism. If citric acid is added to the electrolyte, alkaline earth cations will participate to a different degree, determined by the respective stability constants, in a complexation equilibrium with the citrate ligand. The net effect of such interactions is a decreased concentration of free cation and a corresponding slowing down of the apparent rate of migration. A pair of cations with originally identical migration times, as, for example, calcium and strontium, are slowed down differently, and their full resolution in capillary electrophoresis is achieved. The resolution of a comigration peak pair of sodium and magnesium is achieved by a deceleration of

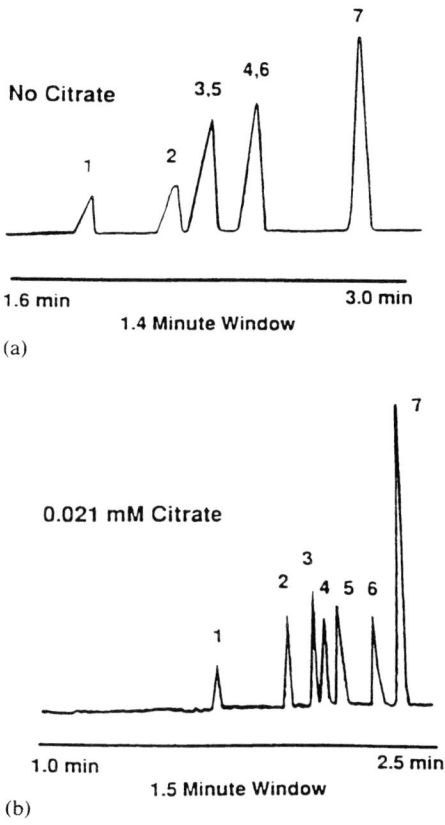

Figure 5.22 Two electropherograms of a mixture of IA and IIA cations. A: Carrier electrolyte: 5 mM UV CAT1 (Waters), pH 5.2. Capillary: 75 μm × 60 × 52 cm, fused silica. Indirect UV detection at 214 nm. Separation voltage: +25 kV. Hydrostatic sample introduction for 30 sec at 9.8 cm. B: Same conditions as A, except for the changed composition (5 mM UV CAT1, 0.021 mM citric acid) and pH (5.5) of the carrier electrolyte. Peak identities and concentrations: (1) potassium, 0.5 ppm; (2) barium, 1.0 ppm; (3) strontium, 0.6 ppm; (4) sodium, 0.5 ppm; (5) calcium, 0.4 ppm; (6) magnesium, 0.2 ppm; (7) lithium, 0.5 ppm. Reproduced with permission.[28,29]

magnesium, while the migration behavior of sodium remains relatively unchanged. The resolving effect of the addition of citrate is illustrated in Figure 5.22B. The longer analytical run times resulting from the complexing deceleration of analyte cations can be prevented by a slight increase in electrolyte pH, leading to a compensating increase of the rate of electroosmotic flow. The effect of complexing equilibrium on migration rates, together with corresponding electrophoretic and electroosmotic vectors, is shown in Figure 5.23.

Mobility of Cations in the Capillary

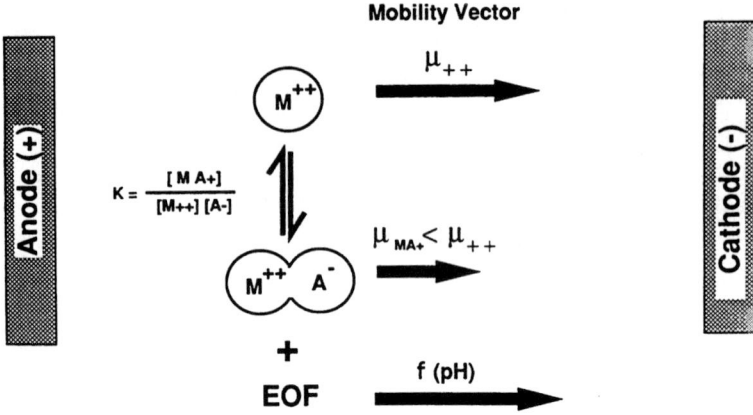

Figure 5.23 Modification of ionic mobility (μ) by complexation. The mobility of cation M is reduced by complexation with complexing ligand A. Reproduced with permission.[5]

A brief comment on the role of coion in carrier electrolyte is necessary at this point. The UV CAT 1 reagent is a UV-absorbing cation making possible the indirect UV detection of analyte cations. It was found preferable over MES, used in the initial studies, because of its relatively higher molar absorptivity at 214 nm. Its electrophoretic mobility is clearly too slow for the separation in Figure 5.22A. Section 2.7.3 provides a discussion of corresponding optimization rules. After an addition of complexing agent, the asymmetry of peak shapes (fronting) was considerably improved. This was a desirable side effect of the deceleration of apparent mobilities due to complexation. The lower apparent mobilities of partially complexed cations are a better match with the electrophoretic mobility of carrier electrolyte coion UV CAT 1.

An alternative way of adjusting selectivity of the separation of IA and IIA cations was reported by Beck and Engelhardt.[30] The resolution problem of sodium and magnesium illustrated in Figure 5.22A can be overcome by employing imidazole as a carrier electrolyte coion. The calcium/strontium peak pair is not discussed by the authors. With a shorter capillary and a relatively high voltage, the method of Beck and Engelhardt also achieves an extremely short analytical run time.

For mixtures of alkali and alkaline earth cations, also containing ammonium ion, the choice of a different electrolyte coion or the addition of citrate

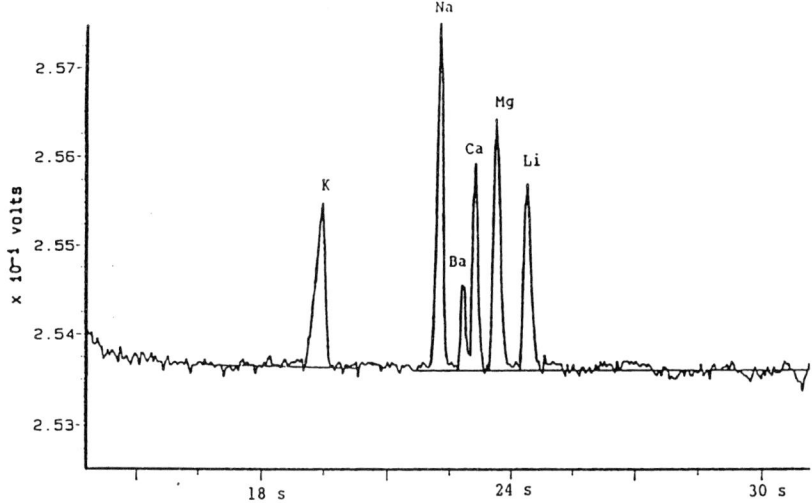

Figure 5.24 Separation of a cation mixture in less than 30 sec. Carrier electrolyte: 3 mM imidazole, pH 6.0. Capillary: 75 μm × 23 × 15 cm, fused silica. Separation voltage: +30 kV. Indirect UV detection at 214 nm. Hydrostatic injection for 30 sec at 9.8 cm. All sample components were at 1 ppm. Reproduced with permission.[30]

ligand alone does not lead to the resolution of all analyte peaks. Potassium and ammonium will still comigrate in carrier electrolytes from Figure 5.22B or 5.24.

To resolve the two peaks, it is necessary to resort to an additional auxiliary mechanism consisting in a complexation by ionic size recognition with the help of a suitable crown ether. Applications of crown ethers for achieving auxiliary separation modes are discussed in general terms in Section 2.4. The improved separation of ammonium from potassium is shown in Figure 5.25.

The electropherogram in Figure 5.26 was generated with an electrolyte containing HIBA instead of citric acid as a complexing agent. This ligand has been used successfully for many years in HPLC eluents for separations of lanthanides and transition metals.[32] Its application to the separation of lanthanides by capillary electrophoresis was also highly successful.

Employment of auxiliary mechanisms, such as dissociation equilibria, complexation, or size-specific association, has opened new possibilities for applying CE to the analysis of cations in various types of samples. The new methodology represents a useful and cost-effective alternative to the currently prevalent spectroscopic techniques.

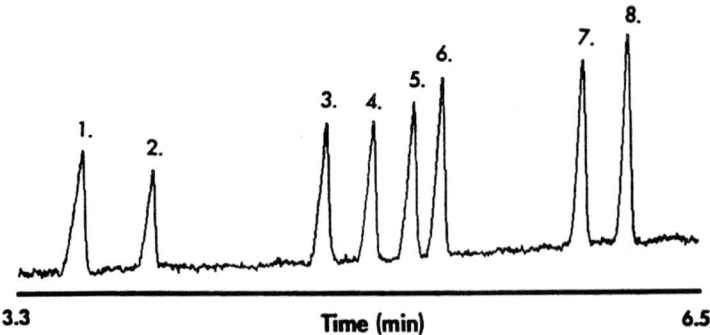

Figure 5.25 Separation of ammonium in a mixture with IA and IIA cations. Carrier electrolyte: 5 mM UV CAT1 (Waters), 6.5 mM α-hydroxyisobutyric acid (HIBA), 2 mM 18-crown-6-ether. Capillary: 75 μm × 60 × 52 cm, fused silica. Separation voltage: +20 kV. Hydrostatic injection for 30 sec at 9.8 cm. Indirect UV detection at 185 nm. Peak identities and concentrations: (1) ammonium, 1 ppm; (2) potassium, 1 ppm; (3) calcium, 0.7 ppm; (4) sodium, 0.6 ppm; (5) magnesium, 0.4 ppm; (6) strontium, 1.5 ppm; (7) barium, 2.0 ppm; (8) lithium, 0.2 ppm. Reproduced with permission.[31]

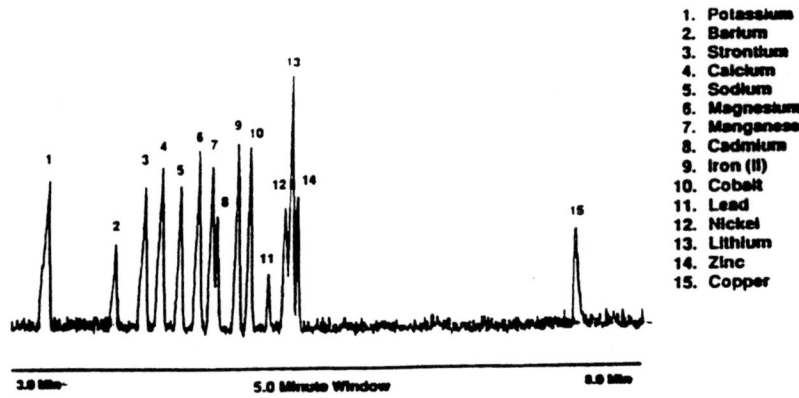

Figure 5.26 Simultaneous separation of alkali, alkaline earth, and transition metal cations by capillary electrophoresis. Carrier electrolyte: 5 mM UV CAT1 (Waters), 6.5 mM HIBA, pH 4.4. Capillary: 75 μm × 37 × 32 cm, fused silica. Separation voltage: +20 kV. Hydrostatic injection: 30 sec at 9.8 cm. Indirect UV detection at 214 nm. Peak identities and concentrations: (1) potassium, 0.8 ppm; (2) barium, 1.5 ppm; (3) strontium, 1.5 ppm; (4) calcium, 0.7 ppm; (5) sodium, 0.6 ppm; (6) magnesium, 0.4 ppm; (7) manganese, 0.8 ppm; (8) cadmium, 0.8 ppm; (9) iron(II), 1.0 ppm; (10) cobalt, 0.8 ppm; (11) lead, 1.0 ppm; (12) nickel, 0.6 ppm; (13) lithium, 0.2 ppm; (14) zinc, 0.4 ppm; (15) copper, 0.6 ppm. Reproduced with permission.[28]

References

1. W. R. Jones and P. Jandik, *Amer. Lab.* **22**, No. 9, 51 (1990).
2. W. R. Jones, P. Jandik, and M. T. Swartz, *J. Chromatogr.* **473**, 171 (1989).
3. H. Shintani and P. K. Dasgupta, *Anal. Chem.* **59**, 802 (1987).
4. W. R. Jones, P. Jandik, and A. L. Heckenberg, *Anal. Chem.* **60**, 1977 (1988).
5. P. Jandik, W. R. Jones, A. Weston, and P. R. Brown, *LC GC* **9**, 634 (1991).
6. W. R. Jones and P. Jandik, *J. Chromatogr.* **546**, 445 (1991).
7. W. Buchberger and P. R. Haddad, *J. Chromatogr.* **608**, 59 (1992).
8. F. E. P. Mikkers, F. M. Everaerts, and Th. P. M. Verheggen, *J Chromatogr.* **169**, 18 (1979).
9. P. Gebauer, M. Deml, P. Bocek, and J. Janak, *J. Chromatogr.* **267**, 455 (1983).
10. F. Foret, M. Deml, V. Kahle, and P. Bocek, *Electrophoresis* **7**, 430 (1986).
11. J. L. Beckers, Th. P. E. M. Verheggen, and F. M. Everaerts, *J. Chromatogr.* **452**, 591 (1988).
12. X. Huang, M. J. Gordon, and R. N. Zare, *J. Chromatogr.* **480**, 285 (1989).
13. M. Aquilar, X. Huang, and R. N. Zare, *J. Chromatogr.* **480**, 427 (1989).
14. S. Hjerten, K. Elenbring, F. Kilar, J. Liao, A. J. C. Chen, C. J. Siebert, and M. Zhu, *J. Chromatogr.* **403**, 47 (1987).
15. F. Foret, S. Fanali, L. Ossicini, and P. Bocek, *J. Chromatogr.* **470**, 299 (1989).
16. W. G. Kuhr and E. S. Yeung, *Anal. Chem.* **60**, 2642 (1988).
17. L. Gross and E. S. Yeung, *J. Chromatogr.* **480**, 169 (1989).
18. P. D. Grossman, H. H. Lauer, S. E. Moring, D. E. Mead, M. F. Oldham, J. H. Nickel, J. R. P. Goudberg, A. Krever, D. H. Ransom, and J. C. Colburn, *American Biotechnology* **8**, 32 (1990).
19. G. Bondoux, P. Jandik, and W. R. Jones, *J. Chromatogr.* **602**, 79 (1992).
20. P. Jandik and W. R. Jones, *J. Chromatogr.* **546**, 431 (1991).
21. G. Bondoux, personal communication.
22. D. T. Gjerde, D. J. Cox, P. Jandik, and J. B. Li, *J. Chromatogr.* **546**, 151 (1991).
23. W. R. Jones and P. Jandik, *J. Chromatogr.* **608**, 385 (1992).
24. T. Tsuda, K. Nomura, and G. Nakagawa, *J. Chromatogr.* **264**, 385 (1983).
25. X. Huang, T. K. J. Pang, M. J. Gordon, and R. N. Zare, *Anal. Chem.* **59**, 2747 (1987).
26. J. L. Beckers, T. P. E. M. Verheggen, and F. M. Everaerts, *J. Chromatogr.* **452**, 591 (1988).
27. A. Nardi, S. Fanali, and F. Foret, *Electrophoresis* **11**, 774 (1990).
28. A. Weston, P. R. Brown, P. Jandik, W. R. Jones, and A. L. Heckenberg, *J. Chromatogr.* **593**, 289 (1992).

29. A. Weston, personal communication, 17 August 1990.
30. W. Beck and H. Engelhardt, *Chromatographia* **33,** 313 (1992).
31. A. Weston, P. R. Brown, and P. Jandik, in preparation.
32. S. Elchuk and R. Cassidy, *Anal. Chem.* **51,** 1434 (1979).

Index

Absorbance, 122
Absorbance noise
  see noise
Acetate, 18, 30, 43, 259, 260, 271
Acetate, carrier electrolyte, 81, 168
Acetonitrile, carrier electrolyte, 51, 53, 268
Additives, 16, 29, 36, 41
Adenosine-5'-monophosphoric acid, 81
Adipate, 43
Adjusted migration index, 204
Air bubbles, 7
Alanine, 150
Alkali, 49
Alkaline earths, 49
Alkanolamines, 161
Alkyl ammonium salts, 28, 31, 161
Alkylamines, 161
Alkylsulfonates, 71
Amines, 161
Aminobenzoate derivatives of
  carbohydrates, 215–25
  electrophoretic mobilities, 230, 243
Aminobenzonitrile derivatives of
  carbohydrates, 231
Aminobutyric acid, carrier electrolyte, 81
Aminopenicillanic (6-aminopenicillanic) acid, 125

Aminopyridine derivatives of
  carbohydrates, 214
  electrophoretic mobilities, 243
Ammonium, 45, 283, 288
Amount sampled, Q, 84, 89, 94
Amoxicillin, 125
Amperometric detection, 168
Amphetamine, 53, 127
Analog-to-digital (A/D) conversion, 198
Analyte zone, 82. *See also* spreading of peaks in CE.
Anions, 19, 160
  inorganic, 15, 160
  detection limits, 270, 272
Anisate, carrier electrolyte, 146
Apparatus for chemical modification of capillaries, 34
Apparent mobility. *See* observed mobility.
Arabinose, 215, 226, 229, 232, 234
Arginine, 168
Arsenate, 147, 261
Asparagine, 150
Aspartate, 150, 260
Atomic vapor lamp, 120
Auxiliary electrode, 170
Azide, 18, 260

Barium, 29, 285, 288
Beer's Law, 110, 122, 129

Benzenetetracarboxylate, 146
Benzo[ghi]perylene, 51
Benzoate, 18, 43, 127, 147, 260, 280
Benzoate, carrier electrolyte, 71, 146
Benzoic acid. *See* benzoate.
Benzoylecgonine, 127
Benzyl-DL-aspartate, 43
Bias between different samples, 90
Bias of electrokinetic injection, 89
Biogenic amines, 171
Biomass degradation products, 249–250
Bismuth, 47
Borate, carrier electrolyte, 39, 53, 125, 127, 213, 214, 219, 229, 232, 246, 249, 251, 252
Boric acid, 73
Bromate, 261
5-Bromo-2,4-dihydroxybenzoic acid, 81
Bromide, 18, 105, 142, 147, 259, 260
Bubble cell, 11
Buffer. *See* carrier electrolyte.
Bulk property detectors, 119
Butalbital, 187
Butanesulfonate, 18, 71, 260, 268
Butyrate, 18, 30, 259, 260

Cadmium, 288
Cadmium in carrier electrolyte, 47
Cadmium lamp, 120
Caffeic acid, 245–6
Caffeine, 187
Calcium, 19, 29, 161, 285, 288
Calcium ion, carrier electrolyte, 267
Calibration plot, 140, 240, 272
Cannabidiol, 53
Capacity factor in MECC, 53
Capillaries. *See also* selection of capillaries.
  fused silica, 14, 29, 107
  polymeric, 14, 29, 57, 81, 107
Capillary gel electrophoresis, 38
Capillary ion analysis, 38, 258–88
Capillary wall, 14, 24, 26
Carbohydrate-borate complexes, 212–17
Carbohydrates, 212–14
  electrophoretic mobilities, 220, 243
  detection limits, 224, 244
Carbonate, 18, 105, 259, 260, 271
Carboxybenzaldehyde, 280
Carboxymethyl-β-cyclodextrine, carrier electrolyte, 56
Carrier electrolyte, 6
Catecholamines, 171
Cations, 17

CBQA tagged amino acids, 155
CE methods for carbohydrates, comparison, 241–2
Cellibiose, 232, 234
Cellobiose, 219, 229
Cerium, 49
Cetyltrimethylammonium bromide (CTAB) additive, 29, 30
CGE. *See* capillary gel electrophoresis.
Charge association, 50
Chemical suppression, 167
Chiral compounds, 24, 251
Chlorate, 18, 41, 260
Chloride, 18, 43, 47, 105, 142, 160, 259, 260, 271, 278
Chloride, carrier electrolyte, 41, 176
Chlorite, 18, 260
Chloroacetate, 260
Chloropropionate, 43
Chromate, 41, 147, 261
Chromate, carrier electrolyte, 18, 73, 105, 139, 142, 259, 260, 266, 271, 279
CIA. *See* capillary ion analysis.
CIA OFM BT Anion (Waters), TTAB. *See* modifier of electroosmotic flow.
Citrate, 18, 260
Citrate, carrier electrolyte, 106, 285
Cloxacillin, 125
CMCE. See countermigration capillary electrophoresis
Cobalt, 288
Cocaine, 127
Codeine, 53, 127, 187
Coelectroosmotic capillary electrophoresis, 44, 99, 257–288
Colloid science, 23
Commercial CE instruments, 180–2, 185–9
Complexing interactions in CE, 46, 267, 285–6
Components of CE systems, 6, 177–182
Concentration in isotachophoretic zones, 100
Conditioning of capillaries, 6, 27
Conductance, equivalent, 21, 87
  limiting ionic equivalent, 13, 18, 87, 101, 282–4
Conductive joint, 169
Conductivity detection, 30, 162
Constant current mode, 193
Copper, 47, 288
Corona discharge, 8
Correlation, 19, 146
Coulombic efficiency, 169

INDEX

Counterelectroosmotic capillary electrophoresis, 42, 99, 211–253
Countermigration capillary electrophoresis, 38, 40
Coupling of capillaries, 113
Cresol isomers, 56
Cresol red, 158
Critical micelle concentration, 32, 51
Crotonate, 260
Crown ether, carrier electrolyte, 54, 287
CTAB. See cetyltrimethylammonium bromide additive.
Cutting of fused silica capillaries, 116
Cyanide, 261
Cyclodextrines, carrier electrolyte, 54
Cylindrical cell, 129
Cysteic acid, 150
Cystine, 150

Dansyl amino acids, 150, 162
Data acquisition rate, 10, 198–200
Data management, 10
Degassing of carrier electrolytes, 7
Delayed data acquisition, 10
Deoxy-D-ribose, 215, 219, 229, 239
Detection, 10, 118–179
  sensitivity, 118–9, 132, 139–40, 143
  indirect, 18, 29, 134–148, 157–162, 168
Detection cells, 129–134
Detection limits. See anions, detection limits, carbohydrates, detection limits.
Detection windows, 114
Detector, 7, 9
Deuterium lamp, 10, 120, 123
Dextrin, 214
Diazepam, 53, 127
Dichloroacetate, 261
Dichromate, 160
Dicloxacillin, 125
Dielectric constant, 23
(Diethylamino)ethyldextran, carrier electrolyte, 48
Diffraction, 130
Diffraction grating, 122
Diffusion coefficient, 60, 68
Disparate levels of anions, 277
Dodecanesulfonate, 268
Dopamine, 171
Doxepin, 106
Dual slope, integrating A/D converter, 199
Dynamic range, 143
Dynamic reserve, 139, 141, 143, 159
Dysprosium, 49

EDTA, carrier electrolyte, 41
Effect of electrolyte concentration on migration times, 263
Effect of EOF modifier concentration on migration times, 264
Effect of organic solvents on migration times, 264
Effect of pH on migration times, 262
Efficiency of separation, 58, 96, 105, 112, 240
Electric control of electroosmosis. See external field, influence on electroosmosis.
Electric field, 15, 17, 23
Electric field distribution in capillary, 193
Electric sample splitter, 94
Electrochemical potential, 61
Electrokinetic experiments, 21–2
Electrokinetic injection. See sample introduction, electromigrative.
Electrolyte. See carrier electrolyte.
Electrolyte density, 83
Electromigrative sampling. See sample introduction, electromigrative.
Electroosmosis. See electroosmotic flow.
Electroosmotic flow, 11, 13, 21, 27
  reversal of, 32, 36. See also modifier of electroosmotic flow.
Electroosmotic Flow Condition, CIA, 258
Electroosmotic mobility, 14
Electropherogram, 10–11
Electrophoretic capillary ion analysis. See capillary ion analysis.
Electrophoretic migration, 12
Electrospray ionization (ESI), 173
Electrostacking, 79, 80–82, 96–98
Electrostacking Condition, CIA, 259
Enanthate, 43
Epinephrine, 171
Erbium, 49
Ethanesulfonate, 18, 71, 260
Ethidium bromide additive, 41
Ethyl-p-aminobenzoate derivatives of carbohydrates, 225–231
  electrophoretic mobilities, 243
Ethylene glycol, carrier electrolyte, 109
Europium, 49
Exponentially modified Gaussian function (EMG), 200
External field, influence on electroosmosis, 36

Fast atom bombardment ionization (FAB), 173

Fentanyl, 53
Ferullic acid, 245–6
Field enhancement ratio, 82, 96
First Coion Condition, CIA, 259
Fixed-wavelength detectors, 120–4
Flavonoids, 244–247
Flow profile, 25
Flunitrazepam, 127
Fluorescein, 158
Fluorescein, carrier electrolyte, 158
Fluorescence detection, 149–162
  indirect, 157–162
  laser induced, 149–162
Fluorescent derivatives of oligosaccharides, 213–14
Fluoride, 18, 105, 142, 259, 260, 271, 278
Forensic drug screen, 53
Formamide, 50
Formate, 18, 30, 147, 259, 260, 267, 271
Fronting of peaks, 70
Fructose, 232, 234, 241
Fucose, 215, 229, 232, 234
Fumarate, 261
Fused Silica Manipulator, 112

Gadolinium, 49
Galactarate, 18, 260
Galactose, 213, 215, 226, 229, 232, 234, 239
Galacturonic acid, 215, 219, 226, 229, 232, 239, 260
Gating, 95
Gaussian peaks, 70
Giddings theory, 172
Gluconate, 18, 239, 260
Glucose, 213, 215, 226, 229, 232, 234, 239, 241
Glucuronic acid, 215, 219, 226, 229, 232, 239, 261
Glutamate, 43, 260
Glutamic acid, 150
Glutarate, 18, 260
Glycerate, 261
Glycine, 150
Glycolate, 261, 267

Heat dissipation, 64–70, 185
Heating power, 66
Helmholtz layer(s), 23
Hemicellulose, 226
Hemimicelles, 32
Heptanesulfonate, 71, 261, 268
Heroin, 53, 127
Hesperidine, 245–6
Hexanesulfonate, 71, 261, 268

Hexanesulfonate, carrier electrolyte, 280
Hexanoate, 30
HIS. See histidine; histidine, carrier electrolyte.
Histidine, carrier electrolyte, 30, 90, 96, 165
Holmium, 49
Home-made CE instruments, 8, 179
HPLC, comparison with CE, 9, 11, 12, 30, 60, 79, 124, 136, 240, 244
Hydrated ionic radii. See Stokes radii.
Hydroxybenzoate, carrier electrolyte, 266
Hydroxybutyrate, 261
Hydroxyethylcellulose additive, 29, 41, 43, 81
Hydroxyisobutyric acid, carrier electrolyte, 49, 288
Hydroxymethylbenzoate, 280
Hydroxypropylcellulose additive, 48
Hydroxypropylmethylcellulose additive, 29
Hydroxyvalerate, 261
Hyperosid, 245–6
Hysteresis of electrophoretic mobilities, 28

IEEKC. See ion exchange electrokinetic chromatography
Imidazole, carrier electrolyte, 248, 287
Immersed flow cell, 151
Imprecision, 10
Inadvertent hydrodynamic flow, 84–85
Inadvertent sample introduction, 68
Inclusion complexation in CE, 54
Indirect UV detection, 134–148
  of carbohydrates, 235–241
Inherent ionic mobility, 12
Injection. See sample introduction.
Injection artifacts, 67
Injection by syringe, 92–93
Integrators, 10
Internal standard calibration, 10
Iodate, 160
Iodide, 147, 261
Ion chromatography, 136, 166
  limitations, 257–258
  comparison with CE, 269, 274, 277
Ion exchange electrokinetic chromatography, 38, 46
Ion exchange interactions in CE, 46
Ion exchange separation, 33, 136
Ion mobility detection, 178
Iron, 288
Isocitrate, 261
Isoleucine, 150
Isopropanol, carrier electrolyte, 171

Isoquercitrin, 245-6
Isotachophoresis, 6, 99, 103
Isotachophoretic steady state, 99, 102

Joule heat, 64-67
 dissipation, 26, 64
 as a cause of peak spreading, 68-70

Ketoglutarate, 261
Kohlrausch Condition, 259
Kohlrausch Regulating Function, 100-2, 269

Lactate, 261
Lactic acid, carrier electrolyte, 47
Lactose, 219, 229, 234
Lambert-Beer's law. See Beer's law.
Lanthanides, 49
Lanthanum, 49
Laser-based refractive index detection, 178
Laser-induced capillary vibration detection, 178
Latex particles, 58
Lead, 288
Leading electrolyte, 99
Length of sample zone. See sample size.
Linearity of calibration plots, 144
Lipophilic interaction, 33
Lithium, 29, 45, 49, 161, 165, 283, 285, 288
Logarithmic linearity test, 144
Lorazepam, 53
LSD, 53
Luteolin, 245-6
Lysine, 150
Lyxose, 215, 232, 234

Magnesium, 29, 49, 161, 285, 288
Maleate, 261
Malonate, 41, 261
Maltose, 215, 234
Maltotriose, 215, 229, 232, 234
Manganese, 288
Mannose, 213, 234, 239
Mannuronic acid, 229, 239
Marfey's reagent, 251
Mass Spectrometric Detection, 172
Mass-to-charge ratio (m/Z), 175
Matrix elimination, 102
Maximum value of current in CE, 8
Maximum value of voltage in CE, 8
MECC. See Micellar electrokinetic capillary chromatography.
MEKC. See Micellar electrokinetic capillary chromatography.

Melibiose, 219, 232, 234
Mercury lamp, 10, 120
Mercury-xenon lamp, 149
MES, 2-(N-morpholino)ethanesulfonic acid, carrier electrolyte, 30, 43, 45, 90, 96, 157, 165, 171, 283
Mesityl oxide, 51
Methadone, 127
Methamphetamine, 127
Methanesulfonate, 71, 261
Methaqualone, 53, 123
Methionine, 150
Methylanthracene, 51
Micellar electrokinetic capillary chromatography, 38, 51-54, 232-5
Micellar electrokinetic chromatography. See Micellar electrokinetic capillary chromatography.
Micellar interactions in CE, 51
Micelles, 32
 anionic, 52
Microcystin, 154
Microinjector, 95
Migration data, 20, 161
Migration index (MI), 202
Migration rate, 12
Migration time, 11
Minimum detectable concentration, 139
Misalignment of optical elements, 130
Mobility vectors, 16-17, 37
Modified capillaries, 21, 33-36, 160
Modifier of electroosmotic flow, 18, 142, 259, 264, 266, 267, 269, 271, 279, 280
 conversion of, 280
Molar absorptivity, 122, 142
Molecular sieving. See capillary gel electrophoresis, countermigration CE.
Molybdate, 18, 147, 260
Monochloroacetate, 267
Monochromator, 122
Monofluorophosphate, 18, 260
Monosaccharides, 213, 229
 in complex matrices, 225
Morphine, 53, 127
Multireflection cell, 132

N. See theoretical plates.
N-acetylneuraminic acid, 239
Nafcillin, 125
Naphthalene, 51
Naphthalenesulfonate, carrier electrolyte, 268
Naphthalenesulfonates, 48
Naphthalene-2-monosulfonate, 43

Naringin, 245–6
Neodymium, 49
Neuron, sampling from, 172
Nickel, 288
Nitrate, 18, 47, 105, 142, 147, 259, 260, 271
Nitrite, 18, 105, 142, 259, 260
Nitrophenols, 102
Noise, 130, 142
Nonanesulfonate, 268
Nonoptimal conditions in CE, 74–75
Norepinephrine, 171
Normalization, 200–4, 274–7
    of migration times, 202
    of peak areas, 200
2-[(N)-[tris (hydroxymethyl)methyl]-
    ethanesulfonic acid, 157
Nuclear power plant water, 272
Nucleosides, 162
Nucleotides, 39, 162

Observed mobility, 12–13, 18–19
Octanesulfonate, 268
OFM BT Anion (Waters), TTAB. See modifier of electroosmotic flow.
Optical cell. See detection cells.
Optical fibers, 133
Optical filter, 121
Optical path length, 109, 132, 142
Optical throughput, 123
Optical tracing diagram, 134
Optimization of efficiency, 58–75
Optimization of selectivity, 37–58
Orange G, 158
Orange juice, 240–1
Orthosilicic acid, 27
Oxacillin, 125
Oxalacetate, 261
Oxalate, 261, 271
Oxazepam, 127

Particles, CE separation of, 56–57
PCP, 53
Peak areas, change with migration rates, 194–198
Peak capacity, 172
Peak deformations, 70–75
Peak widths, 10
Pen-ray lamps, 123
Pentanesulfonate, 18, 71, 260, 268
Pentanoate, 30
Perchlorate, 160, 261
Permanganate, 160
Personal computer, 10, 179
Perylene, 51

Phenobarbital, 53
Phenylcarboxylic acids, 246
Phosphate, 18, 105, 142, 160, 260, 271
Phosphate, carrier electrolyte, 48, 56, 58, 109, 125, 150, 154, 158, 234
Phosphite, 18, 260
Phosphonium salts, 176
Photodiode, 121
Photodiode array detectors. See scanning detection.
Photomultiplier, 121
Phthalate, 147, 260
Phthalate, carrier electrolyte, 146, 176, 266, 267
Poly(uridine 5'-phosphate), 39
Poly(vinylalcohol) additive, 29
Polyacrylamide gel capillary, 39, 214
Polyimide coating of capillaries, 24, 112
Polysaccharides, 226
Polyvinylpyrrolidone (PVA) additive, 29
Potassium, 29, 45, 49, 165, 283, 285, 288
Potentiometric detection, 178
Power supply, 8, 177
Praseodymium, 49
Precision, 182, 272
Precolumn derivatization, 217, 227, 232
Preconcentration by isotachophoresis, 98–104
Preconcentration capillaries, 104–7
Pressure differential, sampling, 83–85
Pressurized water reactors (PWR), 273
Primary circuit water, 274
Propanesulfonate, 18, 71, 260
Propionate, 18, 30, 43, 146, 259, 260, 271
Proteins, 162
Pseudostationary phase, 52
Psilocin, 53
Psilocybin, 53
PTH-arginine, 97
PTH-histidine, 97
Pure water analysis, 102
Purging of capillaries, 84
Pyrazole-3,5-dicarboxylate, 43
Pyrene, 51
Pyridoxine, 109
Pyruvate, 261

Quercitrin, 245–6
Quinine sulfate, carrier electrolyte, 161

Racemic mixtures, 252
Radioisotope detection, 178
Raffinose, 239
Recorders, 10

Rectangular profile of capillaries, 67, 107
Red blood cell, 57
Reduced elution speed (RES), 175
Reductive amination of glucose, 232
Reference beam, 122
Reference electrode, 170
Reference ratios, 89
Reference spectra, 128
Reflection, 130
Reproducibility. *See* precision.
Reproducible sample introduction, 7, 180–2
Resistance of zones in the capillary, 192
Resolution, 58
Retention time, 11, 20
Reverse electrostacking, 191, 279
Reverse phase separation, 33
Reversed optics, 125
Rhamnose, 215, 229, 232, 234, 239
Ribose, 215, 229, 232, 234
Rotary injector for CE, 93
Rubidium, 165
Rutin, 245–6

Saccharose, 241
Salicylate, carrier electrolyte, 146, 160
Samarium, 49
Sample conductivity, 189–194
Sample introduction, 7, 79–107
    hydrostatic, 11, 83–85
    electromigrative, 11, 18, 39, 85–91
Sample introduction methods, comparison, 98
Sample preconcentration, 95–107, 268–77
Sample size, 79, 85
Sample solvent zone. *See* solvent peak.
Sampling. *See* sample introduction.
Sampling potential. *See* sampling voltage.
Sampling rate. *See* data acquisition rate.
Sampling voltage, 86, 88
Saxitoxin, 154
Scanning detection, 121, 126–8
Scanning monochromator detectors. *See* scanning detection.
SDS, carrier electrolyte, 53, 102, 127, 155, 234, 252
Second Coion Condition, CIA, 259
Secondary circuit water, 274
Selected ion monitoring (SIM), 175
Selection of capillaries, 107–18
Selectivity in CE, 63
Selectivity of anion separations, 260–8
Selenate, 261
Selenite, 261

Separation modes, 37–58
    main, 57
    auxillary, 45–57, 62, 287–8
Serine, 150
Serotonin, 171
Sheath flow, 152
Sheathing liquid, 173
Shikimate, 261
Silanol, 26, 32
Silicon dioxide, 26
Siloxane, 26
Siphoning, 83
Smoluchowski equation, 23
Snell's laws, 133
Sodium, 29, 45, 49, 161, 165, 283, 285, 288
Sodium dodecyl sulfate. *See* SDS, carrier electrolyte.
Sodium hydroxide, carrier electrolyte, 58
Solute property detection, 119
Solvent peak, 81, 85, 137–8
Solvent zone. *See* solvent peak.
Solvophobic association, 50
Sorbate, 261
Sorbate, carrier electrolyte, 235, 239, 241
Sorbose, 232, 234
Split-flow injection, 92, 109
Spreading of peaks in CE, 67, 71–74
Stokes law, 17
Stokes radii, 17, 19
Streaming potential, 21, 28
Stripper apparatus for polyimide, 115
Strontium, 29, 285, 288
Succinate, 261
Sulfachloropyridazine, 248
Sulfadiazine, 248
Sulfadimethoxine, 248
Sulfadimidine, 248
Sulfamerazine, 248
Sulfamethoxazole, 248
Sulfanilic acid, 81
Sulfaquanidine, 248
Sulfaquinoxaline, 248
Sulfate, 18, 43, 47, 105, 142, 259, 260, 271
Sulfathiazole, 248
Sulfatroxazole, 248
Sulfide, 261
Sulfite, 261
Sulfonamides, 247–8
Sulfuric acid, carrier electrolyte, 161
Syringe-to-capillary adaptor, 6
System performance, 177–204

Tailing of peaks, 70
Tannins, 250–1

Tartrate, 261
Temperature control, 185–9
Temperature differential. See temperature gradients.
Temperature effect, 184, 213, 222
Temperature fluctuations, 183
Temperature gradients, 65–66, 68, 108
Temperature profile, 64
Terbium, 49
Terephthalate, 280
Terminating electrolyte, 99
Tetraborate. See borate, carrier electrolyte.
Tetradecyltrimethylammonium bromide additive, 30. See also OFM BT Anion (Waters).
Tetrahexylammonium perchlorate, carrier electrolyte, 51
Tetramethylammonium ion, 168
Tetrodoxin, 154
Theoretical limit for N in CE, 60
Theoretical plates, 58–60
Theory, 11–20, 58–67
Thermal conductivities, 65, 68, 108
Thermooptical absorbance detection, 178
Thiocyanate, 261
Thiosulfate, 18, 147, 260
Three-dimensional electropherograms, 127
Three-electrode amperometric system, 170
Thulium, 49
Ticarcillin, 125
Time constant, 10
Toluate, 280
Total ion current, 176
Toxins, 154
Trace levels of anions, 102
Transacotinate, 261
Transfer ratio, 139, 236
Transference number, 86
Transmittance, 122
Triethylamine, 168
Trifluoroacetate, 261

Trimesate, 147, 261
Trimethoprim, 248
Triton-X-100, additive, 29
Tryptic digest, 162
TTAB (tetradecyltrimethylammonium bromide) additive. See modifier of electroosmotic flow.
Tungstate, 18, 147, 260
Tungsten lamp, 122
Two-dimensional separations, 94–172

Undecanesulfonate, 268
Universal calibration, 140–1, 273
Urea, carrier electrolyte, 39, 214
UV CAT 1 (4-methylbenzylamine), carrier electrolyte, 29, 49, 285, 286
UV spectra, 147
UV/visible detection, 119–34

Valence number, 17
Valerate, 18, 260
Vanadate, 147, 261
VanDeemter curve, 60
Variable wavelength detectors, 120, 124
Viscosity, 17, 23, 83
Voltage-to-frequency (VFC) conversion, 198

Water plug, 96
Wavelength kit, 122
Working electrode, 170

Xylose, 213, 215, 226, 229, 234

Ytterbium, 49

Z-shaped cell, 110
Zeta potential, 21, 24, 27, 29, 31
Zinc, 288
Zinc lamp, 120
Zone broadening. See peak spreading.